Argument-Driven Inquiry
in
PHYSICS
VOLUME 1

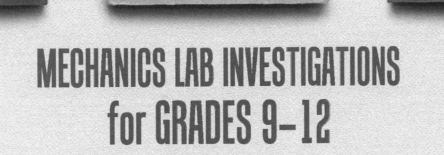

MECHANICS LAB INVESTIGATIONS
for GRADES 9–12

Victor Sampson, Todd L. Hutner, Daniel FitzPatrick,
Adam LaMee, and Jonathon Grooms

National Science Teachers Association
Arlington, Virginia

Claire Reinburg, Director
Rachel Ledbetter, Managing Editor
Amanda Van Beuren, Associate Editor
Donna Yudkin, Book Acquisitions Manager

ART AND DESIGN
Will Thomas Jr., Director

PRINTING AND PRODUCTION
Catherine Lorrain, Director

NATIONAL SCIENCE TEACHERS ASSOCIATION
David L. Evans, Executive Director
David Beacom, Publisher

1840 Wilson Blvd., Arlington, VA 22201
www.nsta.org/store
For customer service inquiries, please call 800-277-5300.

NSTA is committed to publishing material that promotes the best in inquiry-based science education. However, conditions of actual use may vary, and the safety procedures and practices described in this book are intended to serve only as a guide. Additional precautionary measures may be required. NSTA and the authors do not warrant or represent that the procedures and practices in this book meet any safety code or standard of federal, state, or local regulations. NSTA and the authors disclaim any liability for personal injury or damage to property arising out of or relating to the use of this book, including any of the recommendations, instructions, or materials contained therein.

Library of Congress Cataloging-in-Publication Data

Names: Sampson, Victor, 1974- author. | Hutner, Todd, 1981- author. | FitzPatrick, Daniel (Clinical assistant professor of mathematics), author. | LaMee, Adam, author. | Grooms, Jonathon, 1981- author. | National Science Teachers Association.
Title: Argument-driven inquiry in physics. Volume 1, Mechanics lab investigations for grade 9-12 / Victor Sampson, Todd L. Hutner, Daniel FitzPatrick, Adam LaMee, and Jonathon Grooms.
Other titles: Mechanics lab investigations for grade 9-12
Description: Arlington, Virginia : NSTA Press, National Science Teachers Association, [2017] | Includes bibliographical references and index.
Identifiers: LCCN 2017011353 (print) | LCCN 2017021496 (ebook) | ISBN 9781681403762 (e-book) | ISBN 9781681405131 | ISBN 9781681405131(print) | ISBN 168140513Xq(print)
Subjects: LCSH: Mechanics--Study and teaching (Secondary)--Handbooks, manuals, etc. | Mechanics--Experiments. | Physics--Study and teaching (Secondary)--Handbooks, manuals, etc.
Classification: LCC QC129.5 (ebook) | LCC QC129.5 .S36 2017 (print) | DDC 530.078--dc23
LC record available at *https://lccn.loc.gov/2017011353*

Argument-Driven Inquiry
in
PHYSICS
VOLUME 1

MECHANICS LAB INVESTIGATIONS
for GRADES 9–12

CONTENTS

SECTION 3
Forces and Motion: Dynamics

INTRODUCTION LABS

APPLICATION LABS

SECTION 4

Forces and Motion: Circular Motion and Rotation

INTRODUCTION LABS

APPLICATION LAB

SECTION 5

Forces and Motion: Oscillations

INTRODUCTION LABS

APPLICATION LAB

SECTION 6

Forces and Motion: Systems of Particles and Linear Momentum

INTRODUCTION LABS

APPLICATION LABS

SECTION 7

Energy, Work, and Power

INTRODUCTION LABS

APPLICATION LAB

SECTION 8

Appendixes

PREFACE

A Framework for K–12 Science Education (NRC 2012; henceforth referred to as the *Framework*) and the *Next Generation Science Standards* (NGSS Lead States 2013; henceforth referred to as the *NGSS*) call for a different way of thinking about why we teach science and what we expect students to know by the time they graduate high school. As to why we teach science, these documents emphasize that schools need to

> ensure by the end of 12th grade, *all* students have some appreciation of the beauty and wonder of science; possess sufficient knowledge of science and engineering to engage in public discussions on related issues; are careful consumers of scientific and technological information related to their everyday lives; are able to continue to learn about science outside school; and have the skills to enter careers of their choice, including (but not limited to) careers in science, engineering, and technology. (NRC 2012, p. 1)

The *Framework* and the *NGSS* are based on the idea that students need to learn science because it helps them understand how the natural world works, because citizens are required to use scientific ideas to inform both individual choices and collective choices as members of a modern democratic society, and because economic opportunity is increasingly tied to the ability to use scientific ideas, processes, and habits of mind. From this perspective, it is important to learn science because it enables people to figure things out or solve problems.

These two documents also call for a reappraisal of what students need to know and be able to do by time they graduate from high school. Instead of teaching with the goal of helping students remember facts, concepts, and terms, science teachers are now charged with the goal of helping their students become *proficient* in science. To be considered proficient in science, the *Framework* suggests that students need to understand four disciplinary core ideas (DCIs) in the physical sciences,[1] be able to use seven crosscutting concepts (CCs) that span the various disciplines of science, and learn how to participate in eight fundamental scientific and engineering practices (SEPs; called science and engineering practices in the *NGSS*). The DCIs are key organizing principles that have broad explanatory power within a discipline. Scientists use these ideas to explain the natural world. The CCs are ideas that are used across disciplines. These concepts provide a framework or a lens that people can use to explore natural phenomena. These concepts, as a result, often influence what people focus on or pay attention to when they attempt to understand how something works or why something happens. The SEPs are the different activities that scientists and engineers engage in as they attempt to generate new concepts, models, theories, or laws that are both valid and reliable. All three of these dimensions of science are important. Students need to not only know about the DCIs, CCs, and SEPs but also

---

1 Throughout this book, we use the term *physical sciences* when referring to the disciplinary core ideas of the *Framework* (in this context the term refers to a broad collection of scientific fields), but we use the term *physics* when referring to courses at the high school level (as in the title of the book).

must be able to use all three dimensions at the same time to figure things out or to solve problems. These important DCIs, CCs, and SEPs are summarized in Figure 1.

FIGURE 1 _____

The three dimensions of science in *A Framework for K–12 Science Education* and the *Next Generation Science Standards*

Science and engineering practices	Crosscutting concepts
1. Asking Questions and Defining Problems	1. Patterns
2. Developing and Using Models	2. Cause and Effect: Mechanism and Explanation
3. Planning and Carrying Out Investigations	3. Scale, Proportion, and Quantity
4. Analyzing and Interpreting Data	4. Systems and System Models
5. Using Mathematics and Computational Thinking	5. Energy and Matter: Flows, Cycles, and Conservation
6. Constructing Explanations and Designing Solutions	6. Structure and Function
7. Engaging in Argument From Evidence	7. Stability and Change
8. Obtaining, Evaluating, and Communicating Information	

Disciplinary core ideas for the physical sciences*
• PS1: Matter and Its Interactions
• PS2: Motion and Stability: Forces and Interactions
• PS3: Energy
• PS4: Waves and Their Applications in Technologies for Information Transfer

* These disciplinary core ideas represent one of the four subject areas in the *Framework* and the *NGSS*; the other subject areas are life sciences, earth and space sciences, and engineering, technology, and applications of science.

Source: Adapted from NRC 2012 and NGSS Lead States 2013.

To help students become proficient in science in ways described by the National Research Council in the *Framework*, teachers will need to use new instructional approaches that give students an opportunity to use the three dimensions of science to explain natural phenomena or develop novel solutions to problems. This is important because traditional instructional approaches, which were designed to help students "learn about" the concepts, theories, and laws of science rather than how to "figure out" how or why things work, were not created to foster the development of science proficiency inside the classroom. To help teachers make this instructional shift, this book provides 23 laboratory investigations designed using an innovative approach to lab instruction called argument-driven inquiry (ADI). This approach is designed to promote and support three-dimensional instruction inside classrooms because it gives students an opportunity to use DCIs, CCs, and SEPs to construct and critique claims about how things work or why things happen. The lab activities described in this book will also enable students to develop the disciplinary-based literacy skills outlined in the *Common Core State Standards* for English language arts (NGAC and CCSSO 2010) because ADI gives students an opportunity to give presentations to their peers, respond to audience questions and critiques, and then write, evaluate, and revise reports as part of each lab. Use of these labs, as a result, can help teachers align their teaching with current recommendations for improving classroom instruction in science and for making physics more meaningful for students.

The labs included in this book all focus on the topic of mechanics. Thus, these labs focus on only two of the four physical sciences DCIs from the *NGSS* that are outlined in Figure 1. These two DCIs are Motion and Stability: Forces and Interactions (PS2) and Energy (PS3). The other two DCIs for physical sciences from the *NGSS* are a focus of other books in the ADI series. All the labs, however, are well aligned with at least two of the seven CCs and seven of the eight SEPs. In addition, the labs in this book are well aligned with the big ideas and science practices for Advanced Placement (AP) Physics 1, 2, and C: Mechanics (see Figure 2). These labs, as a result, can be used in a wide range of physics courses, including, but not limited to, a conceptual physics course for 9th or 10th graders that is aligned with the *NGSS*, an introductory physics course for juniors or seniors, or even an AP Physics 1, 2, or C: Mechanics course.

FIGURE 2

Selected big ideas and science practices for AP Physics 1 and 2 and the content areas and laboratory objectives for AP Physics C: Mechanics

AP Physics 1 and 2 big ideas	AP Physics 1 and 2 science practices
• Objects and systems have properties such as mass and charge. Systems may have internal structure.	1. Use representations and models to communicate scientific phenomena and solve scientific problems.
• Fields existing in space can be used to explain interactions.	2. Use mathematics appropriately.
• The interactions of an object with other objects can be described by forces.	3. Engage in scientific questioning to extend thinking or to guide investigations.
• Interactions between systems can result in changes in those systems.	4. Plan and implement data collection strategies in relation to a particular scientific question.
• Changes that occur as a result of interactions are constrained by conservation laws.	5. Perform data analysis and evaluation of evidence.
	6. Work with scientific explanations and theories.
	7. Connect and relate knowledge across various scales, concepts, and representations in and across domains.

Continued

FIGURE 2 (*continued*)

AP Physics C: Mechanics content areas	AP Physics C: Mechanics laboratory objectives
• Kinematics	1. Design experiments
• Newton's laws of motion	2. Observe and measure real phenomena
• Work, energy, and power	3. Analyze data
• Systems of particles and linear momentum	4. Analyze errors
• Circular motion and rotation	5. Communicate results
• Oscillations and gravitation	

Source: Adapted from *http://apcentral.collegeboard.com/apc/public/courses/teachers_corner/2262.html* (for AP Physics 1); *http://apcentral.collegeboard.com/apc/public/courses/teachers_corner/225113.html* (for AP Physics 2); *http://apcentral.collegeboard.com/apc/public/courses/teachers_corner/2264.html* (for AP Physics C: Mechanics).

References

National Governors Association Center for Best Practices and Council of Chief State School Officers (NGAC and CCSSO). 2010. *Common core state standards.* Washington, DC: NGAC and CCSSO.

National Research Council (NRC). 2012. *A framework for K–12 science education: Practices, crosscutting concepts, and core ideas.* Washington, DC: National Academies Press.

NGSS Lead States. 2013. *Next Generation Science Standards: For states, by states.* Washington, DC: National Academies Press. *www.nextgenscience.org/next-generation-science-standards*

ABOUT THE AUTHORS

Victor Sampson is an associate professor of STEM (science, technology, engineering, and mathematics) education and the director of the Center for STEM Education (see *http://stemcenter.utexas.edu*) at The University of Texas at Austin (UT-Austin). He received a BA in zoology from the University of Washington, an MIT from Seattle University, and a PhD in curriculum and instruction with a specialization in science education from Arizona State University. Victor also taught high school biology and chemistry for nine years. He specializes in argumentation in science education, teacher learning, and assessment. To learn more about his work in science education, go to *www.vicsampson.com*.

Todd L. Hutner is the assistant director for teacher education and center development for the Center of STEM Education at UT-Austin. He received a BS and an MS in science education from Florida State University (FSU) and a PhD in curriculum and instruction from UT-Austin. Todd's classroom teaching experience includes teaching chemistry, physics, and Advanced Placement (AP) physics in Texas and earth science and astronomy in Florida. His current research focuses on the impact of both teacher education and education policy on the teaching practice of secondary science teachers.

Daniel FitzPatrick is a clinical assistant professor and master teacher in the UTeach program at the UT-Austin. He received a BS and an MA in mathematics from UT Austin and is currently a doctoral student in STEM education. Prior to his work in higher education, Daniel taught both middle and high school mathematics in public and charter schools. His interests include argumentation in mathematics education, teacher preparation, and the use of dynamic software in teaching and learning mathematics.

Adam LaMee has taught high school physics in Florida for 12 years and is currently the PhysTEC teacher-in-residence at the University of Central Florida (*http://sciences. ucf.edu/physics/phystec*). He received a BS in physics and a BS in anthropology from FSU. Adam is a Quarknet Teaching and Learning Fellow. He also contributed to the development of Florida's state science education standards and teacher certification exams, worked on the CERN's CMS (Compact Muon Solenoid) experiment, and has researched game-based assessment and performance assessment alternatives to large-scale testing. Learn more about his work at *www.adamlamee.com*.

Jonathon Grooms is an assistant professor of curriculum and pedagogy in the Graduate School of Education and Human Development at The George Washington University. He received a BS in secondary science and mathematics teaching with a focus in chemistry and physics from FSU. Upon graduation, Jonathon joined FSU's

Office of Science Teaching, where he directed the physical science outreach program Science on the Move. He also earned a PhD in science education from FSU. To learn more about his work in science education, go to *www.jgrooms.com*.

INTRODUCTION

The Importance of Helping Students Become Proficient in Science

The current aim of science education in the United States is for *all* students to become proficient in science by the time they finish high school. *Science proficiency,* as defined by Duschl, Schweingruber, and Shouse (2007), consists of four interrelated aspects. First, it requires an individual to know important scientific explanations about the natural world, to be able to use these explanations to solve problems, and to be able to understand new explanations when they are introduced to the individual. Second, it requires an individual to be able to generate and evaluate scientific explanations and scientific arguments. Third, it requires an individual to understand the nature of scientific knowledge and how scientific knowledge develops over time. Finally, and perhaps most important, an individual who is proficient in science should be able to participate in scientific practices (such as planning and carrying out investigations, analyzing and interpreting data, and arguing from evidence) and communicate in a manner that is consistent with the norms of the scientific community. These four aspects of science proficiency include the knowledge and skills that all people need to have in order to be able to pursue a degree in science, be prepared for a science-related career, and participate in a democracy as an informed citizen.

This view of science proficiency serves as the foundation for the *Framework* (NRC 2012) and the *NGSS* (NGSS Lead States 2013). Unfortunately, our educational system was not designed to help students become proficient in science. As noted in the *Framework,*

> K–12 science education in the United States fails to [promote the development of science proficiency], in part because it is not organized systematically across multiple years of school, emphasizes discrete facts with a focus on breadth over depth, and does not provide students with engaging opportunities to experience how science is actually done. (p. 1)

Our current science education system, in other words, was never designed to give students an opportunity to learn how to use scientific explanations to solve problems, generate or evaluate scientific explanations and arguments, or participate in the practices of science. Our current system was designed to help students learn facts, vocabulary, and basic process skills because many people think that students need a strong foundation in the basics to be successful later in school or in a future career. This vision of science education defines *rigor* as covering more topics and *learning* as the simple acquisition of new ideas or skills.

Our views about what counts as rigor, therefore, must change to promote and support the development of science proficiency. Instead of using the number of different topics covered in a course as a way to measure rigor in our schools, we must start to measure rigor in terms of the number of opportunities students have to use the ideas of science as a way to make sense of the world around them. Students, in other words, should be expected to learn how to use the core ideas of science as conceptual tools to plan and carry out investigations, develop and evaluate explanations, and question how we know what we know. A rigorous course, as result, would be one where students are expected to do science, not just learn about science.

Our views about what learning is and how it happens must also change to promote and support the development of science proficiency. Rather then viewing learning as a simple process where people accumulate more information over time, learning needs to be viewed as a personal and social process that involves "people entering into a different way of thinking about and explaining the natural world; becoming socialized to a greater or lesser extent into the practices of the scientific community with its particular purposes, ways of seeing, and ways of supporting its knowledge claims" (Driver et al. 1994, p. 8). Learning, from this perspective, requires a person to be exposed to the language, the concepts, and the practices of science that make science different from other ways of knowing. This process requires input and guidance about "what counts" from people who are familiar with the goals of science, the norms of science, and the ways things are done in science. Thus, learning is dependent on supportive and informative interactions with others.

Over time, people will begin to appropriate and use the language, the concepts, and the practices of science as their own when they see how valuable they are as a way to accomplish their own goals. Learning thus involves seeing new ideas and ways of doing things, trying out these new ideas and practices, and adopting them when they are useful. This entire process, however, can only happen if teachers provide students with multiple opportunities to use scientific ideas to solve problems, to generate or evaluate scientific explanations and arguments, and to participate in the practices of science inside the classroom. This is important because students must have a supportive and educative environment to try out new ideas and practices, make mistakes, and refine what they know and what they do before they are able to adopt the language, the concepts, and the practices of science as their own.

A New Approach to Teaching Science

We need to use different instructional approaches to create a supportive and educative environment that will enable students to learn the knowledge and skills they need to become proficient in science. These new instructional approaches will need to give students an opportunity to learn how to "figure out" how things work or why things happen. Rather than simply encouraging students to learn about the facts, concepts, theories, and laws of science, we need to give students more opportunities to develop explanations for natural phenomena and design solutions to problems.

This emphasis on "figuring things out" instead of "learning about things" represents a big change in the way we will need to teach science at all grade levels. To figure out how things work or why things happen in a way that is consistent with how science is actually done, students must do more than hands-on activities. Students must learn how to use disciplinary core ideas (DCIs), crosscutting concepts (CCs), and science and engineering practices (SEPs) to develop explanations and solve problems (NGSS Lead States 2013; NRC 2012).

A DCI is a scientific idea that is central to understanding a variety of natural phenomena. An example of a DCI in physics is the force of gravity, which is a type of interaction that occurs between objects with mass. This DCI not only explains the motion of planets around the Sun but also the motion of a rock dropped from a bridge.

CCs are those concepts that are important across the disciplines of science; there are similarities and differences in the treatment of the CC in each discipline. The CCs can be used as a lens to help people think about what to focus on or pay attention to during an investigation. For example, one of the CCs from the *Framework* is Energy and Matter: Flows, Cycles, and Conservation. This CC is important in many different fields of study, including but not limited to mechanics, thermodynamics, electricity and magnetism, and nuclear physics. This CC is equally important in biology; biologists use this CC to explore topics such as cellular processes, growth and development, and ecosystems. It is important to highlight the centrality of this idea, and other CCs, for students as we teach the subject-specific DCIs.

SEPs describe what scientists do to investigate the natural world. The practices outlined in the *Framework* and the *NGSS* explain and extend what is meant by *inquiry* in science and the wide range of activities that scientists engage in as they attempt to generate and validate new ideas. Students engage in practices to build, deepen, and apply their knowledge of DCIs and CCs. The SEPs include familiar aspects of inquiry, including such activities as Asking Questions and Defining Problems, Planning and Carrying Out Investigations, and Analyzing and Interpreting Data. More important, however, the SEPs include other activities that are at the core of doing science. These activities include Developing and Using Models, Constructing Explanations and Designing Solutions, Engaging in Argument From Evidence, and Obtaining, Evaluating, and Communicating Information. All of these SEPs are important to learn, because there is no single scientific method that all scientists must follow; scientists engage in different practices, at different times, and in different orders depending on what they are studying and what they are trying to accomplish at that point in time.

This focus on students using DCIs, CCs, and SEPs during a lesson is called *three-dimensional instruction* because students have an opportunity to use all three dimensions of science to understand how something works, to explain why something happens, or to develop a novel solution to a problem. When teachers use three-dimensional instruction inside their classrooms, they encourage students to develop or use conceptual models, design investigations, develop explanations, share and critique ideas, and argue from evidence, all of which allow students to develop the knowledge and skills they need to be proficient in science (NRC 2012). Current research suggests that all students benefit from three-dimensional instruction because it gives all students more voice and choice during a lesson and it makes the learning process inside the classroom more active and inclusive (NRC 2012).

We think the school science laboratory is the perfect place to integrate three-dimensional instruction into the science curriculum. Well-designed lab activities can provide opportunities for students to participate in an extended investigation where they can not only use one or more DCIs to understand how something works, to explain why something happens, or to develop a novel solution to a problem but also use several different CCs and SEPs during the same lesson. For example, a teacher can give students an opportunity to explore the motion of several different balls as they roll down a ramp. The teacher can then encourage them to use Newton's second law of motion (part of the DCI Motion and Stability: Forces and Interactions) and their understanding of (a) Cause and Effect: Mechanism and Explanation and (b) Scale, Proportion, and Quantity (two different CCs) to plan and carry out an investigation to figure out how the net force acting on a ball of a given mass affects its acceleration. During this investigation they must collect, analyze, and interpret data; use mathematics; construct explanations; argue from evidence; and obtain, evaluate, and communicate information (tasks that involve six different SEPs). Using multiple DCIs, CCs, and SEPs at the same time is important because it creates a classroom experience that parallels how science is done. This, in turn, gives all students who participate in a school science laboratory an opportunity to deepen their understanding of what it means to do science and to develop science-related identities. In the following section, we will describe how to promote and support the development of science proficiency during school science laboratories through three-dimensional instruction.

How School Science Laboratories Can Help Foster the Development of Science Proficiency Through Three-Dimensional Instruction

School science laboratory experiences[1] tend to follow a similar format in most U.S. science classrooms (Hofstein and Lunetta 2004; NRC 2005). This format begins with the teacher introducing students to an important concept or principle through direct instruction, usually by giving a lecture about it or by assigning a chapter from a

---

1 *School science laboratory experiences* are defined as "an opportunity for students to interact directly with the material world using the tools, data collection techniques, models, and theories of science" (NRC 2005, p. 3).

textbook to read. This portion of instruction often takes several class periods. Next, the students will complete a hands-on laboratory activity. The purpose of the hands-on activity is help students understand a concept or principle that was introduced to the students earlier. To ensure that students "get the right result" during the lab and that the lab actually illustrates, confirms, or verifies the target concept or principle, the teacher usually provides students with a step-by-step procedure to follow and a data table to fill out. Students are then asked to answer a set of analysis questions to ensure that everyone "reaches the right conclusion" based on the data they collected during the lab. The laboratory experience then ends with the teacher going over what the students should have done during the lab, what they should have observed, and what answers they should have given in response to the analysis questions, to ensure that the students "learned what they were supposed have learned" from the hands-on activity. This final step of the laboratory experience is usually done, once again, through whole-class direct instruction.

Recent research, however, suggests that this type of approach to laboratory instruction does little to help students learn key concepts. The National Research Council (2005, p. 5), for example, conducted a synthesis of several different studies that examined what students learn from laboratory instruction and found that "research focused on the goal of student mastery of subject matter indicates that typical laboratory experiences are no more or less effective than other forms of science instruction (such as reading, lectures, or discussion)." This finding is troubling because, as noted earlier, the main goal of this type of lab experience is to help students understand an important concept or principle by giving them a hands-on and concrete experience with it. In addition, this type of laboratory experience does little to help students learn how plan and carry out investigations or analyze and interpret data, because students have no voice or choice during the activity. Students are expected to simply follow a set of directions rather than having to think about what data they will collect, how they will collect it, and what they will need to do to analyze it once they have it. These types of activities also can lead to misunderstanding about the nature of scientific knowledge and how this knowledge is developed over time, because of the emphasis on following procedure and getting the right results. These "cookbook" labs, as a result, do not reflect how science is done at all.

Over the last decade, many teachers have changed their labs to be inquiry-based in order to address the many shortcomings of typical cookbook lab activities. Inquiry-based lab experiences that are consistent with the definition of *inquiry* found in the *National Science Education Standards* (NRC 1996) and *Inquiry and the National Science Education Standards* (NRC 2000) share five key features:

1. These labs are designed so students need to answer a scientifically oriented

question by conducting an investigation.

2. These labs give students an opportunity to develop their own method to collect data during the investigation.

3. Students are expected to formulate an answer to the question based on their analysis of the data they collected.

4. Students connect their answer to some theory, model, or law; this is the aspect of inquiry-based instruction that students tend to struggle with most.

5. Students are given an opportunity to communicate and justify their answer to the question.

These types of labs tend to be used by teachers as a way to introduce students to important content and to give students an opportunity to learn how to collect and analyze data in science (NRC 2012).

Although inquiry-based approaches give students much more voice and choice during a lab, especially when compared with more typical cookbook approaches, they do not do as much as they could do to promote the development of science proficiency. Teachers tend to use inquiry-based labs as a way to help students learn about new ideas rather than as a way to help students learn how to figure out how things work or why they happen. Students, as a result, rarely have an opportunity to learn how to use DCIs, CCs, and SEPs to develop explanations or solve prob-lems. In addition, inquiry-based approaches rarely give students an opportunity to participate in the full range of scientific practices. Inquiry-based labs tend to be designed so students have many opportunities to learn how to ask questions, plan and carry out investigations, and analyze and interpret data but few opportunities to learn how to participate in the practices that focus on how new ideas are developed, shared, refined, and eventually validated within the scientific community. These important practices include developing models, constructing explanations, arguing from evidence, and obtaining, evaluating, and communicating information (Duschl, Schweingruber, and Shouse 2007; NRC 2005). Inquiry-based labs also do not give students an opportunity to improve their science-specific literacy skills. Students are rarely expected to read, write, and speak in a scientific manner because the focus of these labs is learning about content and how to collect and analyze data in science, not how to propose, critique, and revise ideas.

Changing the focus and nature of inquiry-based labs so they are more consistent with three-dimensional instruction can help address these issues. To implement such a change, teachers will not only have to focus on using DCIs, CCs and SEPs but will also need to emphasize "how we know" in the physical sciences (i.e., how new knowledge is generated and validated) equally with "what we know" about forces, motion, and energy (i.e., the theories, laws, and unifying concepts). We have found that this shift in focus is best accomplished by making the practice of arguing from evidence or scientific argumentation the central feature of all laboratory activities. We define *scientific argumentation* as the process of proposing, supporting, evaluating, and refining claims based on evidence (Sampson, Grooms, and Walker, 2011). The *Framework* (NRC 2012) provides a good description of the role argumentation plays in science:

> Scientists and engineers use evidence-based argumentation to make the case for their ideas, whether involving new theories or designs, novel ways of collecting data, or interpretations of evidence. They and their peers then attempt to identify weaknesses and limitations in the argument, with the ultimate goal of refining and improving the explanation or design. (p. 46)

When teachers make the practice of arguing from evidence the central focus of lab activities, students have more opportunities to learn how to construct and support scientific knowledge claims through argument (NRC 2012). Students also have more opportunities to learn how to evaluate the claims and arguments made by others. Students, as a result, learn how to read, write, and speak in a scientific manner because they need to be able to propose and support their claims when they share them and evaluate, challenge, and refine the claims made by others.

We developed the argument-driven inquiry (ADI) instructional model (Sampson and Gleim 2009; Sampson, Grooms, and Walker 2009, 2011) as a way to change the focus and nature of labs so they are consistent with three-dimensional instruction. ADI gives students an opportunity to learn how to use DCIs, CCs, and SEPs to figure out how things work or why things happen. This instructional approach also places scientific argumentation as the central feature of all laboratory activities. ADI lab investigations, as a result, make lab activities more authentic and educative for students and thus help teachers promote and support the development of science proficiency. This instructional model reflects current research about how people learn science (NRC 1999, 2005, 2008, 2012) and is also based on what is known about how to engage students in argumentation and other important scientific practices (Erduran and Jimenez-Aleixandre 2008; McNeill and Krajcik 2008; Osborne, Erduran, and Simon 2004; Sampson and Clark 2008). We will explain the stages of ADI and how each stage works in Chapter 1.

How to Use This Book

The intended audience of this book is primarily practicing high school physics teachers. We recognize that physics teachers teach many different types of physics courses. Some courses are conceptual in nature, some are algebra based, and some are calculus based. We understand how teaching these different types of physics courses results in different challenges and needs. We have therefore designed the laboratory investigations included in this book to meet the needs of teachers who teach a wide range of courses. Some labs, for example, require students to determine a general relationship or trend and do not require a lot of mathematics. These labs can be used in a physics course that is more conceptual in nature. Other labs, in contrast, require students to develop a mathematical model that students can use to explain and predict the motion of an object over time. These labs are intended for students in Advanced Placement (AP) Physics C: Mechanics who are concurrently enrolled in or have successfully completed an introductory calculus course. The majority of the labs, however, were written for an algebra-based physics course. These labs require some algebra, such as determining a mathematical relationship between two variables (which is often, but not always, a linear relationship). All of the labs were designed to give students an opportunity to learn how to use DCIs, CCS, and SEPs to figure things out.

As we wrote the labs for this book, we kept in mind the fact that physics is often a two-year program of study in many school districts. Students, as a result, often take Physics I in their 11th-grade year along with Algebra II and then take either AP Physics 1, AP Physics 2, or AP Physics C: Mechanics along with either AP Statistics or AP Calculus. We have therefore aligned the labs with the *NGSS* performance expectations (where applicable) and the AP Physics 1 and AP Physics C: Mechanics learning objectives so teachers can use these labs in either an introductory physics course or an AP physics course. We believe that it is important to focus on three-dimensional instruction in both contexts because students need to learn how to use DCIs, CCs, and SEPs to figure out how things work or why things happen even when they are taking AP physics. Lab instruction is also a major component of the AP physics curriculum. In AP Physics 1 and AP Physics C: Mechanics, for example, the College Board recommends that at least 25% of instructional time be devoted to laboratory experiences. These experiences should therefore do more than demonstrate, illustrate, or verify a target concept; they should promote and support the development of the four aspects of science proficiency.

One of the recent advances in physics education has been the development of physics-specific equipment that students can use during investigations; probeware and video cameras for collecting data; and data analysis software, including video analysis software, which enables students to explore the data they collect during an investigation. We recognize that while some physics teachers work in settings where this equipment is readily available and funds are easily accessed to purchase additional equipment, many others do not work in such settings. To address this concern, we have designed many of the labs in this book so they can be conducted in lower-tech ways, by using stopwatches and metersticks. Sometimes, however, a lab may not be worth doing if students do not have access to specific equipment such as a video camera and video analysis software. When materials are optional, we indicate in the Lab Handout that students "may also consider using" optional equipment. If this equipment is not available to you, when introducing the lab just let students know that they do not have the option to use that equipment. We also recognize that the initial cost to purchase the necessary equipment may be high, especially when compared with equipment needed for a chemistry or biology course. However, the replacement costs for these labs are minimal because the equipment should last several years; in contrast, biology or chemistry courses may require annual replacement of chemicals or specimens.

Finally, we want to make clear that we do not expect teachers to use every lab in this book over the course of an academic year. We wrote this book to support the teaching of mechanics, which is a topic found in the first-semester curriculum of a physics course. Concepts included under the topic of mechanics include kinematics, dynamics, circular motion and rotation, oscillations, momentum, and energy. We suggest that teachers who use this book choose one (or two, at most) labs for each of the six major concepts.

There are two types of labs included in the book: (1) introduction labs and (2) application labs. An introduction lab should be used at the beginning of a unit. These labs often require little formal knowledge of the target concept before students begin the investigation. For example, the lab on free fall (Lab 2) is an introduction lab and does not require students to know about the acceleration due to gravity, but students are still expected to use a DCI (Motion and Stability: Forces and Interactions) and two CCs (Patterns, Stability and Change) to figure out the relationship between the mass of an object and its acceleration during free fall. After students complete the lab, teachers can move forward to formalize the laws and formulas related to acceleration due to gravity through other means of instruction.

Application labs, on the other hand, are designed to come at the end of a unit. The intent of these labs is give student an opportunity to apply their knowledge of a specific concept they learned about earlier in the course, along with their knowledge of DCIs and CCs, to a novel situation. For example, Lab 23 requires students to use their knowledge about conservation of energy and the relationship between energy and power to determine the horsepower of a remote control toy car.

Organization of This Book

This book is divided into eight sections. Section 1 includes two chapters: the first chapter describes the ADI instructional model, and the second chapter describes the development of the ADI lab investigations and an overview of what is included with each investigation. Sections 2–7 contain the 23 lab investigations. Each investigation includes three components:

- Teacher Notes, which provide information about the purpose of the lab and what teachers need to do to guide students through it.

- Lab Handout, which can be photocopied and given to students at the beginning of the lab. It provides the students with a phenomenon to investigate, a guiding question to answer, and an overview of the DCIs and CCs that students can use during the investigation.

- Checkout Questions, which can be photocopied and given to students at the conclusion of the lab activity as an optional assessment. The Checkout Questions consist of items that target students' understanding of the DCIs, the CCs, and the nature of scientific knowledge (NOSK) and the nature of scientific inquiry (NOSI) concepts addressed during the lab.

Section 8 consists of five appendixes:

- Appendix 1 contains several standards alignment matrixes that can be used to assist with curriculum or lesson planning.

- Appendix 2 provides an overview of the CCs and the NOSK and NOSI concepts that are a focus of the lab investigations. This information about the CCs and the NOSK and NOSI are included as a reference for teachers.

- Appendix 3 provides several options (in tabular format) for implementing an ADI investigation over multiple 50-minute class periods.

- Appendix 4 provides options for investigation proposals, which students can use as graphic organizers to plan an investigation. The proposals can be photocopied and given to students during the lab.

- Appendix 5 provides a peer-review guide and teacher scoring rubric, which can also be photocopied and given to students.

Safety Practices in the Science Laboratory

It is important for all of us to do what we can to make school science laboratory experiences safer for everyone in the classroom. We recommend four important guidelines to follow. First, we need to have proper safety equipment such as, but not limited to, fume hoods, fire extinguishers, eye wash, and showers in the classroom or laboratory. Second, we need to ensure that students use appropriate personal protective equipment (PPE; e.g., sanitized indirectly vented chemical-splash goggles, chemical-resistant aprons and nonlatex gloves) during all components of laboratory activities (i.e., setup, hands-on investigation, and takedown). At a minimum, the PPE we provide for students to use must meet the ANSI/ISEA Z87.1 D.3 standard. Third, we must review and comply with all safety policies and procedures, including but not limited to appropriate chemical management, that have been established by our place of employment. Finally, and perhaps most important, we all need to adopt safety standards and better professional safety practices and enforce them inside the classroom or laboratory.

We provide safety precautions for each investigation and recommend that all teachers follow these safety precautions to provide a safer learning experience inside the classroom. The safety precautions associated with each lab investigation are based, in part, on the use of the recommended materials and instructions, legal safety standards, and better professional safety practices. Selection of alternative materials or procedures for these activities may jeopardize the level of safety and therefore is at the user's own risk.

We also recommend that you encourage students to read the National Science Teacher Association's document *Safety in the Science Classroom, Laboratory, or Field Sites* before allowing them to work in the laboratory for the first time. This document is available online at *www.nsta.org/docs/SafetyInTheScienceClassroomLabAndField.pdf*. Your students and their parent(s) or guardian(s) should then sign the document to acknowledge that they understand the safety procedures that must be followed during a school science laboratory experience.

Remember that a lab includes three parts: (1) setting up the lab and preparing the materials, (2) conducting the actual investigation, and (3) the cleanup, also called the takedown. The safety procedures and PPE we recommend for each investigation apply to all three parts.

References

Duschl, R. A., H. A. Schweingruber, and A. W. Shouse, eds. 2007. *Taking science to school: Learning and teaching science in grades K–8*. Washington, DC: National Academies Press.

Driver, R., Asoko, H., Leach, J., Mortimer, E., and Scott, P. 1994. Constructing scientific knowledge in the classroom. *Educational Researcher* 23: 5–12.

Erduran, S., and M. Jimenez-Aleixandre, eds. 2008. *Argumentation in science education: Perspectives from classroom-based research*. Dordrecht, The Netherlands: Springer.

Hofstein, A., and V. Lunetta. 2004. The laboratory in science education: Foundations for the twenty-first century. *Science Education* 88: 28–54.

McNeill, K., and J. Krajcik. 2008. Assessing middle school students' content knowledge and reasoning through written scientific explanations. In *Assessing science learning: Perspectives from research and practice,* eds. J. Coffey, R. Douglas, and C. Stearns, 101–116. Arlington, VA: NSTA Press.

National Research Council (NRC). 1999. *How people learn: Brain, mind, experience, and school.* Washington, DC: National Academies Press.

National Research Council (NRC). 2000. *Inquiry and the National Science Education Standards.* Washington, DC: National Academies Press.

National Research Council (NRC). 2005. *America's lab report: Investigations in high school science.* Washington, DC: National Academies Press.

National Research Council (NRC). 2008. *Ready, set, science: Putting research to work in K–8 science classrooms.* Washington, DC: National Academies Press.

National Research Council (NRC). 2012. *A framework for K–12 science education: Practices, crosscutting concepts, and core ideas.* Washington, DC: National Academies Press.

NGSS Lead States. 2013. *Next Generation Science Standards: For states, by states.* Washington, DC: National Academies Press. *www.nextgenscience.org/next-generation-science-standards.*

Osborne, J., S. Erduran, and S. Simon. 2004. Enhancing the quality of argumentation in science classrooms. *Journal of Research in Science Teaching* 41 (10): 994–1020.

Sampson, V., and D. Clark. 2008. Assessment of the ways students generate arguments in science education: Current perspectives and recommendations for future directions. *Science Education* 92 (3): 447–472.

Sampson, V., and L. Gleim. 2009. Argument-driven inquiry to promote the understanding of important concepts and practices in biology. *American Biology Teacher* 71 (8): 471–477.

Sampson, V., J. Grooms, and J. Walker. 2009. Argument-driven inquiry: A way to promote learning during laboratory activities. *The Science Teacher* 76 (7): 42–47.

Sampson, V., J. Grooms, and J. Walker. 2011. Argument-driven inquiry as a way to help students learn how to participate in scientific argumentation and craft written arguments: An exploratory study. *Science Education* 95 (2): 217–257.

SECTION 1
Using Argument-Driven Inquiry

CHAPTER 1
Argument-Driven Inquiry

Stages of Argument-Driven Inquiry

The argument-driven inquiry (ADI) instructional model was designed to change the focus and nature of labs so they are consistent with the three-dimensional instructional approach. ADI therefore gives students an opportunity to learn how to use disciplinary core ideas (DCIs), crosscutting concepts (CCs), and science and engineering practices (SEPs) (NGSS Lead States 2013; NRC 2012) to figure out how things work or why things happen. This instructional approach also places scientific argumentation as the central feature of all laboratory activities. ADI lab investigations, as a result, make lab activities more authentic (students have an opportunity to engage in all the practices of science) *and* educative (students receive the feedback and explicit guidance that they need to improve on each aspect of science proficiency).

In this chapter, we will explain what happens during each of the eight stages of the ADI instructional model. These eight stages are the same for every ADI laboratory experience. Students, as result, quickly learn what is expected of them during each stage of an ADI lab and can focus on learning how to use DCIs, CCs, and SEPs to develop explanations or solve problems. Figure 3 (p. 4) summarizes the eight stages of the ADI instructional model.

Stage 1: Identify the Task and the Guiding Question

An ADI lab activity begins with the teacher identifying a phenomenon to investigate and offering a guiding question for the students to answer. The goal of the teacher at this stage of the model is to capture the students' interest and provide them with a reason to complete the investigation. To aid in this, teachers should provide each student with a copy of the Lab Handout. This handout includes a brief introduction that provides a description of the puzzling phenomenon or a problem to solve, the DCI and CCs that students can use during the investigation, a reason to investigate, and the task the students will need to complete. This handout also includes information about the nature of the argument they will need to produce, some helpful tips on how to get started, and criteria that will be used to judge argument quality (e.g., the sufficiency of the claim and the quality of the evidence).

Teachers often begin an ADI investigation by selecting a different student to read each section of the Lab Handout out loud while the other students follow along. As the students read, they can annotate the text to identify important or useful ideas and information or terms that may be unfamiliar or confusing. After each section is read, the teacher can pause to clarify expectations, answer questions, and provide additional information as needed.

Teachers can also spark student interest by giving a demonstration or showing a video of the phenomenon.

It is also important for the teacher to hold a "tool talk" during this stage, taking a few minutes to explain how to use the available lab equipment, how to use a computer simulation, or even how to use software to analyze data. Teachers need to hold a tool talk because students are often unfamiliar with specialized lab equipment, simulations, or software. Even if the students are familiar with the available tools, they will often use them incorrectly or in an unsafe manner unless they are reminded about how the tools work and the proper way to use them. The teacher should therefore review specific safety protocols and precautions as part of the tool talk.

FIGURE 3

Stages of the argument-driven inquiry instructional model

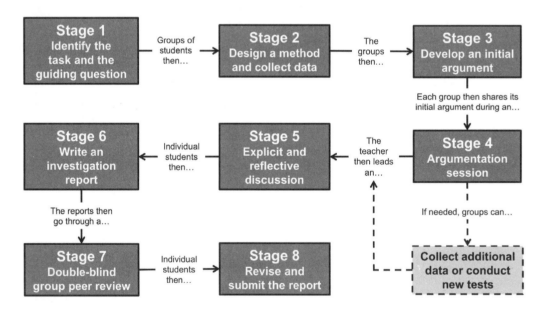

Including a tool talk during the first stage is useful because students often find it difficult to design a method to collect the data needed to answer the guiding question (the task of stage 2) when they do not understand how to use the available materials. We also recommend that teachers give students a few minutes to tinker with the equipment, simulation, or software they will be using to collect data as part of the tool talk. We have found that students can quickly figure out how the equipment, simulation, or software works and what they can and cannot do with it simply by tinkering with the available materials for 5–10 minutes. When students are given an opportunity to tinker with the equipment, simulation, or software as part of the tool talk, they end up designing much better investigations (the task of stage 2) because they understand what they can and cannot do with the tools they will use to collect data.

Once all the students understand the goal of the activity and how to use the available materials, the teacher should divide the students into small groups (we recommend three or four students per group) and move on to the second stage of the instructional model.

Stage 2: Design a Method and Collect Data

In stage 2, small groups of students develop a method to gather the data they need to answer the guiding question and carry out that method. How students complete this stage depends on the nature of the investigation. Some investigations call for groups to answer the guiding question by designing a controlled experiment, whereas others require students to analyze an existing data set (e.g., a database or information sheets). If students need assistance in designing their method, teachers can have students complete an investigation proposal. These proposals guide students through the process of developing a method by encouraging them to think about what type of data they will need to collect, how to collect it, and how to analyze it. We have included six different investigation proposals in Appendix 4 (p. 533) of this book that students can use to design their investigations. Investigation Proposal A (long or short version) can be used when students need to collect systematic observations for a descriptive investigation. Investigation Proposal B (long or short version) or Investigation Proposal C (long or short version) can be used when students need to design a comparative or experimental study to test potential explanations or relationships as part of their investigation. Investigation Proposal B requires students to design a test of two alternative hypotheses, and Investigation Proposal C requires students to design a test of three alternative hypotheses.

The overall intent of this stage is to provide students with an opportunity to interact directly with the natural world (or in some cases with data drawn from the natural world) using appropriate tools and data collection techniques and to learn how to deal with the uncertainties of empirical work. This stage of the model also gives students a chance to learn why some approaches to data collection or analysis work better than others and how the method used during a scientific investigation is based on the nature of the question and the phenomenon under investigation. At the end of this stage, students should have collected all the data they need to answer the guiding question.

Stage 3: Develop an Initial Argument

The next stage of the instructional model calls for students to develop an initial argument in response to the guiding question. To do this, each group needs to be encouraged to first analyze the measurements (e.g., temperature and mass) and/or observations (e.g., appearance and location) they collected during stage 2 of the model. Once the groups have analyzed and interpreted the results of their analysis, they can create an initial argument.

The argument consists of a claim, the evidence they are using to support their claim, and a justification of their evidence. The *claim* is their answer to the guiding question. The *evidence* consists of an analysis of the data they collected and an interpretation of the analysis. The *justification of the evidence* is a statement that defends their choice of evidence by explaining why it is important and relevant, making the concepts or assumptions underlying the analysis and interpretation explicit. The components of a scientific argument are illustrated in Figure 4.

FIGURE 4 _____

The components of a scientific argument and criteria for evaluating its quality

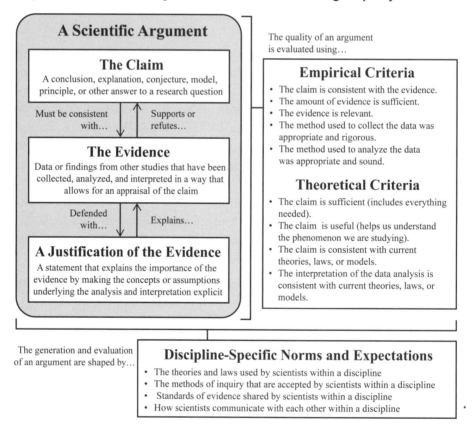

To illustrate each of the three structural components of a scientific argument, consider the following example. This argument was made in response to the guiding question, "What variables affect the period of a pendulum?"

> *Claim:* The only variable that affects the period of the pendulum is the length of the pendulum. The mass of the bob and the angle of release do not affect the period of the pendulum.

Evidence: When we changed the mass of the bob and held the length of the string and the angle of release constant, the period of the pendulum was approximately 1.5 s each time. We then changed the release angle from 5 to 15 degrees (and held the length of the string and the mass of the bob constant), and again the period remained approximately 1.5 s. Finally, we changed the length of the string and held the mass of the bob and the release angle constant. As the length of the string increased from 0.5 m to 1.5 m, the period of the pendulum increased from 1.5 to 2.5 s. The length of the string, as a result, was the only factor that resulted in a meaningful change in the period of a pendulum.

Justification of the evidence: Our evidence is based on four important assumptions. First, the period of a pendulum is defined as the time it takes the bob to travel from one extreme position, to the opposite extreme position, and then back. Second, there are only two forces acting on the pendulum bob. The force of gravity acts downward upon the bob, and a tension force from the string pulls upward on the bob. We can ignore air resistance because its effects are smaller than the uncertainty in our measurements. Third, the pendulum bob accelerates toward the equilibrium point as it moves back and forth because an unbalanced force is acting on it (Newton's second law of motion). Fourth, there is always some uncertainty in measurement, so only large differences are meaningful.

The claim in this argument provides an answer to the guiding question. The author then uses genuine evidence to support the claim by providing an analysis of the data collected (periods of the pendulum under three different conditions) and an interpretation of the analysis (the length of the string was the only factor that resulted in a meaningful change). Finally, the author provides a justification of the evidence in the argument by making explicit the underlying concept and assumptions (the definition of a period, the nature of forces acting on the bob, why the bob accelerates in different directions, and uncertainty in measurement) guiding the analysis of the data and the interpretation of the analysis.

It is important for students to understand that, in science, some arguments are better than others. An important aspect of science and scientific argumentation involves the evaluation of the various components of the arguments put forward by others. Therefore, the framework provided in Figure 4 also highlights two types of criteria that students can and should be encouraged to use to evaluate an argument in science: empirical criteria and theoretical criteria. *Empirical criteria* include

- how well the claim fits with all available evidence,
- the sufficiency of the evidence,
- the relevance of the evidence,
- the appropriateness and rigor of the method used to collect the data, and
- the appropriateness and soundness of the method used to analyze the data.

Theoretical criteria, on the other hand, refer to standards that are important in science but are not empirical in nature; examples of these criteria are

- the sufficiency of the claim (i.e., Does it include everything needed?);
- the usefulness of the claim (i.e., Does it help us understand the phenomenon we are studying?);
- how consistent the claim is with accepted theories, laws, or models (i.e., Does it fit with our current understanding of force and interactions or stability and motion?); and
- how consistent the interpretation of the results of the analysis is with accepted theories, laws, or models (i.e., is the interpretation based on what is known about forces and interactions or stability and motion?).

What counts as quality within these different components, however, varies from discipline to discipline (e.g., physics, chemistry, geology, biology) and within the specific fields of each discipline (e.g., astrophysics, biophysics, nuclear physics, quantum mechanics, thermodynamics). This variation is due to differences in the types of phenomena investigated, what counts as an accepted mode of inquiry (e.g., descriptive studies, experimentation, computer modeling), and the theory-laden nature of scientific inquiry. It is important to keep in mind that "what counts" as a quality argument in science is discipline and field dependent.

To allow for the critique and refinement of the tentative argument during the next stage of ADI, each group of students should create their initial argument in a medium that can easily be viewed by the other groups. We recommend using a 2′ x 3′ whiteboard. Students should include the guiding question of the lab and the three main components of the argument on the board. Figure 5 shows the general layout for a presentation of an argument, and Figure 6 provides an example of argument crafted by students. Students can also create their initial arguments using presentation software such as Microsoft's PowerPoint or Apple's Keynote and devote one slide to each component of an argument. The choice of medium is not important as long as students are able to easily modify the content of their argument as they work and it enables others to easily view their argument.

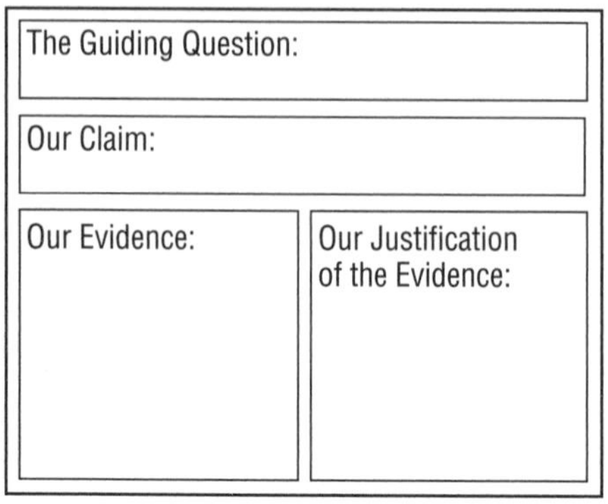

The intention of this stage of the model is to provide the student groups with an opportunity to make sense of what they are seeing or doing during the investigation. As students work together to create an initial argument, they must talk with each other and determine if their analysis is useful or not and how

FIGURE 5 _____

The components of an argument that should be included on a whiteboard (outline)

The Guiding Question:	
Our Claim:	
Our Evidence:	Our Justification of the Evidence:

to best interpret the trends, differences, or relationships that they identify as a result of the analysis. They must also decide if the evidence (data that have been analyzed and interpreted) that they chose to include in their argument is relevant, sufficient, and an acceptable way to support their claim. This process, in turn, enables the groups of students to evaluate competing ideas and weed out any claim that is inaccurate, does not fit with all the available data, or contains contradictions.

This stage of the model is challenging for students because they are rarely asked to make sense of a phenomenon based on raw data, so it is important for teachers to actively work to support their "sense-making." In this stage, teachers should circulate from group to group to act as a resource person for the students, asking questions urging them to think about what they are doing and why. To help students remember the goal of the activity, you can ask questions such as "What are you trying to figure out?" You can ask them questions such as to "Why is that information important?" or "Why is that analysis useful?" to encourage them to think about whether or not the data they are analyzing is relevant or the analysis is informative. To help them remember to use rigorous criteria to evaluate the merits of a tentative claim, you can ask, "Does that fit with all the data?" or "Is that consistent with what we know about forces and motion?"

It is important to remember that at the beginning of the school year, students will struggle to develop arguments and will often rely on inappropriate criteria such as plausibility (e.g., "That sounds good to me") or fit with personal experience (e.g., "But that is what I saw on TV once") as they attempt to make sense of their data. However, as students learn why it is useful to use evidence in an argument, what makes evidence valid or acceptable in science, and why it is important to justify why they used a particular type of evidence through practice, *students will improve their ability to argue from evidence* (Grooms, Enderle, and Sampson 2015). This is an important principle underlying the ADI instructional model.

FIGURE 6

An example of a student-generated argument on a whiteboard

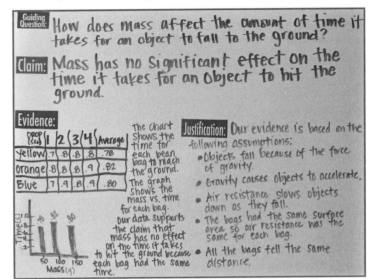

Stage 4: Argumentation Session

The fourth stage of ADI is the argumentation session. In this stage, each group is given an opportunity to share, evaluate, and revise their initial arguments by interacting with members from the other groups (see Figure 7). This stage is included in the model for three reasons:

1. Scientific argumentation (i.e., arguing from evidence) is an important practice in science because critique and revision lead to better outcomes.

2. Research indicates that students learn more about the content and develop better critical-thinking skills when they are exposed to the alternative ideas, respond to the questions and challenges of other students, and evaluate the merits of competing ideas (Duschl, Schweingruber, and Shouse 2007; NRC 2012).

3. Students learn how to distinguish between ideas using rigorous scientific criteria and are able to develop scientific habits of mind (such as treating ideas with initial skepticism, insisting that the reasoning and assumptions be made explicit, and insisting that claims be supported by valid evidence) during the argumentation sessions.

FIGURE 7

A student presents her group's argument to students from other groups during the argumentation session.

This stage, as a result, provides the students with an opportunity to learn from and about scientific argumentation.

It is important to note, however, that supporting and promoting productive interactions between students in the classroom can be difficult because the practice of arguing from evidence is foreign to most students when they first begin participating in ADI. To aid these interactions, students are required to generate their arguments in a medium that can be seen by others. By looking at whiteboards, paper, or slides, students tend to focus their attention on evaluating evidence and the DCI that was used to justify the evidence rather than attacking the source of the ideas. As a result, this strategy often makes the discussion more productive and makes it easier for student to identify and weed out faulty ideas. It is also important for the students to view the argumentation session as an opportunity to learn. The teacher, therefore, should describe the argumentation session as an opportunity for students to collaborate with their peers and as a chance to give each other feedback so the quality of all the arguments can be improved, rather than as an opportunity determine who is right or wrong.

FIGURE 8

A modified gallery walk format is used during the argumentation session to allow multiple groups to share their arguments at the same time.

To ensure that all students remain engaged during the argumentation session, we recommend that teachers use a modified "gallery walk" format rather than a whole-class presentation format. In the modified gallery walk format, one or two members of the group stay at their workstation to share their groups' ideas while the other group members go to different groups one at a time to listen to and critique the arguments developed by their classmates (see Figure 8). This type of format ensures that all ideas are heard and more students are actively involved in the process. We recommend that the students who are responsible for critiquing arguments visit at least three different groups during the argumentation session. We also recommend that the presenters keep a record of the critiques made by their classmates and any suggestions for improvement. The students who are responsible for critiquing the arguments should also be encouraged to keep a record of good ideas or potential ways to improve their own arguments as they travel from group to group.

Just as is the case in earlier stages of ADI, it is important for the classroom teacher to be involved in (without leading) the discussions during the argumentation session. Once again, the teacher should move from group to group to keep students on task and model good scientific argumentation. The teacher can ask the presenter(s) questions such as "Why did you decide to analyze the available data like that?" or "Were there any data that did not fit with your claim?" to encourage students to use empirical criteria to evaluate the quality of the arguments. The teacher can also ask the presenter(s) to explain how the claim they are presenting fits with the theories, laws, or models of science or to explain why the evidence they used is important. In addition, the teacher can ask the students who are listening to the presentation questions such as "Do you think their analysis is accurate?" or "Do you think their interpretation is sound?" or even "Do you think their claim fits with what we know about forces and motion?" These questions can serve to remind students to use empirical and theoretical criteria to evaluate an argument during the discussions. Overall, it is the goal of the teacher at this stage of the lesson to encourage students to think about how they know what they know and why some claims are more valid or acceptable in science. This stage of the model, however, is not the time to tell the students that they are right or wrong.

At the end of the argumentation session, it is important to give the students time to meet with their original group so they can discuss what they learned by interacting with individuals from the other groups and they can revise their initial arguments. This process can begin with the presenter(s) sharing the critiques and the suggestions for improvement that they heard during the argumentation session. The students who visited the other groups during the argumentation can then share their ideas for making the arguments better based on what they observed and discussed at other stations. Students often realize that the way they collected or analyzed data was flawed in some way at this point in the process. The teacher should therefore encourage students to collect new data or reanalyze the data they collected as needed. Teachers can also give students time to conduct additional tests of ideas or claims. At the end of this stage, each group should have a final argument that is much better than their initial one.

Stage 5: Explicit and Reflective Discussion

The teacher should lead a whole-class explicit and reflective discussion during stage 5 of ADI. The intent of this discussion is to give students an opportunity to think about and share what they know and how they know it. This stage enables the classroom teacher to ensure that all students understand the DCI and CCs they used during the investigation. It is also encourages students to think about ways to improve their participation in scientific practices such as planning and carrying out investigations, analyzing and interpreting data, and arguing from evidence. At this point in the instructional sequence, the teacher should also encourage students to think about one or two nature of scientific knowledge or nature of scientific inquiry concepts. It is important to stress that an explicit and reflective discussion is not a lecture; it is an opportunity for students to think about important ideas and practices and to share what they know or do not understand. The more students talk during this stage, the more meaningful the experience will be for them and the more a teacher can learn about student thinking.

Teachers should begin the discussion by asking students to share what they know about the DCI and the CCs they used to figure things out during the lab (the DCI and CCs can be found in the "Your Task" section of the Lab Handout). The teacher can give several images as prompts and then ask students questions to encourage students to think about how these ideas or concepts helped them explain the phenomenon under investigation and how they used these ideas or concepts to provide a justification of the evidence in their arguments. The teacher should not tell the students what results they should have obtained or what information should be included in each argument. Instead, the teachers should focus on the students' thoughts about the DCI and CCs by providing a context for students to share their views and explain their thinking. Remember, this stage of ADI is a *discussion*, not a lecture. We provide recommendations about what teachers can do and the types of questions that teachers can ask to facilitate a productive discussion about the DCI and CCs during this stage as part of the teacher notes for each lab investigation.

Next, the teacher should encourage the students to think about what they learned about the practices of science and how to design better investigations in the future. This is important because students are expected to design their own investigations, decide how to analyze and interpret data, and support their claims with evidence in every ADI lab investigation. These practices are complex, and students cannot be expected to master them without being given opportunities to try, fail, and then learn from their mistakes. To encourage students to learn from their mistakes during a lab, students must have an opportunity to reflect on what went well and what went wrong during their investigation. The teacher should therefore encourage the students to think about what they did during their investigation, how they choose to analyze and interpret data, how they decide to argue from evidence, and what they could do better. The teacher can then use the students' ideas to highlight what does and does not count as quality or rigor in science and to offer advice about ways to improve in the future. Over time, students will gradually improve their abilities to participate in the practices of science as they learn what works and what does not. To help facilitate this process, in the Teacher Notes for each ADI lab investigation we provide questions that teachers can ask students to help elicit their ideas about the practices of science and set goals for future investigations.

The teacher should end this stage with an explicit discussion of one or two aspects of the nature of scientific knowledge (NOSK) or nature of scientific inquiry (NOSI), using what the students did during the investigation to help illustrate these important concepts (NGSS Lead States 2013). This stage provides a golden opportunity for explicit instruction about NOSK and how this knowledge develops over time in a context that is meaningful to the students. For example, teachers can use the lab as a way to illustrate the differences between

- observations and inferences,
- data and evidence, and
- theories and laws.

Teachers can also use the lab investigation as a way to illustrate NOSI. For example, teachers might discuss

- how the culture of science, societal needs, and current events influence the work of scientists;
- the wide range of methods that scientists can use to collect data;
- what does and does not count as an experiment in science; and
- the role that creativity and imagination play during an investigation.

Recent research suggests that students only develop an appropriate understanding of the nature of science when teachers discuss these concepts in an *explicit* fashion (Abd-El-Khalick and Lederman 2000; Lederman and Lederman 2004; Schwartz, Lederman, and Crawford 2004). In addition, by embedding a discussion of NOSK and NOSI into each lab investigation, teachers can highlight these important concepts over and over again throughout the school year rather than just focusing on them during a single unit. This type of approach makes it easier for students to learn these abstract and sometimes counterintuitive concepts. As part of the Teacher Notes for each lab investigation, we provide recommendations about which concepts to focus on and examples of questions that teachers can ask to facilitate a productive discussion about these concepts during this stage of the instructional sequence.

Stage 6: Write an Investigation Report

Stage 6 is included in the ADI model because writing is an important part of doing science. Scientists must be able to read and understand the writing of others as well as evaluate its worth. They also must be able to share the results of their own research through writing. In addition, writing helps students learn how to articulate their thinking in a clear and concise manner, encourages metacognition, and improves student understanding of the content (Wallace, Hand, and Prain 2004). Finally, and perhaps most important, writing makes each student's thinking visible to the teacher (which facilitates assessment) and enables the teacher to provide students with the educative feedback they need to improve.

In stage 6, each student is required to write an individual investigation report using his or her group's argument. The report should be centered on three fundamental questions:

1. What question were you trying to answer and why?
2. What did you do to answer your question and why?
3. What is your argument?

Teachers should encourage students to use tables or graphs to help organize their evidence and require them to reference this information in the body of the report. Stage 6 is important because it allows them to learn how to construct an explanation, argue from evidence, and communicate information. It also enables students to master the disciplinary-based writing skills outlined in the *Common Core State Standards* in English Language Arts (*CCSS ELA;* NGAC and CCSSO 2010). The report can be written during class or can be assigned as homework.

The format of the report is designed to emphasize the persuasive nature of science writing and to help students learn how to communicate in multiple modes (words, figures, tables, and equations). The three-question format is well aligned with the components of a traditional laboratory report (i.e., introduction, procedure, results and discussion) but allows students to see the important role argument plays in science. We strongly recommend that teachers *limit the length of the investigation* report to two double-spaced pages or one single-spaced page. This limitation encourages students to write in a clear and concise manner, because there is little room for extraneous information. This limitation is less intimidating than a more lengthy report requirement, and it lessens the work required in the subsequent stages.

Stage 7: Double-Blind Group Peer Review

During stage 7, each student is required to submit to the teacher one or more copies of his or her investigation report. We recommend that students bring in multiple copies of their report to make it easier for a group of students to review it at the same time; however, this is not a requirement if students are unable to bring in multiple copies of their reports. Instead of reading multiple copies of the same report as they review it, the group of reviewers can simply share a single copy of a report. Students should not place their names on the report before they turn it in to the teacher at the beginning of this stage; instead they should use an identification number to maintain anonymity—to ensure that reviews are based on the ideas presented and not the person presenting the ideas.

We recommend that teachers place students into groups of three to review the reports (these groups can be different from the groups that students worked in during stages 1–4). The teacher then gives each group a report written by a single student (or the multiple copies of the report submitted by a single student) and a peer-review guide and teacher scoring rubric (PRG/TSR). We included two versions of the PRG/TSR in Appendix 5 (p. 543): one version is designed for a high school course, and one is designed for an Advanced Placement (AP) course. The students in the group are then asked to review the report (or copies of the report) as a team using the PRG/TSR (see Figure 9, p. 16). The PRG/TSR contains specific criteria that are to be used by the group as they evaluate the quality of each section of investigation report as well as quality of the writing. There is also space for the reviewers to provide the author with feedback about how to improve the report. Once a group finishes reviewing a report as a team, they are given another report to review. When students are grouped together in threes, they only need to review three different reports. Be sure to give students only 15 minutes to review each set of reports (we recommend

FIGURE 9

A group of students reviewing a report written by a classmate using the peer-review guide and teacher scoring rubric

setting a timer to help manage time). When students are grouped into three and given 15 minutes to complete each review, the entire peer-review process can be completed in one 50-minute class period (3 different reports × 15 minutes = 45 minutes).

Reviewing each report as a group using the PRG/TSR is an important component of the peer-review process because it provides students with a forum to discuss "what counts" as high quality or acceptable and, in so doing, forces them to reach a consensus during the process. This method also helps prevent students from checking off "yes" for each criterion on the PRG/TSR without thorough consideration of the merits of the paper. It is also important for students to provide constructive and specific feedback to the author when areas of the paper are found to not meet the standards established by the PRG/TSR. The peer-review process provides students with an opportunity to read good and bad examples of the reports. This helps the students learn new ways to organize and present information, which in turn will help them write better on subsequent reports.

This stage of the model also gives students more opportunities to develop reading skills that are needed to be successful in science. Students must be able to determine the central ideas or conclusions of a text and determine the meaning of symbols, key terms, and other domain-specific words. In addition, students must be able to assess the reasoning and evidence that an author includes in a text to support his or her claim and compare or contrast findings presented in a text with those from other sources when they read a scientific text. Students can develop all these skills, as well as the other discipline-based reading standards found in the *CCSS ELA*, when they are required to read and critically review reports written by their classmates.

Stage 8: Revise and Submit the Report

The final stage in the ADI instructional model is to revise the report based on the suggestions given during the peer review. If the report met all the criteria, the student may simply submit the paper to the teacher with the original peer-reviewed "rough draft" and PRG/TSR attached, ensuring that his or her name replaces the identification number. Students whose reports are found by the peer-review group to be acceptable can maintain the option to revise it if they so desire after reviewing the work of other students. If a report was found to be unacceptable by the reviewers during the peer-review stage, the author is required to rewrite his or her report using the reviewers' comments and suggestions as a guideline. The author is also required to explain what he or she did to improve each section of the report in response to the reviewers' suggestions (or explain why he or she decided to ignore the reviewers' suggestions) in the author response section of the PRG/TSR.

Once the report is revised, it is turned in to the teacher for evaluation with the original rough draft and the PRG/TSR attached. The teacher can then provide a score on the PRG/TSR in the column labeled "Teacher Score" and use these ratings to assign an overall grade for the report. This approach provides students with a chance to improve their writing mechanics and develop their reasoning and understanding of the content. This process also offers students the added benefit of reducing academic pressure by providing support in obtaining the highest possible grade for their final product.

The PRG/TSR is designed to be used with any ADI lab investigation. Teachers, as a result, can use the same scoring rubric throughout the entire school year. This is beneficial for several reasons. First, the criteria for what counts as a high-quality report do not change from lab to lab. Students therefore quickly learn what is expected from them when they write a report and teachers do not have to spend valuable class time explaining the various components of the PRG/TSR each time they assign a report. Second, the PRG/TSR makes it clear what components of a report need to be improved next time, because the grade is not based on a holistic evaluation of the report. Students, as a result, can see which aspects of their writing are strong and which aspects need improvement. Finally, and perhaps most important, PRG/TSR provides teachers with a standardized measure of student performance that can be compared over multiple reports across semesters. Teachers can therefore track improvement over time.

The Role of the Teacher During Argument-Driven Inquiry

If the ADI instructional model is to be successful and student learning is to be optimized, the role of the teacher during a lab activity designed using this model must be different than the teacher's role during a more traditional lab. The teacher *must* act as a resource for the students, rather than as a director, as students work through each stage of the activity; the teacher must encourage students to think about *what they are doing* and *why they made that decision* throughout the process. This encouragement should take the form of probing questions that teachers ask as they walk around the classroom, such as "Why do you want to set up your equipment that way?" or "What type of data will you need to collect to be able to answer that question?"

Teachers must restrain themselves from telling or showing students how to "properly" conduct the investigation. However, teachers must emphasize the need to maintain high standards for a scientific investigation by requiring students to use rigorous standards for "what counts" as a good method or a strong argument in the context of science.

Finally, and perhaps most important for the success of an ADI activity, teachers must be willing to let students try and fail, and then help them learn from their mistakes. Teachers should not try to make the lab investigations included in this book "student-proof" by providing additional directions to ensure that students do everything right the first time. We have found that students often learn more from an ADI lab activity when they design a flawed method to collect data during stage 2 or analyze their results in an inappropriate manner during stage 3, because their classmates quickly point out these mistakes during the argumentation session (stage 4) and it leads to more teachable moments.

Because the teacher's role in an ADI lab is different from what typically happens in laboratories, we've provided a chart describing teacher behaviors that are consistent and inconsistent with each stage of the instructional model (see Table 1). This table is organized by stage because what the students and the teacher need to accomplish during each stage is different. It might be helpful to keep this table handy as a guide when you are first attempting to implement the lab activities found in the book.

TABLE 1

Teacher behaviors during the stages of the ADI instructional model

Stage	What the teacher does that is …	
	Consistent with ADI model	**Inconsistent with ADI model**
1: Identify the task and the guiding question	• Sparks students' curiosity • "Creates a need" for students to design and carry out an investigation • Organizes students into collaborative groups of three or four • Supplies students with the materials they will need • Holds a "tool talk" to show students how to use equipment or to illustrate proper technique • Reviews relevant safety precautions and protocols • Provides students with hints • Allows students to tinker with the equipment they will be using later	• Provides students with possible answers to the research question • Tells students that there is one correct answer • Provides a list of vocabulary terms or explicitly describes the content addressed in the lab
2: Design a method and collect data	• Encourages students to ask questions as they design their investigations • Asks groups questions about their method (e.g., "Why did you do it this way?") and the type of data they expect from that design • Reminds students of the importance of specificity when completing their investigation proposals	• Gives students a procedure to follow • Does not question students about the method they design or the type of data they expect to collect • Approves vague or incomplete investigation proposals
3: Develop an initial argument	• Reminds students of the research question and what counts as appropriate evidence in science • Requires students to generate an argument that provides and supports a claim with genuine evidence • Asks students what opposing ideas or rebuttals they might anticipate • Provides related theories and reference materials as tools	• Requires only one student to be prepared to discuss the argument • Moves to groups to check on progress without asking students questions about why they are doing what they are doing • Does not interact with students (uses the time to catch up on other responsibilities) • Tells students the right answer
4: Argumentation session	• Reminds students of appropriate behaviors in the learning community • Encourages students to ask questions of peers • Keeps the discussion focused on the elements of the argument • Encourages students to use appropriate criteria for determining what does and does not count	• Allows students to negatively respond to others • Asks questions about students' claims before other students can ask • Allows students to discuss ideas that are not supported by evidence • Allows students to use inappropriate criteria for determining what does and does not count

Continued

TABLE 1 (*continued*)

5: Explicit and reflective discussion	• Encourages students to discuss what they learned about the content and how they know what they know • Encourages students to discuss what they learned about the nature of scientific knowledge and the nature of scientific inquiry • Encourages students to think of ways to be more productive next time	• Provides a lecture on the content • Skips over the discussion about the nature of scientific knowledge and the nature of scientific inquiry to save time • Tells students "what they should have learned" or "this is what you all should have figured out"
6: Write an investigation report	• Reminds students about the audience, topic, and purpose of the report • Provides the peer-review guide in advance • Provides an example of a good report and an example of a bad report	• Has students write only a portion of the report • Moves on to the next activity/topic without providing feedback
7: Double-blind group peer review	• Reminds students of appropriate behaviors for the review process • Ensures that all groups are giving a quality and fair peer review to the best of their ability • Encourages students to remember that while grammar and punctuation are important, the main goal is an acceptable scientific claim with supporting evidence and justification • Holds the reviewers accountable	• Allows students to make critical comments about the author (e.g., "This person is stupid") rather than their work (e.g., "This claim needs to be supported by evidence") • Allows students to just check off "Yes" on each item without providing a critical evaluation of the report
8: Revise and submit the report	• Requires students to edit their reports based on the reviewers' comments • Requires students to respond to the reviewers' ratings and comments • Has students complete the Checkout Questions after they have turned in their report	• Allows students to turn in a report without a completed peer-review guide • Allows students to turn in a report without revising it first

References

Abd-El-Khalick, F., and N. G. Lederman. 2000. Improving science teachers' conceptions of nature of science: A critical review of the literature. *International Journal of Science Education* 22: 665–701.

Duschl, R. A., H. A. Schweingruber, and A. W. Shouse, eds. 2007. *Taking science to school: Learning and teaching science in grades K–8*. Washington, DC: National Academies Press.

Grooms, J., Enderle, P., and Sampson, V. 2015. Coordinating scientific argumentation and the *Next Generation Science Standards* through argument driven inquiry. *Science Educator* 24 (1): 45–50.

Lederman, N. G., and J. S. Lederman. 2004. Revising instruction to teach the nature of science. *The Science Teacher* 71 (9): 36–39.

National Governors Association Center for Best Practices and Council of Chief State School Officers (NGAC and CCSSO). 2010. *Common core state standards*. Washington, DC: NGAC and CCSSO.

National Research Council (NRC). 2012. *A framework for K–12 science education: Practices, crosscutting concepts, and core ideas*. Washington, DC: National Academies Press.

NGSS Lead States. 2013. *Next Generation Science Standards: For states, by states*. Washington, DC: National Academies Press. *www.nextgenscience.org/next-generation-science-standards*.

Schwartz, R. S., N. Lederman, and B. Crawford. 2004. Developing views of nature of science in an authentic context: An explicit approach to bridging the gap between nature of science and scientific inquiry. *Science Education* 88: 610–645.

Wallace, C., B. Hand, and V. Prain, eds. 2004. *Writing and learning in the science classroom*. Boston: Kluwer Academic Publishers.

CHAPTER 2
Lab Investigations

This book includes 23 physics lab investigations designed around the argument-driven inquiry (ADI) instructional model. Please note that these investigations are not designed to replace an existing curriculum, but to transform the laboratory component of a science course. A teacher can use these investigations as a way to introduce students to a new concept related to a disciplinary core idea (DCI) at the beginning of a unit (introduction labs) or as a way to give students an opportunity to apply a specific concept that they learned about earlier in class in a novel situation (application labs). All of the labs, however, were designed to give students an opportunity to learn how to use DCIs, crosscutting concepts (CCs), and science and engineering practices (SEPs) to figure things out. The 23 lab investigations have been aligned with the following sources to facilitate curriculum and lesson planning:

- *A Framework for K–12 Science Education* (see Standards Matrix A in Appendix 1, p. 513)
- Aspects of the nature of scientific knowledge (NOSK) and the nature of scientific inquiry (NOSI) (see Standards Matrix B in Appendix 1, p. 515)
- The *Next Generation Science Standards* (see Standards Matrix C in Appendix 1, p. 516)
- Advanced Placement (AP) Physics 1 science practices and big ideas (see Standards Matrix D in Appendix 1, p. 518)
- AP Physics C: Mechanics content areas and laboratory objectives (see Standards Matrix E in Appendix 1, p. 520)
- The *Common Core State Standards* in English language arts (*CCSS ELA*) (see Standards Matrix F in Appendix 1, p. 521)
- The *Common Core State Standards* in mathematics (*CCSS Mathematics*) (see Standards Matrix G in Appendix 1, p. 522)

Many of the ideas for the investigations in this book came from existing resources; however, we modified these existing activities to fit with the ADI instructional model. Once the ADI lab investigations were created, several practicing physicists reviewed them for content accuracy. Several different physics teachers then piloted the labs in their courses (including general and honors sections). After piloting, each lab investigation and all related instructional materials were revised based on the feedback of these teachers. The modified labs were then piloted and modified for a second time by other physics teachers at other locations and refined again. The final iteration of each lab is included in this book.

Research that has been conducted on ADI in classrooms indicates that students have much better inquiry and writing skills after participating in at least eight ADI investigations and make substantial gains in their understanding of DCIs, CCs, SEPs, NOSK, and NOSI (Grooms, Enderle, and Sampson 2015; Sampson et al. 2013). To learn more about the research associated with the ADI instructional model, visit *www.argumentdriveninquiry.com*.

Teacher Notes

Each teacher must decide when and how to use a laboratory experience to best support student learning. To help with these decisions, we have included Teacher Notes for each investigation. These notes include information about the purpose of the lab, the time needed to implement each stage of the model for that lab, the materials needed, hints for implementation, and connections to standards. In the following subsections, we will describe the information provided in each section of the Teacher Notes.

Purpose

This section describes the content of the lab and indicates whether the activity is designed to help students think about a new idea or think with a new idea. Labs that are designed to help students *think about* a new idea are called *introduction labs*. Introduction labs require students to explore potential cause-and-effect relationships or how things change over time. These labs are best used at the beginning of a unit of study. Labs that are designed to help students learn to *think with* a new idea are called *application labs*. Application labs require students to think with a new idea they are already familiar with to develop an explanation or to solve a problem. These labs are best used at the end of the unit of study. Please note that because of the nature of the ADI approach, in both cases very little emphasis needs to be placed on making sure the students "learn the vocabulary first" or "know the content" before the lab investigation begins. Instead, with the combination of the information provided in the Lab Handout and your students' evolving understanding of the DCIs, CCs, and SEPs, they will develop a better understanding of the content *as they work through the eight stages of ADI*. The purpose also highlights the NOSK or NOSI concepts that should be emphasized during the explicit and reflective discussion stage of the activity.

Underlying Physics Concepts

This section of the Teacher Notes provides a basic overview of the major concepts that the students will explore and or use during the investigation.

Timeline

Unlike most traditional laboratories, ADI labs typically take four or five days to complete. The amount of time it takes to complete each lab will vary depending on how long it takes to collect data and whether or not the students write in class or at home. The time associated with each ADI lab investigation may be longer in the first few labs your students conduct, but the time will grow shorter as your students become familiar with the practices used in the model (argumentation, designing investigations, writing reports). We therefore provide suggestions about which stages of ADI you should be able to complete in a 50-minute class period.

Materials and Preparation

This section describes the lab supplies (consumables and equipment) and instructional materials (e.g., Lab Handout, Investigation Proposal, and peer-review guide and teacher scoring rubric [PRG/TSR]) needed to implement the lab activity. The lab supplies are designed for one group; however, multiple groups can share if resources are scarce. Although we have included specific suggestions for some lab supplies based on our finding that these supplies worked best during the field tests, substitutions can be made if needed. Always be sure to test all lab supplies before conducting the lab with the students, because using new materials often has unexpected consequences.

This section also describes the setup that needs to be done *before* students can do the investigation. Please note that some of the labs may require preparation up to 24 hours in advance. Make sure to read this section at least two days before you plan to have the students in your class start the investigation.

Safety Precautions

This section provides an overview of potential safety hazards as well as safety protocols that should be followed to make the laboratory safer for students. These are based on legal safety standards and current safety practices. Teachers should also review and follow all local polices and protocols used within their school district and/or school (e.g., the district chemical hygiene plan, Board of Education safety policies).

Laboratory waste disposal is discussed in this section if it is relevant to the lab.

Topics for the Explicit and Reflective Discussion

This section begins with an overview of the DCI and CCs that students use to figure things out during the lab. We provide advice about ways to encourage students to think about how these ideas or concepts helped them explain the phenomenon under investigation and how they used these ideas or concepts to provide a justification of the evidence in their arguments. The section also provides some advice for teachers about how to encourage students to reflect on ways to improve the design of their investigation in the future. This section concludes with an overview of the relevant NOSK and NOSI concepts to discuss during the explicit and reflective discussion and some sample questions that teachers can pose to help students be reflective about what they know about these concepts.

Hints for Implementing the Lab

These labs have been tested by many teachers many times. As a result, we have collected hints from the teachers for each stage of the ADI process. These hints are designed to help you avoid some of the pitfalls earlier teachers have experienced and to make the investigation run smoothly. The section also includes tips for making the investigation safer.

Connections to Standards

This section is designed to inform curriculum and lesson planning by highlighting how the investigation can be used to address performance expectations from the *NGSS* (where applicable); learning objectives from AP Physics 1; learning objectives from AP Physics C: Mechanics; *CCSS ELA*; and *CCSS Mathematics*.

Instructional Materials

The instructional materials included in this book are reproducible copy masters that are designed to support students as they participate in an ADI lab activity. The materials needed for each lab include a Lab Handout, the PRG/TSR, and a set of Checkout Questions. Some labs also require an investigation proposal. In the following subsections, we will provide an overview of these important materials.

Lab Handout

At the beginning of each lab activity, each student should be given a copy of the Lab Handout. This handout provides information about the phenomenon that they will investigate and a guiding question for the students to answer. The handout also provides hints for students to help them design their investigation in the "Getting Started" section, information about what to include in their initial argument, and the requirements for the investigation report. The last part of the Lab Handout provides space for students to keep track of critiques, suggestions for improvement, and good ideas that arise during the argumentation session.

Peer-Review Guide and Teacher Scoring Rubric

The PRG/TSR is designed to make the criteria that are used to judge the quality of an investigation report explicit. We recommend that teachers make one copy for each student and provide it to the students before they begin writing their investigation report. This will ensure that students understand how they will be evaluated. Then during the double-blind group peer-review stage of the model, each group should fill out the peer-review guide as they review the reports of their classmates. (Each group will need to review at least three reports.) The reviewers should rate the report on each criterion and then provide advice to the author about ways to improve. Once the review is complete, the author needs to revise his or her report and respond to the reviewers' rating and comments in the appropriate sections in the PRG/TSR. The PRG/TSR should be submitted to the instructor along with the first draft and the final report for a final evaluation. To score the report, the teacher can simply fill out the "Teacher Score" column of the PRG/TSR and then total the scores.

Checkout Questions

To facilitate formative assessment inside the classroom, we have included a set of Checkout Questions for each lab investigation. The questions target the key ideas and the NOSK and NOSI concepts that are addressed in the lab. Students should complete the Checkout Questions on the same day they turn in their final report. One handout is needed for each student. The students should complete these questions on their own. The teacher can then use the students' responses, along with the report, to determine if the students learned what they needed to during the lab, and then reteach as needed.

Investigation Proposal

To help students design better investigations, we have developed and included three different types of investigation proposals in this book, with a short version and a long version of each type. These investigation proposals are optional, but we have found that students design and carry out much better investigations when they are required to fill out a proposal and then get teacher feedback about their method before they begin. We provide recommendations about which investigation proposal (A, B, or C) to use for a particular lab as part of the Teacher Notes. If a teacher decides to use an investigation proposal as part of a lab activity, we recommend providing one copy for each group. The lab handout for students also has a heading labeled "Investigation Proposal Required?" that is followed by "Yes" and "No" check boxes. The teacher should be sure to have students check the appropriate box on the Lab Handout when introducing the lab activity.

References

Grooms, J., Enderle, P, and Sampson, V. 2015. Coordinating scientific argumentation and the *Next Generation Science Standards* through argument driven inquiry. *Science Educator* 24 (1): 45–50.

National Governors Association Center for Best Practices and Council of Chief State School Officers (NGAC and CCSSO). 2010. *Common core state standards.* Washington, DC: NGAC and CCSSO.

National Research Council (NRC). 2012. *A framework for K–12 science education: Practices, crosscutting concepts, and core ideas.* Washington, DC: National Academies Press.

NGSS Lead States. 2013. *Next Generation Science Standards: For states, by states.* Washington, DC: National Academies Press. *www.nextgenscience.org/next-generation-science-standards.*

Sampson, V., Enderle, P., Grooms, J., and Witte, S. 2013. Writing to learn and learning to write during the school science laboratory: Helping middle and high school students develop argumentative writing skills as they learn core ideas. *Science Education* 97 (5): 643–670.

SECTION 2
Motion and Interactions

Kinematics

Introduction Labs

LAB 1

Teacher Notes

Lab 1. Acceleration and Velocity: How Does the Direction of Acceleration Affect the Velocity of an Object?

Purpose

The purpose of this lab is to *introduce* students to the core idea of forces and motion, part of the disciplinary core idea (DCI) of Motion and Stability: Forces and Interactions from the *NGSS*, by having them explore the relationship between acceleration and velocity. This lab also gives students an opportunity to learn about the crosscutting concepts (CCs) of (a) Patterns and (b) Scale, Proportion, and Quantity from the *NGSS*. In addition, this lab can be used to help students understand two big ideas from AP Physics: (a) the interactions of an object with other objects can be described by forces and (b) interactions between systems can result in changes in those systems. As part of the explicit and reflective discussion, students will also learn about (a) the difference between observations and inferences in science and (b) the nature and role of experiments in science.

Underlying Physics Concepts

Acceleration is the rate of change of velocity with respect to time, and the formula for acceleration is shown in Equation 1.1, where **a** is acceleration; **v** is velocity; and t is time. In SI units, acceleration is measured in meters per second squared (m/s²); velocity is measured in meters per second (m/s); and time is measured in seconds (s). For the instantaneous acceleration at any time, t, the acceleration is the derivative of the velocity with respect to time.

$$\textbf{(Equation 1.1)} \quad \mathbf{a} = \Delta\mathbf{v}/\Delta t$$

As a rate of change, acceleration is visible as the slope of an object's velocity versus time graph. The average acceleration over any time period is shown by the slope of the line between the two points on an acceleration versus time graph. The instantaneous acceleration is also visible on a velocity versus time graph. In this case, the instantaneous acceleration is the slope of the tangent line to the graph at a specific time. An analysis of the graph's slope using the familiar *rise over run* reveals it has units of velocity divided by units of time. Velocity and acceleration are *vectors* and are described by a magnitude and direction. The steepness of the slope represents the *magnitude* of the acceleration, and the sign of the slope represents its *direction*.

The reference frame of the system, chosen by whoever is observing it, dictates the positive and negative directions for all further discussion of the object's motion. An object moving in the positive direction can be said to have *positive velocity*. *Negative velocity* occurs when an object is moving in the opposite direction. It is important to note, however, that the choice of positive direction is arbitrary, so what one observer calls positive velocity may be called negative by a different observer. To avoid confusion, physicists have established sign conventions when studying motion. Typically, motion in the right, upward, north, or east direction is given a positive sign convention. Motion in the left, downward, south, or west direction is given a negative sign convention.

When a velocity graph increases (i.e., when the slope is positive), the acceleration is positive and is said to be pointing in the positive direction. This occurs whenever the velocity is getting more positive. This can occur when the initial velocity is in the positive direction, and we would say the object is "speeding up" or "getting faster." More formally, we say the magnitude of the velocity is increasing. A graph of this situation is shown in Figure 1.1.

A velocity versus time graph also contains information regarding the position of the object with respect to its starting position. From an algebraic standpoint, the change in position is equal to the area between the time axis and the line. In Figure 1.1, the total area between the line and the time axis is positive, so relative to the initial position the object has moved in the positive direction. For students in calculus, the position is the integral of the velocity function with respect to time. From a purely descriptive standpoint, the object represented by Figure 1.1 moves in the positive direction for the duration of the motion.

A positive acceleration can also occur when the object has an initial velocity in the negative direction. In this case, we would say the object is "slowing down." More formally, we say that initially the magnitude of the velocity is decreasing, as shown in Figure 1.2. It is important to point out that in Figure 1.2, the graph passes through the *x*-axis. This means that at some point in time, the object comes to a rest and then moves in the positive direction.

With regard to the position of the object represented in Figure 1.2, the position is again the area between the line and the time axis. Descriptively, the position of the object starts at the initial position and then moves with a decreasing magnitude of the velocity in the negative direction. At the time when the velocity function crosses the time axis, the object comes to rest. The object then reverses direction and moves in the positive direction. However, the displacement of the object remains negative until the object has moved an equal amount in the positive direction. When the area between the line and the time axis is equal to zero, the object is at the initial position. As shown in Figure 1.2, the area under the curve is equal to zero when time is equal to zero and at the end of the motion. We can also use our

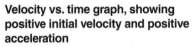

FIGURE 1.1

Velocity vs. time graph, showing positive initial velocity and positive acceleration

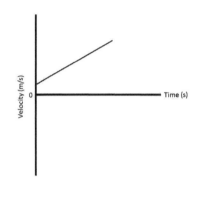

FIGURE 1.2

Velocity vs. time graph, showing negative initial velocity and positive acceleration

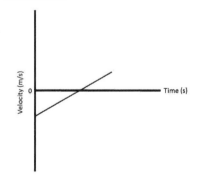

LAB 1

FIGURE 1.3

Position vs. time graph corresponding to the velocity vs. time graph in Figure 1.2

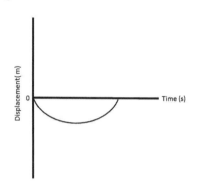

FIGURE 1.4

Velocity vs. time graph, showing positive initial velocity and negative acceleration

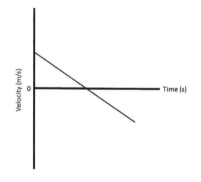

FIGURE 1.5

Velocity vs. time graph, showing negative initial velocity and negative acceleration

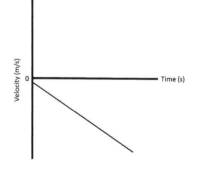

information from Figure 1.2 to graph a position versus time graph, shown in Figure 1.3.

Notice how the displacement is negative for the entirety of the motion shown in Figure 1.3, despite the velocity having a positive value for part of the time. For this object, the object begins with an initial velocity in the negative direction; moves in the negative direction while it slows down; comes to rest; reverses direction and moves in the positive direction (i.e., positive velocity) while its position is still negative with respect to its initial position; and finally returns to the initial position with a positive velocity.

Similarly, a negative acceleration occurs when the velocity is positive and slowing down *or* when the velocity is negative and speeding up (i.e., becoming more negative). These two cases are shown in Figures 1.4 and 1.5, respectively.

Acceleration can also be observed qualitatively on a position versus time graph. Since the slope of the graph represents velocity, accelerating motion yields a curved position graph. Positive acceleration (i.e., increasing velocity) corresponds to an increasing slope. This can appear as a section of an upward-opening parabola and is referred to as "concave up" position graph. A "concave down" position graph depicts negative acceleration and can resemble a section of a downward-opening parabola.

Timeline

The instructional time needed to complete this lab investigation is 220–280 minutes. Appendix 3 (p. 531) provides options for implementing this lab investigation over several class periods. Option A (280 minutes) should be used if students are unfamiliar with scientific writing, because this option provides extra instructional time for scaffolding the writing process. You can scaffold the writing process by modeling, providing examples, and providing hints as students write each section of the report. Option A should also be used if students are unfamiliar with any of the data collection and analysis tools. Option B (220 minutes) should be used if students are familiar with scientific writing and have developed the skills needed to write an investigation report on their own. In option B, students complete stage 6 (writing the investigation report) and stage 8 (revising the investigation report) as homework.

Materials and Preparation

The materials needed to implement this investigation are listed in Table 1.1. The equipment can be purchased from a science supply company such as Carolina, Flinn Scientific, PASCO, Vernier, or Ward's Science. The data collection and analysis software and/or the video analysis software can be purchased from Vernier (Logger *Pro*) or PASCO (SPARKvue or Capstone). These companies also have apps that can be used on Apple- or Android-based tablets and cell phones. We recommend consulting with your school's information technology coordinator to determine the best option for your students.

TABLE 1.1

Materials list for Lab 1

Item	Quantity
Safety glasses or goggles	1 per student
Kinematic cart with fan attachment	1 per group
Track for cart	1 per group
Motion detector/sensor	1 per group
Interface for motion detector/sensor (if used)	1 per group
Video camera	1 per group
Computer or tablet with data collection and analysis software (for use with the motion detector/sensor) or video analysis software (for use with the video camera)	1 per group
Meterstick (for use with the video camera)	1 per group
Investigation Proposal C (optional)	1 per group
Whiteboard, 2' × 3'*	1 per group
Lab Handout	1 per student
Peer-review guide and teacher scoring rubric	1 per student
Checkout Questions	1 per student

* As an alternative, students can use computer and presentation software such as Microsoft PowerPoint or Apple Keynote to create their arguments.

Be sure to use a set routine for distributing and collecting the materials during the lab investigation. One option is to set up the materials for each group at each group's lab station before class begins. This option works well when there is a dedicated section of the classroom for lab work and the materials are large and difficult to move (such as a dynamics track). A second option is to have all the materials on a table or cart at a central

LAB 1

location. You can then assign a member of each group to be the "materials manager." This individual is responsible for collecting all the materials his or her group needs from the table or cart during class and for returning all the materials at the end of the class. This option works well when the materials are small and easy to move (such as stopwatches, metersticks, or beanbags). It also makes it easy to inventory the materials at the end of the class before students leave for the day.

Safety Precautions

Remind students to follow all normal lab safety rules. In addition, tell students to take the following safety precautions:

1. Wear sanitized safety glasses or goggles during lab setup, hands-on activity, and takedown.
2. Keep fingers and toes out of the way of moving objects.
3. Do not place fingers into the fan.
4. Keep hair, clothing, and jewelry away from the fan attachment while it is switched on.
5. Wash hands with soap and water after completing the lab.

Topics for the Explicit and Reflective Discussion
Reflecting on the Use of Core Ideas and Crosscutting Concepts During the Investigation

Teachers should begin the explicit and reflective discussion by asking students to discuss what they know about the core idea they used during the investigation. The following are some important concepts related to the core idea of forces and motion that students need to use to determine a mathematical relationship between velocity and acceleration:

- Quantities such as position, displacement, distance, velocity, speed, and acceleration are used to describe the motion of an object.
- Displacement, velocity, and acceleration are vector quantities.
- *Displacement* is a change in position. *Velocity* is the rate of change of position over a period of time. *Acceleration* is the rate of change in velocity over a period of time.
- A reference frame is needed to determine the direction and the magnitude of the displacement, velocity, and acceleration of an object.

To help students reflect on what they know about forces and motion, we recommend showing them two or three images using presentation software that help illustrate these important ideas. You can then ask the students the following questions to encourage them to share how they are thinking about these important concepts:

1. What do we see going on in this image?

2. Does anyone have anything else to add?

3. What might be going on that we can't see?

4. What are some things that we are not sure about here?

You can then encourage students to think about how CCs played a role in their investigation. There are at least two CCs that students need to use to determine a mathematical relationship between velocity and acceleration: (a) Patterns and (b) Scale, Proportion, and Quantity (see Appendix 2 [p. 527] for a brief description of these CCs). To help students reflect on what they know about these CCs, we recommend asking them the following questions:

1. Why is it important to look for patterns during an investigation?

2. What patterns did you identify during your investigation? What did the identification of these patterns allow you to do?

3. Why is it important keep track of changes in a system quantitatively during an investigation?

4. What did you keep track of quantitatively during your investigation? What did that allow you to do?

You can then encourage the students to think about how they used all these different concepts to help answer the guiding question and why it is important to use these ideas to help justify their evidence for their final arguments. Be sure to remind your students to explain why they included the evidence in their arguments and make the assumptions underlying their analysis and interpretation of the data explicit in order to provide an adequate justification of their evidence.

Reflecting on Ways to Design Better Investigations

It is important for students to reflect on the strengths and weaknesses of the investigation they designed during the explicit and reflective discussion. Students should therefore be encouraged to discuss ways to eliminate potential flaws, measurement errors, or sources of uncertainty in their investigations. To help students be more reflective about the design of their investigation and what they can do to make their investigations more rigorous in the future, you can ask them the following questions:

1. What were some of the strengths of the way you planned and carried out your investigation? In other words, what made it scientific?

2. What were some of the weaknesses of the way you planned and carried out your investigation? In other words, what made it less scientific?

LAB 1

3. What rules can we make, as a class, to ensure that our next investigation is more scientific?

Reflecting on the Nature of Scientific Knowledge and Scientific Inquiry

This investigation can be used to illustrate two important concepts related to the nature of scientific knowledge and the nature of scientific inquiry: (a) the difference between observations and inferences in science and (b) the nature and role of experiments in science (see Appendix 2 for a brief description of these two concepts). Be sure to review these concepts during and at the end of the explicit and reflective discussion. To help students think about these concepts in relation to what they did during the lab, you can ask them the following questions:

1. You had to make observations and inferences during your investigation. Can you give me some examples of these observations and inferences?

2. Can you work with your group to come up with a rule that you can use to decide if a piece of information is an observation or an inference? Be ready to share in a few minutes.

3. I asked you to design and carry out an experiment as part of your investigation. Can you give me some examples of what experiments are used for in science?

4. Can you work with your group to come up with a rule that you can use to decide if an investigation is an experiment or not? Be ready to share in a few minutes.

You can also use presentation software or other techniques to encourage your students to think about these concepts. You can show examples of information from the investigation that are either observations or inferences and ask students to classify each example and explain their thinking. You can also show images of different types of investigations (such as a person collecting data in the field as part of a descriptive study, a person working in the library doing a literature review, a person working on a computer to analyze an existing data set, and an actual experiment) and ask students to indicate if they think each image represents an experiment and why or why not.

Be sure to remind your students that, to become proficient in science, it is important that they understand what counts as scientific knowledge and how that knowledge develops over time.

Hints for Implementing the Lab

- Allowing students to design their own procedures for collecting data gives students an opportunity to try, to fail, and to learn from their mistakes. However, you can scaffold students as they develop their procedures by having them fill out an investigation proposal. These proposals provide a way for you to offer students hints and suggestions without telling them how to do it. You can also check the proposals quickly during a class period. For this lab we suggest using Investigation Proposal C.

- Learn how to use the cart and fan attachment before the lab begins. It is important for you to know how to use the equipment so you can help students when technical issues arise.

- Allow the students to become familiar with the cart and fan attachment as part of the tool talk before they begin to design their investigation. Giving them 5–10 minutes to examine the equipment will let them see what they can and cannot do with it.

- Be sure to allow students to go back and re-collect data at the end of the argumentation session. Students often realize that they made numerous mistakes when they were collecting data as a result of their discussions during the argumentation session. The students, as a result, will want a chance to re-collect data, and the re-collection of data should be encouraged when time allows. This also offers an opportunity to discuss what scientists do when they realize a mistake is made inside the lab.

If students use a motion detector/sensor

- Learn how to use the motion detector/sensor and the data collection and analysis software before the lab begins. It is important for you to know how to use the equipment and software so you can help students when technical issues arise.

- Allow the students to become familiar with the motion detector/sensor and data collection and analysis software as part of the tool talk before they begin to design their investigation. Giving them 5–10 minutes to examine the equipment and software will let them see what they can and cannot do with it.

If students use video analysis

- We suggest allowing students to familiarize themselves with the video analysis software before they finalize the procedure for the investigation, especially if they have not used such software previously. This gives students an opportunity to learn how to work with the software and to improve the quality of the video they take.

- Remind students to hold the video camera as still as possible. Any movement of the camera will introduce error into their analysis. If using actual camcorders, we recommend using a tripod to hold the camera steady. If students are using a camera on a cell phone or tablet, we recommend using a table to help steady the camera.

LAB 1

- Remind students to place a meterstick in the same field of view as the motion they are capturing with the video camera. Also, the meterstick should be approximately the same distance from the camera as the motion. Most video analysis software requires the user to define a scale in the video (this allows the software to establish distances and, subsequently, other variables dependent on distance and displacement).

Connections to Standards

Table 1.2 highlights how the investigation can be used to address learning objectives from AP Physics 1; learning objectives from AP Physics C: Mechanics; *Common Core State Standards,* in English language arts (*CCSS ELA*); and *Common Core State Standards,* Mathematics (*CCSS Mathematics*).

TABLE 1.2 _____

Lab 1 alignment with standards

***NGSS* performance expectations**	• None
AP Physics 1 learning objectives	• 3.A.1.1: The student is able to express the motion of an object using narrative, mathematical, and graphical representations. • 3.A.1.2: The student is able to design an experimental investigation of the motion of an object. • 3.A.1.3: The student is able to analyze experimental data describing the motion of an object and is able to express the results of the analysis using narrative, mathematical, and graphical representations. • 3.A.3.1: The student is able to analyze a scenario and make claims (develop arguments, justify assertions) about the forces exerted on an object by other objects for different types of forces or components of forces. • 3.B.1.2: The student is able to design a plan to collect and analyze data for motion (static, constant, or accelerating) from force measurements and carry out an analysis to determine the relationship between the net force and the vector sum of the individual forces.
AP Physics C: Mechanics learning objectives	• I.A.1.a: Students should understand the general relationships among position, velocity, and acceleration for the motion of a particle along a straight line. • I.A.1.b(1): Write down expressions for velocity and position as functions of time, and identify or sketch graphs of these quantities.
Literacy connections (*CCSS ELA*)	• *Reading:* Key ideas and details, craft and structure, integration of knowledge and ideas • *Writing:* Text types and purposes, production and distribution of writing, research to build and present knowledge, range of writing • *Speaking and listening:* Comprehension and collaboration, presentation of knowledge and ideas
Mathematics connections (*CCSS Mathematics*)	• *Mathematical practices:* Make sense of problems and persevere in solving them, reason abstractly and quantitatively, construct viable arguments and critique the reasoning of others, model with mathematics, use appropriate tools strategically, attend to precision, look for and express regularity in repeated reasoning • *Number and quantity:* Reason quantitatively and use units to solve problems, represent and model with vector quantities, perform operations on vectors • *Algebra:* Interpret the structure of expressions, understand solving equations as a process of reasoning and explain the reasoning, solve equations and inequalities in one variable, represent and solve equations and inequalities graphically • *Functions:* Understand the concept of a function and use function notation; interpret functions that arise in applications in terms of the context; analyze functions using different representations; construct and compare linear and exponential models and solve problems; interpret expressions for functions in terms of the situation they model • *Statistics and probability:* Summarize, represent, and interpret data on two categorical and quantitative variables; interpret linear models; make inferences and justify conclusions from sample surveys, experiments, and observational studies

LAB 1

Lab Handout

Lab 1. Acceleration and Velocity: How Does the Direction of Acceleration Affect the Velocity of an Object?

Introduction

The ability to describe motion is the basis of much of physics. To explain why objects move the way they do, we must first be able to describe how an object moves. *Velocity* and *acceleration* are terms often used to describe motion. These two terms, however, have different meanings in science. The rate of change in an object's position is called its velocity. An object's velocity describes how fast it travels and its direction of motion. The rate of change of velocity is called acceleration. Like position, velocity and acceleration are vector quantities with a magnitude (i.e., how much) and a direction (i.e., which way). Choosing a frame of reference when you study an object's motion is up to you. Your reference frame will specify which direction is the positive direction and which one is negative.

Graphs that show a change in position of an object over time (called a position vs. time graph) and the change in velocity of an object over time (called a velocity vs. time graph) can help us describe the motion of an object. On these types of graphs, the rate of change for the object of interest corresponds to the slope of the line. For example, the slope of the line on a position vs. time graph at a specific time is the velocity of an object at that time, because velocity is the rate of change of an object's position with respect to time. For a graph with a curved line, such as those shown in Figures L1.1 and L1.2, the slope of the line at a given point in time on the graph is defined as the slope of the tangent line to the curve at that specific point.

Not all rates of change remain constant, however. For example, it is possible for an object to increase or decrease in velocity. We can also represent this change on a motion graph. If the slope of the tangent line gets steeper or gets less steep downward as time increases, this is called a "concave up" graph (see Figure L1.1). A concave up position versus time graph indicates that an object's velocity is increasing over time. When

FIGURE L1.1

A concave up position vs. time graph indicates that the velocity of an object is increasing.

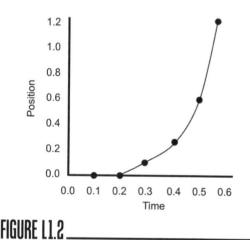

FIGURE L1.2

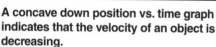

A concave down position vs. time graph indicates that the velocity of an object is decreasing.

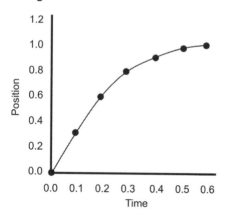

the slopes of the tangent lines decrease, in contrast, it is called a "concave down" graph. A concave down position versus time graph can either be getting less steep upward or becoming more steep downward (see Figure L1.2). Either way, a concave down position versus time graph indicates that the velocity of the object is decreasing over time.

Physicists use the term acceleration to describe any change in the velocity of an object. The acceleration of an object can be either positive or negative. The direction of that acceleration, however, is important because it may cause an object's velocity to change differently. In this investigation you will have an opportunity to explore the motion of a cart on a track in order to explain how the movement of that cart changes when it accelerates in different directions.

Your Task

You will observe the motion of a cart and then use what you know about vectors and graphs, patterns, and scale, proportion, and quantity to determine a mathematical relationship between velocity and acceleration.

The guiding question of this investigation is, *How does the direction of acceleration affect the velocity of an object?*

Materials

You may use any of the following materials during your investigation (some items may not be available):

- Safety glasses or goggles (required)
- Cart with fan attachment
- Track for cart
- Video camera
- Motion detector/sensor and interface
- Computer or tablet with data collection and analysis software and/or video analysis software
- Meterstick

Safety Precautions

Follow all normal lab safety rules. In addition, take the following safety precautions:

1. Wear sanitized safety glasses or goggles during lab setup, hands-on activity, and takedown.

2. Keep fingers and toes out of the way of moving objects.

3. Do not place fingers into the fan.

4. Keep hair, clothing, and jewelry away from the cart while the fan attachment is switched on.

5. Wash hands with soap and water after completing the lab.

LAB 1

Investigation Proposal Required?   ☐ Yes   ☐ No

Getting Started

To answer the guiding question, you will need to design and carry out an experiment using a fan cart and a track. The cart can be placed on the track so it will only move in two directions (i.e., left or right). You can use the fan to apply a force to the cart so it will accelerate in different directions. You have two options for tracking the motion of the cart on the track. The first option is to use a motion detector/sensor attached to the end of the track. The motion detector/sensor will allow you to record the exact position of the cart on the track at a given point in time and how the position of the cart relative to the motion detector/sensor changes on the track over time. Figure L1.3 shows how you can set up the equipment for this investigation if you use a motion detector/sensor to track the motion of the cart. The second option for tracking the motion of a cart is to use video analysis software. In this case, you will need to use a video camera and a meterstick. Place the meterstick in front of the track and then video record the cart as it moves. You can then upload the video to a computer or tablet and use video analysis software to examine the motion of the cart. The meterstick is important because it will provide a reference point in the video so you can measure the cart's displacement over time. Figure L.1.4 shows how you can set up the equipment for this investigation if you decide to use video analysis software to track the motion of the cart.

FIGURE L1.3 _____

How to examine the motion of a cart using a motion sensor

FIGURE L1.4 _____

How to examine the motion of a cart using a video camera

Before you can begin to design an experiment using this equipment, however, you will need to determine what type of data you need to collect, how you will collect it, and how you will analyze it. To determine what type of data you need to collect, think about the following questions:

- What are the boundaries and components of the system you are studying?
- Which factor(s) might control the rate of change in this system?
- How could you keep track of changes in this system quantitatively?

- What information do you need to determine the cart's velocity and acceleration?
- What will be the reference frame?
- Are your variables vector quantities or scalar quantities?

To determine *how you will collect the data*, think about the following questions:

- How will you change the direction of acceleration?
- What comparisons will you need to make?
- What measurement scale or scales should you use to collect data?
- How will you make sure that your data are of high quality (i.e., how will you reduce error)?
- For any vector quantities, which directions are positive and which directions are negative?

To determine *how you will analyze the data*, think about the following questions:

- How will you identify both the magnitude and direction of the vector quantities?
- What type of calculations will you need to make?
- How could you use mathematics to describe a change over time?
- How could you use mathematics to describe a relationship between variables?
- What types of patterns could you look for in your data?

Connections to the Nature of Scientific Knowledge and Scientific Inquiry

As you work through your investigation, you may want to consider

- the difference between observations and inferences in science, and
- the nature and role of experiments in science.

Initial Argument

Once your group has finished collecting and analyzing your data, your group will need to develop an initial argument. Your initial argument needs to include a claim, evidence supporting your claim, and a justification of the evidence. The claim is your group's answer to the guiding question. The evidence is an analysis and interpretation of your data. Finally, the justification of the evidence is why your group thinks the evidence matters. The justification of the evidence is important because scientists can use different kinds of evidence to support their claims. Your group will create your initial argument on a whiteboard. Your whiteboard should include all the information shown in Figure L1.5 (p. 46).

LAB 1

Argumentation Session

The argumentation session allows all of the groups to share their arguments. One or two members of each group will stay at the lab station to share that group's argument, while the other members of the group go to the other lab stations to listen to and critique the other arguments. This is similar to what scientists do when they propose, support, evaluate, and refine new ideas during a poster session at a conference. If you are presenting your group's argument, your goal is to share your ideas and answer questions. You should also keep a record of the critiques and suggestions made by your classmates so you can use this feedback to make your initial argument stronger. You can keep track of specific critiques and suggestions for improvement that your classmates mention in the space below.

Argument presentation on a whiteboard

The Guiding Question:	
Our Claim:	
Our Evidence:	Our Justification of the Evidence:

Critiques about our initial argument and suggestions for improvement:

If you are critiquing your classmates' arguments, your goal is to look for mistakes in their arguments and offer suggestions for improvement so these mistakes can be fixed. You should look for ways to make your initial argument stronger by looking for things that the other groups did well. You can keep track of interesting ideas that you see and hear during the argumentation in the space below. You can also use this space to keep track of any questions that you will need to discuss with your team.

Interesting ideas from other groups or questions to take back to my group:

Once the argumentation session is complete, you will have a chance to meet with your group and revise your initial argument. Your group might need to gather more data or design a way to test one or more alternative claims as part of this process. Remember, your goal at this stage of the investigation is to develop the best argument possible.

Report

Once you have completed your research, you will need to prepare an investigation report that consists of three sections. Each section should provide an answer to the following questions:

1. What question were you trying to answer and why?

2. What did you do to answer your question and why?

3. What is your argument?

Your report should answer these questions in two pages or less. This report must be typed, and any diagrams, figures, or tables should be embedded into the document. Be sure to write in a persuasive style; you are trying to convince others that your claim is acceptable or valid!

LAB 1

Lab 1. Acceleration and Velocity: How Does the Direction of Acceleration Affect the Velocity of an Object?

1. Two carts undergo positive acceleration, but their velocities are in opposite directions.

    a. Sketch each cart and label each one with arrows representing the directions of its velocity and its acceleration

    b. Sketch a single velocity versus time graph representing each cart.

    c. Sketch a single position versus time graph representing each cart.

2. Positive acceleration will cause an object to speed up.

   a. I agree with this statement.
   b. I disagree with this statement.

   Explain your answer, using an example from your investigation about the direction of acceleration and velocity.

3. *Observations* and *inferences* are terms that have the same meaning in science.

   a. I agree with this statement.
   b. I disagree with this statement.

   Explain your answer, using an example from your investigation about the direction of acceleration and velocity.

4. Scientists always design and carry out an experiment to answer scientific questions.

   a. I agree with this statement.
   b. I disagree with this statement.

   Explain your answer, using an example from your investigation about the direction of acceleration and velocity.

5. Why is it useful to identify patterns during an investigation? In your answer, be sure to include examples from at least two different investigations.

6. How are vector quantities and scalar quantities different in science? In your answer, be sure to include examples from at least two different investigations.

LAB 2

Teacher Notes

Lab 2. Acceleration and Gravity: What Is the Relationship Between the Mass of an Object and Its Acceleration During Free Fall?

Purpose

The purpose of this lab is to *introduce* students to the core idea of forces and motion, part of the disciplinary core idea (DCI) of Motion and Stability: Forces and Interactions from the *NGSS,* by having them explore the relationship between the mass of an object and its acceleration during free fall. This lab also gives students an opportunity to learn about the crosscutting concepts (CCs) of (a) Patterns and (b) Stability and Change from the *NGSS.* In addition, this lab can be used to help students understand three big ideas from AP Physics: (a) fields existing in space can be used to explain interactions, (b) interactions of an object with other objects can be described by forces, and (c) interactions between systems can result in changes in those systems. As part of the explicit and reflective discussion, students will also learn about (a) the difference between laws and theories in science and (b) the difference between data and evidence in science.

Underlying Physics Concepts

Acceleration is defined as the rate of change of velocity with respect to time. For objects in free fall, we can relate the distance an object falls, the time it takes to fall, and its acceleration using equation 2.1, below, where $\mathbf{y}$ is vertical displacement; $\mathbf{v}_0$ is initial velocity, t is time, and $\mathbf{a}$ is acceleration. In SI units, displacement is measured in meters (m); velocity in meters per second (m/s), time in seconds (s); and acceleration in meters per second squared (m/s2).

$$\textbf{(Equation 2.1)} \quad \mathbf{y} = \mathbf{v}_0 t + \tfrac{1}{2}\mathbf{a}t^2$$

When an object is dropped from rest, the initial velocity is zero (i.e., $\mathbf{v}_0 = 0$), so the equation simplifies to $\mathbf{y} = \tfrac{1}{2}\mathbf{a}t^2$. If we are to solve for the acceleration, we get Equation 2.2.

$$\textbf{(Equation 2.2)} \quad \mathbf{a} = 2\mathbf{y}/t^2$$

For those who use this lab in a calculus-based physics course, *acceleration* can also be defined as the second derivative of the position function with respect to time.

Regardless of an object's mass, the acceleration due to Earth's gravitational pull near the surface is constant (neglecting other forces, such as air resistance). Physicists use the lowercase letter $\mathbf{g}$ (instead of $\mathbf{a}$) to represent this acceleration in formulas (as in $\mathbf{g} = -9.8$ m/s^2).

Our discussion thus far does not mean that the gravitational force between Earth and any one object is constant. Using Newton's second law, we know that the force acting on an object is equal to the mass of that object times the object's acceleration. Therefore, the force between Earth and an object is a function of both the acceleration due to gravity (-9.8 m/s^2) and the falling object's mass. The equation for this is $\mathbf{F}_g = m\mathbf{g}$, where $\mathbf{F}_g$ is the gravitational force, m is the mass of the falling object, and $\mathbf{g}$ is the acceleration due to gravity (-9.8 m/s^2).

Because forces come in pairs, it is also important to note that when an object is in free fall, the object accelerates toward Earth *and* Earth accelerates toward the object. However, because Earth is so massive and because the greatest gravitational influence on Earth is the Sun, we can ignore the movement of Earth. (*Note:* We often ignore many factors as we explore different topics in high school physics courses. For example, we often ignore the role of air resistance on falling objects because this effect is small for small objects when compared with the effect of Earth's gravity. Or, we ignore the effect of the electric force on a charged sphere from a single electron placed near it and focus on what the charged sphere does to the electron.)

Finally, it is important to note that the gravitational acceleration constant, $\mathbf{g} = -9.8$ m/s^2, only applies to objects on Earth. The value for $\mathbf{g}$ on other planets is different. This is because $\mathbf{g}$ is derived from Newton's law of universal gravity, $\mathbf{g} = GM/\mathbf{r}^2$, where $\mathbf{g}$ is the acceleration due to gravity on a body (planet, star, moon, etc.), G is the universal gravity constant (6.67×10^{-11} N·m^2/kg^2), M is the mass of the large body (planet, star, moon, etc.), and $\mathbf{r}$ is the radius of the large body. Thus, $\mathbf{g}$ on the Moon is -1.63 m/s^2 and on Mars is -3.75 m/s^2.

Timeline

The instructional time needed to complete this lab investigation is 220–280 minutes. Appendix 3 (p. 531) provides options for implementing this lab investigation over several class periods. Option A (280 minutes) should be used if students are unfamiliar with scientific writing, because this option provides extra instructional time for scaffolding the writing process. You can scaffold the writing process by modeling, providing examples, and providing hints as students write each section of the report. Option A can also be used if you are introducing students to the video analysis programs. Option B (220 minutes) should be used if students are familiar with scientific writing and have developed the skills needed to write an investigation report on their own. In option B, students complete stage 6 (writing the investigation report) and stage 8 (revising the investigation report) as homework.

Materials and Preparation

The materials needed to implement this investigation are listed in Table 2.1 (p. 54). Most of the equipment can be purchased from a science supply company such as Carolina, Flinn Scientific, PASCO, Vernier, or Ward's Science. Video analysis software can be purchased from Vernier (Logger *Pro*) or PASCO (SPARKvue or Capstone). These companies also have apps that can be used on Apple- or Android-based tablets and cell phones. We recommend consulting with your school's information technology coordinator to determine the best option for your students.

You will need to prepare your own beanbags before the first class period for this investigation. The beanbags must all be the same size, but each beanbag must be a different mass. The bags should be about 25 cm × 25 cm in size. You can make your own beanbags by sewing two pieces of fabric together or by using large sealable plastic bags. We recommend filling the beanbags with either plastic beads or rice. The plastic beads can be purchased from a retail craft store such as Michaels or Wal-Mart. Prepare the beanbags for this investigation as follows:

- Beanbag A: Add 120 ml (½ cup) of plastic beads (or rice) and seal the bag.
- Beanbag B: Add 240 ml (1 cup) of plastic beads (or rice) and seal the bag.
- Beanbag C: Add 360 ml (1½ cups) of plastic beads (or rice) and seal the bag.
- Beanbag D: Add 480 ml (2 cups) of plastic beads (or rice) and seal the bag.
- Beanbag E: Add 600 ml (2½ cups) of plastic beads (or rice) and seal the bag.

TABLE 2.1

Materials list for Lab 2

Item	Quantity
Safety glasses or goggles	1 per student
Beanbags A–E	1 of each type per group
Meterstick	1 per group
Stopwatch	1 per group
Electronic or triple beam balance	1 per group
Masking tape	As needed
Investigation Proposal C (optional)	1 per group
Whiteboard, 2' × 3'*	1 per group
Lab Handout	1 per student
Peer-review guide and teacher scoring rubric	1 per student
Checkout Questions	1 per student
Equipment for video analysis (optional)	
Video camera	1 per group
Computer or tablet with video analysis software	1 per group

*As an alternative, students can use computer and presentation software such as Microsoft PowerPoint or Apple Keynote to create their arguments.

Be sure to use a set routine for distributing and collecting the materials during the lab investigation. One option is to set up the materials for each group at each group's lab station before class begins. This option works well when there is a dedicated section of the classroom for lab work and the materials are large and difficult to move (such as a dynamics track). A second option is to have all the materials on a table or cart at a central location. You can then assign a member of each group to be the "materials manager." This individual is responsible for collecting all the materials his or her group needs from the table or cart during class and for returning all the materials at the end of the class. This option works well when the materials are small and easy to move (such as stopwatches, metersticks, or beanbags). It also makes it easy to inventory the materials at the end of the class before students leave for the day.

Safety Precautions

Remind students to follow all normal lab safety rules. In addition, tell the students to take the following safety precautions:

1. Wear sanitized safety glasses or goggles during lab setup, the hands-on activity, and takedown.

2. Do not throw the beanbags.

3. Do not stand on tables or chairs.

4. Wash hands with soap and water after completing the lab.

Topics for the Explicit and Reflective Discussion

Reflecting on the Use of Core Ideas and Crosscutting Concepts During the Investigation

Teachers should begin the explicit and reflective discussion by asking students to discuss what they know about the core idea they used during the investigation. The following are some important concepts related to the core idea of forces and motion that students need to use to determine the relationship between mass and acceleration due to gravity:

- Gravity is an attractive force between two objects that have mass.

- Objects accelerate toward the center of Earth when in free fall because of the gravitational force between the object and Earth.

- *Displacement* is a change in position. *Velocity* is the rate of change of position over a period of time. *Acceleration* is the rate of change in velocity over a period of time.

- *Air resistance* is the result of an object moving through a layer of air and colliding with air molecules. The force of air resistance acting on a falling object is therefore dependent on the velocity and the cross-sectional surface area of the falling object.

To help students reflect on what they know about forces and motion, we recommend showing them two or three images using presentation software that help illustrate these important ideas. You can then ask the students the following questions to encourage them to share how they are thinking about these important concepts:

1. What do we see going on in this image?

2. Does anyone have anything else to add?

3. What might be going on that we can't see?

4. What are some things that we are not sure about here?

You can then encourage students to think about how CCs played a role in their investigation. There are at least two CCs that students need to use to determine a mathematical relationship between velocity and acceleration: (a) Patterns and (b) Stability and Change (see Appendix 2 [p. 527] for a brief description of these CCs). To help students reflect on what they know about these CCs, we recommend asking them the following questions:

1. Why is it important to look for patterns during an investigation?

2. What patterns did you identify during your investigation? What did the identification of these patterns allow you to do?

3. Why is it important to think about what controls or affects the rate of change in a system?

4. Which factor(s) might have controlled the rate of change in the velocity (i.e., acceleration) of a falling object in your investigation? What did exploring these factors allow you to do?

You can then encourage the students to think about how they used all these different concepts to help answer the guiding question and why it is important to use these ideas to help justify their evidence for their final arguments. Be sure to remind your students to explain why they included the evidence in their arguments and make the assumptions underlying their analysis and interpretation of the data explicit in order to provide an adequate justification of their evidence.

Reflecting on Ways to Design Better Investigations

It is important for students to reflect on the strengths and weaknesses of the investigation they designed during the explicit and reflective discussion. Students should therefore be encouraged to discuss ways to eliminate potential flaws, measurement errors, or sources of uncertainty in their investigations. To help students be more reflective about the design of their investigation and what they can do to make their investigations more rigorous in the future, you can ask them the following questions:

1. What were some of the strengths of the way you planned and carried out your investigation? In other words, what made it scientific?

2. What were some of the weaknesses of the way you planned and carried out your investigation? In other words, what made it less scientific?

3. What rules can we make, as a class, to ensure that our next investigation is more scientific?

Reflecting on the Nature of Scientific Knowledge and Scientific Inquiry

This investigation can be used to illustrate two important concepts related to the nature of scientific knowledge and the nature of scientific inquiry: (a) the difference between laws and theories in science and (b) the difference between data and evidence in science (see Appendix 2 for a brief description of these two concepts). Be sure to review these concepts during and at the end of the explicit and reflective discussion. To help students think about these concepts in relation to what they did during the lab, you can ask them the following questions:

1. Laws and theories are different in science. Is the finding that the acceleration due to gravity is constant near Earth's surface (i.e., all objects experience the same acceleration due to gravity during free fall regardless of their mass) an example of a law or a theory? Why?

2. Can you work with your group to come up with a rule that you can use to decide if something is a law or a theory? Be ready to share in a few minutes.

3. You had to talk about data and evidence during your investigation. Can you give me some examples of data and evidence from your investigation?

4. Can you work with your group to come up with a rule that you can use to decide if a piece of information is data or evidence? Be ready to share in a few minutes.

You can also use presentation software or other techniques to encourage your students to think about these concepts. You can show examples of either a law (such as $\mathbf{g} = GM/\mathbf{r}^2$) or a theory (such as *gravity is the curvature of four-dimensional space-time due to the presence of mass*) and ask students to indicate if they think it is a law or a theory and why. You can also show examples of information from the investigation that are either data or evidence and ask students to classify each example and explain their thinking.

Be sure to remind your students that, to become proficient in science, it is important that they understand what counts as scientific knowledge and how that knowledge develops over time.

LAB 2

Hints for Implementing the Lab

- As one of the early labs during the semester, this lab allows students to develop lab skills related to making accurate measurements. Because most concepts in physics ask how physical systems develop or change over time, learning how to keep accurate time is particularly important.

- This lab is unique in that it is designed for students to get a null result. That is, the answer to the guiding question is that there is no relationship between the mass of an object and its acceleration during free fall. This lab intentionally presents students with the null result as it teaches them that sometimes in science an experiment leads to no relationship.

- Your students may not be used to investigations that produce a null result. As such, they may try to force a relationship when there isn't one, or they will claim there must be a problem with their experimental design or the equipment that leads to the null result. This is acceptable. Chances are, one group or a few groups will accept the null result and then, during the argumentation session, other groups will "come around" to this view as well.

- Allowing students to design their own procedures for collecting data gives students an opportunity to try, to fail, and to learn from their mistakes. However, you can scaffold students as they develop their procedure by having them fill out an investigation proposal. These proposals provide a way for you to offer students hints and suggestions without telling them how to do it. You can also check the proposals quickly during a class period. For this lab we suggest using Investigation Proposal C.

- Allow the students to become familiar with the equipment and materials as part of the tool talk before they begin to design their investigation. Giving them 5–10 minutes to examine the equipment and materials will let them see what they can and cannot do with them.

- If students calculate acceleration by measuring the time with a stopwatch, they will get a value for **g** much larger than −9.8 m/s^2. This is acceptable, because they will still get a very similar value for each object. The purpose of the lab is for students to realize that mass does not affect the acceleration of an object due to gravity. This also provides a great opportunity to discuss the nature of scientific knowledge and the nature of scientific inquiry. One important topic for students to think about related to the nature of scientific inquiry is the limits to the precision of our equipment and the constraints that these limits place on resultant data. Furthermore, technological change, such as computers and video cameras, allows us to become more and more precise in our measurements.

- Be sure to allow students to go back and re-collect data at the end of the argumentation session. Students often realize that they made numerous mistakes when they were collecting data as a result of their discussions during the argumentation session. The students, as a result, will want a chance to re-collect data, and the re-collection of data should be encouraged when time allows. This also offers an opportunity to discuss what scientists do when they realize a mistake is made inside the lab.

If students use video analysis

- We suggest allowing students to familiarize themselves with the video analysis software before they finalize the procedure for the investigation, especially if they have not used such software previously. This gives students an opportunity to learn how to work with the software and to improve the quality of the video they take.

- Remind students to hold the video camera as still as possible. Any movement of the camera will introduce error into their analysis. If using actual camcorders, we recommend using a tripod to hold the camera steady. If students are using a camera on a cell phone or tablet, we recommend using a table to help steady the camera.

- Remind students to place a meterstick in the same field of view as the motion they are capturing with the video camera. Also, the meterstick should be approximately the same distance from the camera as the motion. Most video analysis software requires the user to define a scale in the video (this allows the software to establish distances and, subsequently, other variables dependent on distance and displacement).

Connections to Standards

Table 2.2 (p. 60) highlights how the investigation can be used to address learning objectives from AP Physics 1; learning objectives from AP Physics C: Mechanics, *Common Core State Standards,* in English language arts (*CCSS ELA*); and *Common Core State Standards,* Mathematics (*CCSS Mathematics*).

LAB 2

TABLE 2.2 _____

Lab 2 alignment with standards

NGSS performance expectations	• None
AP Physics 1 learning objectives	• 1.C.1.1: The student is able to design an experiment for collecting data to determine the relationship between the net force exerted on an object, its inertial mass, and its acceleration. • 3.A.1.1: The student is able to express the motion of an object using narrative, mathematical, and graphical representations. • 3.A.1.2: The student is able to design an experimental investigation of the motion of an object. • 3.A.1.3: The student is able to analyze experimental data describing the motion of an object and is able to express the results of the analysis using narrative, mathematical, and graphical representations. • 3.A.3.1: The student is able to analyze a scenario and make claims (develop arguments, justify assertions) about the forces exerted on an object by other objects for different types of forces or components of forces. • 3.B.1.2: The student is able to design a plan to collect and analyze data for motion (static, constant, or accelerating) from force measurements and carry out an analysis to determine the relationship between the net force and the vector sum of the individual forces. • 4.A.2.1: The student is able to make predictions about the motion of a system based on the fact that acceleration is equal to the change in velocity per unit time, and velocity is equal to the change in position per unit time.
AP Physics C: Mechanics learning objectives	• I.A.1.b(1): Write down expressions for velocity and position as functions of time, and identify or sketch graphs of these quantities. • I.A.1.b(2): Use the equations $\mathbf{v} = \mathbf{v}_0 + \mathbf{a}t$; $x = x_0 + \mathbf{v}_0 t + \frac{1}{2}\mathbf{a}t^2$ and $\mathbf{v}^2 = \mathbf{v}_0^2 + 2\mathbf{a}(x - x_0)$ to solve problems involving one-dimensional motion with constant acceleration.
Literacy connections (*CCSS ELA*)	• *Reading:* Key ideas and details, craft and structure, integration of knowledge and ideas • *Writing:* Text types and purposes, production and distribution of writing, research to build and present knowledge, range of writing • *Speaking and listening:* Comprehension and collaboration, presentation of knowledge and ideas

Mathematics connections (*CCSS Mathematics*)	• *Mathematical practices:* Make sense of problems and persevere in solving them, reason abstractly and quantitatively, construct viable arguments and critique the reasoning of others, use appropriate tools strategically, attend to precision, look for and express regularity in repeated reasoning • *Number and quantity:* Reason quantitatively and use units to solve problems, represent and model with vector quantities, perform operations on vectors • *Algebra:* Interpret the structure of expressions, understand solving equations as a process of reasoning and explain the reasoning, solve equations and inequalities in one variable, represent and solve equations and inequalities graphically • *Functions:* Understand the concept of a function and use function notation, interpret functions that arise in applications in terms of the context, analyze functions using different representations, interpret expressions for functions in terms of the situation they model • *Statistics and probability:* Summarize, represent, and interpret data on two categorical and quantitative variables; understand and evaluate random processes underlying statistical experiments; make inferences and justify conclusions from sample surveys, experiments, and observational studies

Lab Handout

Lab 2. Acceleration and Gravity: What Is the Relationship Between the Mass of an Object and Its Acceleration During Free Fall?

Introduction

The motion of an object is the result of all the different forces that are acting on the object. If you push a toy car across the floor, it moves in the direction you pushed it. If the car then hits a wall, the force of the wall causes the car to stop. Applying a push or a pull to an object is an example of a contact force, where one object applies a force to another object through direct contact. There are other types of forces that can act on objects that do not involve objects touching. For example, a strong magnet can pull on a paper clip and make it move without ever actually touching the paper clip. Another example is static electricity. Static electricity in a rubber balloon can cause a person's hair to stand up without the balloon actually touching any of his or her hair. Magnetic forces and electrical forces are therefore called non-contact forces. Perhaps the most common non-contact force is gravity. Gravity is a force of attraction between two objects; the force due to gravity always works to bring objects closer together.

Any two objects that have mass (and, remember, all matter has mass) will also experience a gravitational force of attraction between them. Consider the Sun, the Earth, and the Moon as examples. The Earth and the Moon are very large and have a lot of mass; the force of gravity between the Earth and the Moon is strong enough to keep the Moon orbiting the Earth even though they are very far apart. Similarly, the force of gravity between the Sun and the Earth is strong enough to keep the Earth in orbit around the Sun, despite the Earth and the Sun being millions of miles apart. The force of gravity between two objects depends on the amount of mass of each object and how far apart they are. Objects that are more massive produce a greater gravitational force. The force of gravity between two objects also weakens as the distance between the two objects increases. So even though the Earth and the Sun are very far apart from each other (which means less gravity), the fact that they are both very massive (which means more gravity) results in a gravitational force that is strong enough to keep the Earth in orbit.

The gravitational force that acts between any two objects, as noted earlier, can cause one of those objects to move. For example, a cell phone in free fall (see Figure L2.1) moves toward the center of the Earth because of gravity. Scientists describe the motion of an object in free fall by describing its velocity and acceleration. Velocity is the speed (distance in a specific amount of time) of an object in a given direction. Acceleration is the rate of change in velocity per unit time, most often the rate of change in velocity per second. The amount of force required to produce a specific acceleration in the motion of an object, such as a cell phone, depends on the mass of that object. Therefore, as the mass of an object increases, so does the amount of force that is needed to produce a specific acceleration.

In this investigation you will have an opportunity to explore the relationship between the mass of an object and its acceleration due to gravity during free fall. Many people think that heavier objects accelerate toward the ground faster than lighter ones because gravity will act on heavier objects with more force. Heavier objects will therefore have a greater acceleration because of the force of gravity. Other people, however, think that heavier objects have more inertia (the tendency of an object to resist changes in its motion) so heavier objects will be less responsive to the force of gravity and have a smaller acceleration. Still others think that mass of a falling object has no effect on acceleration due to gravity because the magnitude of the force of gravity acting on a falling object is dependent on the mass of that object. In this case, the greater force of gravity for a more massive object is countered by the greater inertia of the massive object, thereby resulting in an acceleration due to gravity that is unaffected by the falling object's mass.

FIGURE L2.1

A cell phone in free fall

Unfortunately, it is challenging to determine which of these three explanations is the most valid because objects encounter air resistance as they fall. Air resistance is the result of an object moving through a layer of air and colliding with air molecules. The more air molecules that an object collides with, the greater the air resistance force. Air resistance is therefore dependent on the velocity of the falling object and the cross-sectional surface area of the falling object. Since heavier objects are often larger than lighter ones (consider a bowling ball and a marble as an example), it is often difficult to design a fair test of these three explanations. To determine the relationship between mass and acceleration due to gravity, you will therefore need to design an experiment that will allow you to control for the influence of air resistance.

Your Task

Use what you know about forces and motion, patterns, and rates of change to design and carry out an experiment to determine the relationship between mass and acceleration due to gravity.

The guiding question of this investigation is, *What is the relationship between the mass of an object and its acceleration during free fall?*

LAB 2

Materials

You may use any of the following materials during your investigation:

- Safety glasses or goggles (required)
- Beanbags A, B, C, D, and/or E
- Meterstick
- Stopwatch
- Electronic or triple beam balance
- Masking tape

If you have access to the following equipment, you may also consider using a video camera and a computer or tablet with video analysis software.

Safety Precautions

Follow all normal lab safety rules. In addition, take the following safety precautions:

1. Wear sanitized safety glasses or goggles during lab setup, hands-on activity, and takedown.

2. Do not throw the beanbags.

3. Do not stand on tables or chairs.

4. Wash hands with soap and water after completing the lab.

Investigation Proposal Required?   ☐ Yes   ☐ No

Getting Started

To answer the guiding question, you will need to design and conduct an experiment as part of your investigation. To accomplish this task, you must determine what type of data you need to collect, how you will collect it, and how you will analyze it.

To determine *what type of data you need to collect*, think about the following questions:

- What information will you need to be able to determine the acceleration of a falling object?

- Which factor(s) might control the rate of change in the velocity (i.e., acceleration) of a falling object?

- What will be the independent variable and the dependent variable for your experiment?

- Will you measure acceleration directly or will you have to calculate it using other measurements?

To determine *how you will collect the data*, think about the following questions:

- What variables will need to be controlled and how will you control them?

- How many tests will you need to run to have reliable data (to make sure it is consistent)?
- How will you make sure that your data are of high quality (i.e., how will you reduce error)?
- How will you keep track of the data you collect and how will you organize it?

To determine *how you will analyze the data*, think about the following questions:

- How will you calculate the acceleration of a falling object?
- What types of patterns might you look for as you analyze the data you collected?
- What type of calculations will you need to make to take into account multiple trials?
- What types of graphs or tables could you create to help make sense of your data?

Connections to the Nature of Scientific Knowledge and Scientific Inquiry

As you work through your investigation, you may want to consider

- the difference between laws and theories in science, and
- the difference between data and evidence in science.

Initial Argument

Once your group has finished collecting and analyzing your data, your group will need to develop an initial argument. Your initial argument needs to include a claim, evidence to support your claim, and a justification of the evidence. The claim is your group's answer to the guiding question. The evidence is an analysis and interpretation of your data. Finally, the justification of the evidence is why your group thinks the evidence matters. The justification of the evidence is important because scientists can use different kinds of evidence to support their claims. Your group will create your initial argument on a whiteboard. Your whiteboard should include all the information shown in Figure L2.2.

FIGURE L2.2

Argument presentation on a whiteboard

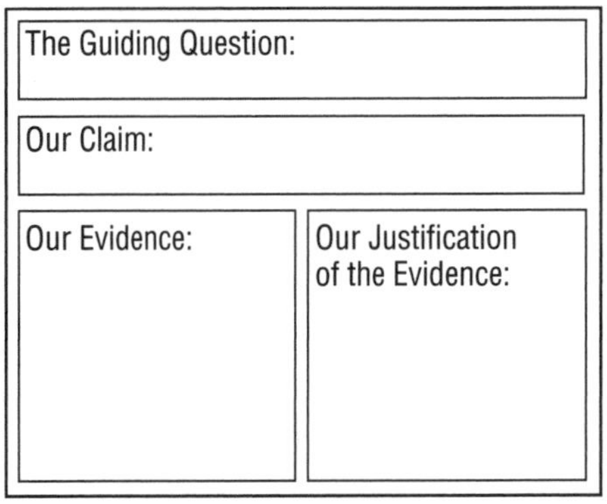

The Guiding Question:

Our Claim:

Our Evidence:

Our Justification of the Evidence:

Argumentation Session

The argumentation session allows all of the groups to share their arguments. One or two members of each group will stay at the lab station to share that group's argument, while the other members of the group go to the other lab stations to listen to and critique the other arguments. This is similar to what scientists do when they propose, support, evaluate, and refine new ideas during a poster session at a conference. If you are presenting your group's argument, your goal is to share your ideas and answer questions. You should also keep a record of the critiques and suggestions made by your classmates so you can use this feedback to make your initial argument stronger. You can keep track of specific critiques and suggestions for improvement that your classmates mention in the space below.

Critiques about our initial argument and suggestions for improvement:

If you are critiquing your classmates' arguments, your goal is to look for mistakes in their arguments and offer suggestions for improvement so these mistakes can be fixed. You should look for ways to make your initial argument stronger by looking for things that the other groups did well. You can keep track of interesting ideas that you see and hear during the argumentation in the space below. You can also use this space to keep track of any questions that you will need to discuss with your team.

Interesting ideas from other groups or questions to take back to my group:

Once the argumentation session is complete, you will have a chance to meet with your group and revise your initial argument. Your group might need to gather more data or design a way to test one or more alternative claims as part of this process. Remember, your goal at this stage of the investigation is to develop the best argument possible.

Report

Once you have completed your research, you will need to prepare an investigation report that consists of three sections. Each section should provide an answer to the following questions:

1. What question were you trying to answer and why?

2. What did you do to answer your question and why?

3. What is your argument?

Your report should answer these questions in two pages or less. This report must be typed, and any diagrams, figures, or tables should be embedded into the document. Be sure to write in a persuasive style; you are trying to convince others that your claim is acceptable or valid!

Checkout Questions

Lab 2. Acceleration and Gravity: What Is the Relationship Between the Mass of an Object and Its Acceleration During Free Fall?

For questions 1–3, assume air resistance can be neglected.

1. The picture at right shows a tall building. A physics class is investigating the time it takes for different objects that are dropped out of a window to fall to the ground. They release four objects from rest. Object A (mass = 5 kg) and object B (mass = 10 kg) are dropped from level 8. Object C (mass = 5 kg) and object D (mass = 10 kg) are dropped from level 4. What is the order in which they hit the ground?

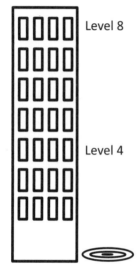

Level 8

Level 4

   a. Object D, then object C, then object B, then object A

   b. Object D, then objects C and B at the same time, then object A.

   c. Object D, then object B, then object C, then object A

   d. Objects C and D at the same time, then objects A and B at the same time

   How do you know?

2. Determine the amount of time that object A would take to fall if level 8 is 50 meters above the ground.

3. How long would it take object C to hit the ground if level 4 is half as high as level 8?

4. A theory turns into a law once it has been proven to be true.

   a. I agree with this statement.

   b. I disagree with this statement.

   Explain your answer, using an example from your investigation about the relationship between the mass of an object and the acceleration due to gravity during free fall.

5. *Data* and *evidence* are terms that have the same meaning in science.

    a. I agree with this statement.

    b. I disagree with this statement.

Explain your answer, using an example from your investigation about the relationship between the mass of an object and the acceleration due to gravity during free fall.

6. Why is it useful to understand the factors that control rates of change during an investigation? In your answer, be sure to include examples from at least two different investigations.

7. Why is it useful to identify patterns during an investigation? In your answer, be sure to include examples from at least two different investigations.

Teacher Notes

Lab 3. Projectile Motion: How Do Changes to the Launch Angle, the Initial Velocity, and the Mass of a Projectile Affect Its Hang Time?

Purpose

The purpose of this lab is to *introduce* students to the core idea of forces and motion, part of the disciplinary core idea (DCI) of Motion and Stability: Forces and Interactions from the *NGSS*, by having them explore projectile motion. This lab also gives students an opportunity to learn about the crosscutting concepts (CCs) of (a) Patterns and (b) Cause and Effect: Mechanism and Explanation from the *NGSS*. In addition, this lab can be used to help students understand two big ideas from AP Physics: (a) fields existing in space can be used to explain interactions and (b) the interactions of an object with other objects can be described by forces. As part of the explicit and reflective discussion, students will also learn about (a) the difference between observations and inferences in science and (b) the nature and role of experiments in science.

Underlying Physics Concepts

The hang time of a projectile launched from ground level can be described by Equation 3.1, where hang time, t, is dependent on the initial velocity, v_0, and the angle of release, θ. In SI units, time is measured in seconds (s); velocity in meters per second (m/s); and the angle of release in either radians (rad) or degrees (°). The constant g corresponds to the acceleration due to gravity and is equal to 9.8 m/s^2. For mathematical purposes, we choose to write g as a positive number in this lab. We do this because using regression techniques to mathematically model the data gathered in this investigation will result in a positive coefficient, which is a function of g.

$$\text{(Equation 3.1)} \qquad t = \frac{2v_0}{g} \sin\theta$$

Although it is unlikely that the introductory physics student will collect data and derive Equation 3.1 from analysis, the relationship that t depends on v_0 and θ is important. If the student holds θ constant, the student will find a linear relationship between t and v_0. Conversely, if the student holds v_0 constant, a linear relationship between t and $\sin\theta$ will be found. It is also important for students to discover that hang time does not depend on the mass of the object because the only force acting on the projectile is the force of gravity (see the Teacher Notes for Lab 2 for a further discussion of the relationship between the mass of an object and its acceleration due to gravity).

Another potential factor that might affect the hang time in the investigation is the height above the ground the projectile is launched from. As the height from which the projectile is launched increases, the hang time of the projectile will also increase. Some students may also explore this relationship, but it is not required given the question for this lab.

For students with strong prior knowledge in mathematics, particularly calculus, the derivation of Equation 3.1 is a useful exercise. Start with the given information that the only force acting on an object in flight is gravity in the vertical direction. Thus, $\mathbf{a} = -\mathbf{g}$ and $\mathbf{v}_{0y} = \mathbf{v}\sin\theta$. The change in the vertical position of the object can be found using Equation 3.2, where $\mathbf{y}$ is the vertical displacement, $\mathbf{a}$ is the acceleration, and t is the time. In SI units, displacement is measured in meters (m).

$$\textbf{(Equation 3.2)} \quad \Delta\mathbf{y} = \mathbf{v}_{0y}t + \frac{1}{2}\mathbf{a}t^2$$

Assuming the projectile is launched and falls back at the same height, $\Delta\mathbf{y} = 0$, so Equation 3.2 can be rewritten as Equation 3.3.

$$\textbf{(Equation 3.3)} \quad 0 = \mathbf{v}_0(\sin\theta)t + \frac{1}{2}(-\mathbf{g})t^2$$

Factoring out t, $0 = t(\mathbf{v}_0\sin\theta - \frac{1}{2}\mathbf{g}t)$. This equation is useful in showing the time when the projectile is at ground level. Thus, the equation is satisfied when t equal 0 seconds (i.e., before the projectile is launched) and when the quantity $\mathbf{v}_0\sin\theta - \frac{1}{2}\mathbf{g}t$ is also equal to zero. Setting $\mathbf{v}_0\sin\theta = \frac{1}{2}\mathbf{g}t$ and solving for t yields our result, namely, $t = \frac{2\mathbf{v}_0}{\mathbf{g}}\sin\theta$.

Timeline

The instructional time needed to complete this lab investigation is 220–280 minutes. Appendix 3 (p. 531) provides options for implementing this lab investigation over several class periods. Option A (280 minutes) should be used if students are unfamiliar with scientific writing, because this option provides extra instructional time for scaffolding the writing process. You can scaffold the writing process by modeling, providing examples, and providing hints as students write each section of the report. Option A can also be used if you are introducing students to the video analysis programs. Option B (220 minutes) should be used if students are familiar with scientific writing and have developed the skills needed to write an investigation report on their own. In option B, students complete stage 6 (writing the investigation report) and stage 8 (revising the investigation report) as homework.

LAB 3

Materials and Preparation

The materials needed to implement this investigation are listed in Table 3.1. The consumables and equipment can be purchased from a science supply company such as Carolina, Flinn Scientific, PASCO, Vernier, or Ward's Science. Video analysis software can be purchased from Vernier (Logger *Pro*) or PASCO (SPARKvue or Capstone). These companies also have apps that can be used on Apple- or Android-based tablets and cell phones. We recommend consulting with your school's information technology coordinator to determine the best option for your students.

TABLE 3.1

Materials list for Lab 3

Item	Quantity
Consumable	
Tape (roll)	1 per group
Equipment	
Safety glasses or goggles	1 per student
Marble launcher	1 per group
Marbles (different masses and sizes)	Several per group
Protractor	1 per group
Tape measure	1 per group
Stopwatch	1 per student
Investigation Proposal C (optional)	1 per group
Whiteboard, 2' × 3'*	1 per group
Lab Handout	1 per student
Peer-review guide and teacher scoring rubric	1 per student
Checkout Questions	1 per student
Equipment for video analysis (optional)	
Video camera	1 per group
Computer or tablet with video analysis software	1 per group

* As an alternative, students can use computer and presentation software such as Microsoft PowerPoint or Apple Keynote to create their arguments.

Be sure to use a set routine for distributing and collecting the materials during the lab investigation. One option is to set up the materials for each group at each group's lab

station before class begins. This option works well when there is a dedicated section of the classroom for lab work and the materials are large and difficult to move (such as a dynamics track). A second option is to have all the materials on a table or cart at a central location. You can then assign a member of each group to be the "materials manager." This individual is responsible for collecting all the materials his or her group needs from the table or cart during class and for returning all the materials at the end of the class. This option works well when the materials are small and easy to move (such as stopwatches, metersticks, or marbles). It also makes it easy to inventory the materials at the end of the class before students leave for the day.

Safety Precautions

Remind students to follow all normal lab safety rules. In addition, tell students to take the following safety precautions:

1. Wear sanitized safety glasses or goggles during lab setup, hands-on activity, and takedown.

2. Do not throw marbles or launch them at people.

3. Do not stand on tables or chairs.

4. Pick up marbles or other materials on the floor immediately to avoid a trip, slip, or fall hazard.

5. Wash hands with soap and water after completing the lab.

Topics for the Explicit and Reflective Discussion
Reflecting on the Use of Core Ideas and Crosscutting Concepts During the Investigation

Teachers should begin the explicit and reflective discussion by asking students to discuss what they know about the core idea they used during the investigation. The following are some important concepts related to the core idea of forces and motion that students need to use to determine which variable has the biggest impact on the hang time of a projectile:

- Quantities such as position, displacement, distance, velocity, speed, and acceleration are used to describe the motion of an object.

- Displacement, velocity, and acceleration are vector quantities.

- *Displacement* is a change in position. *Velocity* is the rate of change of position over a period of time. *Acceleration* is the rate of change in velocity over a period of time.

- The only force acting on a projectile in motion is gravity (if we ignore air resistance).

- Every projectile experiences air resistance, but its effect is negligible in this case so it can be ignored.

To help students reflect on what they know about forces and motion, we recommend showing them two or three images using presentation software that help illustrate these important ideas. You can then ask the students the following questions to encourage them to share how they are thinking about these important concepts:

1. What do we see going on in this image?

2. Does anyone have anything else to add?

3. What might be going on that we can't see?

4. What are some things that we are not sure about here?

You can then encourage students to think about how CCs played a role in their investigation. There are at least two CCs that students need to use to determine which variable has the biggest impact on the hang time of a projectile: (a) Patterns and (b) Cause and Effect: Mechanism and Explanation (see Appendix 2 [p. 527] for a brief description of these CCs). To help students reflect on what they know about these CCs, we recommend asking them the following questions:

1. Why is it important to look for patterns during an investigation?

2. What patterns did you identify during your investigation? What did the identification of these patterns allow you to do?

3. Why is it important to identify causal relationships in science?

4. What did you have to do during your investigation to determine if a factor did or did not cause a change in the hang time of a projectile? Why was that useful to do?

You can then encourage the students to think about how they used all these different concepts to help answer the guiding question and why it is important to use these ideas to help justify their evidence for their final arguments. Be sure to remind your students to explain why they included the evidence in their arguments and make the assumptions underlying their analysis and interpretation of the data explicit in order to provide an adequate justification of their evidence.

Reflecting on Ways to Design Better Investigations

It is important for students to reflect on the strengths and weaknesses of the investigation they designed during the explicit and reflective discussion. Students should therefore be encouraged to discuss ways to eliminate potential flaws, measurement errors, or sources of uncertainty in their investigations. To help students be more reflective about the design of their investigation and what they can do to make their investigations more rigorous in the future, you can ask them the following questions:

1. What were some of the strengths of the way you planned and carried out your investigation? In other words, what made it scientific?

2. What were some of the weaknesses of the way you planned and carried out your investigation? In other words, what made it less scientific?

3. What rules can we make, as a class, to ensure that our next investigation is more scientific?

Reflecting on the Nature of Scientific Knowledge and Scientific Inquiry

This investigation can be used to illustrate two important concepts related to the nature of scientific knowledge and the nature of scientific inquiry: (a) the difference between observations and inferences in science and (b) the nature and role of experiments in science (see Appendix 2 for a brief description of these two concepts). Be sure to review these concepts during and at the end of the explicit and reflective discussion. To help students think about these concepts in relation to what they did during the lab, you can ask them the following questions:

1. You had to make observations and inferences during your investigation. Can you give me some examples of these observations and inferences?

2. Can you work with your group to come up with a rule that you can use to decide if a piece of information is an observation or an inference? Be ready to share in a few minutes.

3. I asked you to design and carry out an experiment during your investigation. Can you give me some examples of what experiments are used for in science?

4. Can you work with your group to come up with a rule that you can use to decide if an investigation is an experiment or not? Be ready to share in a few minutes.

You can also use presentation software or other techniques to encourage your students to think about these concepts. You can show examples of information from the investigation that are either observations or inferences and ask students to classify each example and explain their thinking. You can also show images of different types of investigations (such as a physicist or an astronomer collecting data using a telescope as part of an observational and descriptive study, a person working in the library doing a literature review, a person working on a computer to analyze an existing data set, and an actual experiment) and ask students to indicate if they think that each image represents an experiment and why or why not.

Be sure to remind your students that, to be proficient in science, it is important that they understand what counts as scientific knowledge and how that knowledge develops over time.

Hints for Implementing the Lab

- Allowing students to design their own procedures for collecting data gives students an opportunity to try, to fail, and to learn from their mistakes. However, you can scaffold students as they develop their procedure by having them fill out an investigation proposal. These proposals provide a way for you to offer students hints and suggestions without telling them how to do it. You can also check the proposals quickly during a class period. For this lab we suggest using Investigation Proposal C.

- Learn how to use the marble launcher before the lab begins. It is important for you to know how to use the equipment so you can help students when technical issues arise.

- Allow the students to become familiar with the marble launcher as part of the tool talk before they begin to design their investigation. Giving them 5–10 minutes to examine the equipment will let them see what they can and cannot do with it.

- The horizontal displacement of the marble may be quite long. Students will need a wide berth to launch their marbles. We recommend collecting data outside on a flat, grassy area. This will provide the space necessary for students to launch their marbles. A grassy area will also limit the ability of the marbles to roll after they land. Students can also collect data in a hallway, gymnasium, or the cafeteria.

- Be sure to allow students to go back and re-collect data at the end of the argumentation session. Students often realize that they made numerous mistakes when they were collecting data as a result of their discussions during the argumentation session. The students, as a result, will want a chance to re-collect data, and the re-collection of data should be encouraged when time allows. This also offers an opportunity to discuss what scientists do when they realize a mistake is made inside the lab.

If students use video analysis

- We suggest allowing students to familiarize themselves with the video analysis software before they finalize the procedure for the investigation, especially if they have not used such software previously. This gives students an opportunity to learn how to work with the software and to improve the quality of the video they take.

- Remind students to hold the video camera as still as possible. Any movement of the camera will introduce error into their analysis. If using actual camcorders, we recommend using a tripod to hold the camera steady. If students are using a camera on a cell phone or tablet, we recommend using a table to help steady it.

- Remind students to place a meterstick in the same field of view as the motion they are capturing with the video camera. Also, the meterstick should be approximately the same distance from the camera as the motion. Most video analysis software requires the user to define a scale in the video (this allows the software to establish distances and, subsequently, other variables dependent on distance and displacement).

Connections to Standards

Table 3.2 highlights how the investigation can be used to address learning objectives from AP Physics 1; learning objectives from AP Physics C: Mechanics, *Common Core State Standards*, in English language arts (*CCSS ELA*); and *Common Core State Standards, Mathematics* (*CCSS Mathematics*).

TABLE 3.2 _____

Lab 3 alignment with standards

NGSS performance expectations	• None
AP Physics 1 learning objectives	• 3.A.1.1: The student is able to express the motion of an object using narrative, mathematical, and graphical representations. • 3.A.1.2: The student is able to design an experimental investigation of the motion of an object. • 3.A.1.3: The student is able to analyze experimental data describing the motion of an object and is able to express the results of the analysis using narrative, mathematical, and graphical representations. • 3.A.3.1: The student is able to analyze a scenario and make claims (develop arguments, justify assertions) about the forces exerted on an object by other objects for different types of forces or components of forces. • 3.B.1.2: The student is able to design a plan to collect and analyze data for motion (static, constant, or accelerating) from force measurements and carry out an analysis to determine the relationship between the net force and the vector sum of the individual forces. • 3.B.2.1: The student is able to create and use free-body diagrams to analyze physical situations to solve problems with motion qualitatively and quantitatively.
AP Physics C: Mechanics learning objectives	• I.A.1.b(2): Use the equations $\mathbf{v} = \mathbf{v}_0 + \mathbf{a}t$; $x = x_0 + \mathbf{v}_0 t + \frac{1}{2}\mathbf{a}t^2$ and $\mathbf{v}^2 = \mathbf{v}_0^2 + 2\mathbf{a}(x - x_0)$ to solve problems involving one-dimensional motion with constant acceleration. • I.A.2.c(1): Write down expressions for the horizontal and vertical components of velocity and position as functions of time, and sketch or identify graphs of these components. • I.A.2.c(2): Use these expressions in analyzing the motion of a projectile that is projected with an arbitrary initial velocity.

continued

Table 3.2 (*continued*)

Literacy connections (*CCSS ELA*)	• *Reading:* Key ideas and details, craft and structure, integration of knowledge and ideas • *Writing:* Text types and purposes, production and distribution of writing, research to build and present knowledge, range of writing • *Speaking and listening:* Comprehension and collaboration, presentation of knowledge and ideas
Mathematics connections (*CCSS Mathematics*)	• *Mathematical practices:* Make sense of problems and persevere in solving them, reason abstractly and quantitatively, construct viable arguments and critique the reasoning of others, model with mathematics, use appropriate tools strategically, attend to precision, look for and make use of structure, look for and express regularity in repeated reasoning • *Number and quantity:* Reason quantitatively and use units to solve problems, represent and model with vector quantities, perform operations on vectors • *Algebra:* Interpret the structure of expressions, create equations that describe numbers or relationships, understand solving equations as a process of reasoning and explain the reasoning, solve equations and inequalities in one variable, represent and solve equations and inequalities graphically • *Functions:* Understand the concept of a function and use function notation; interpret functions that arise in applications in terms of the context; analyze functions using different representations; build a function that models a relationship between two quantities; construct and compare linear and exponential models and solve problems; interpret expressions for functions in terms of the situation they model • *Statistics and probability:* Summarize, represent, and interpret data on two categorical and quantitative variables; interpret linear models; understand and evaluate random processes underlying statistical experiments; make inferences and justify conclusions from sample surveys, experiments, and observational studies

Lab Handout

Lab 3. Projectile Motion: How Do Changes to the Launch Angle, the Initial Velocity, and the Mass of a Projectile Affect Its Hang Time?

Introduction

Projectile motion is defined as the flight of an object near the Earth's surface under the action of gravity alone. Understanding the factors that affect the motion of projectiles has led to several major milestones in human history. Take spears as an example. Anthropologists have found evidence that human ancestors used stone-tipped spears for hunting 500,000 years ago (Wilkins et al. 2012). The ability to make and then use a spear to hunt large animals allowed our ancestors to gather more food. Later on, as people developed a better understanding of the factors that govern projectile motion, they were able to build new tools for launching projectiles that could travel farther in the air and hit targets with great accuracy. These tools included such things as bows and arrows, trebuchets, and cannons.

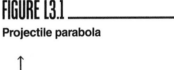

FIGURE L3.1

Projectile parabola

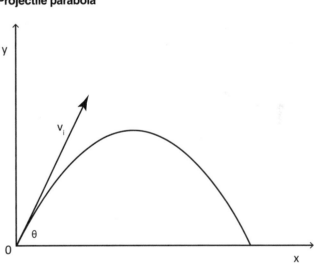

When a projectile is in flight, we assume that gravity is the sole force acting on it. The path of a projectile is a curve called a parabola, as shown in Figure L3.1. The projectile will have both a horizontal and vertical component to its velocity at any given time. Although scientists recognize that air resistance does affect the flight of a projectile, under most circumstances the effect of air resistance can be ignored. When we ignore air resistance, the initial velocity of the projectile governs the horizontal component of the projectile's velocity and the force of gravity governs its vertical component.

People often want to be able to predict how long a projectile will stay in the air (i.e., the hang time) after it is launched. There are a number of variables that may, or may not, affect the hang time of a projectile. These variables include the launch angle (denoted as θ) the initial velocity of the projectile, and the mass of the projectile. Some of these variables may also interact with each other, so the effect of any one variable may differ depending on the value of another variable. People therefore need to understand not only how these three variables affect the motion of a projectile but also how they interact with each other to predict how long a projectile will remain in the air after it is launched.

LAB 3

Your Task

Use what you to know about both linear and projectile motion, the importance of looking for patterns in data, and how to identify causal relationships to design and carry out a series of experiments to determine which variables impact the hang time of a projectile.

The guiding question of this investigation is, *How do changes to the launch angle, the initial velocity, and the mass of a projectile affect its hang time?*

Materials

You may use any of the following materials during your investigation:

Consumable
- Tape

Equipment
- Safety glasses or goggles (required)
- Marble launcher
- Marbles (different sizes and masses)
- Protractor
- Tape measure
- Stopwatch

If you have access to the following equipment, you may also consider using a video camera and a computer or tablet with video analysis software.

Safety Precautions

Follow all normal lab safety rules. In addition, take the following safety precautions:

1. Wear sanitized safety glasses or goggles during lab setup, hands-on activity, and takedown.

2. Do not throw marbles or launch them at people.

3. Do not stand on tables or chairs.

4. Pick up marbles and other materials on the floor immediately to avoid a trip, slip, or fall hazard.

5. Wash hands with soap and water after completing the lab.

Investigation Proposal Required?   ☐ Yes   ☐ No

Getting Started

To answer the guiding question, you will need to design and carry out several different experiments. Each experiment should look at one potential variable that may or may not affect the hang time of a projectile. To accomplish this task, you must determine what type of data you need to collect, how you will collect it, and how you will analyze it.

To determine *what type of data you need to collect,* think about the following questions:

- What are the boundaries and the components of the system you are studying?
- Which variable or variables could cause a change in the hang time of a projectile?
- What information will you need to track changes in these variables?
- What information will you need to be able to track changes in the hang time of a projectile?
- What will be the independent variable and the dependent variable for each experiment?

To determine *how you will collect the data,* think about the following questions:

- What conditions need to be satisfied to establish a cause-and-effect relationship?
- What will you hold constant during each experiment?
- What variables will need to be controlled during each experiment, and how will you control them?
- How will you make sure that your data are of high quality (i.e., how will you reduce error)?

To determine *how you will analyze the data,* think about the following questions:

- What types of patterns might you look for as you analyze your data?
- How could you use mathematics to describe a relationship between variables?
- What type of calculations will you need to make?

Connections to the Nature of Scientific Knowledge and Scientific Inquiry

As you work through your investigation, you may want to consider

- the difference between observations and inferences in science, and
- the nature and role of experiments in science.

Initial Argument

Once your group has finished collecting and analyzing your data, your group will need to develop an initial argument. Your initial argument needs to include a claim, evidence to support your claim, and a justification of the evidence. The *claim* is your group's answer to the guiding question. The *evidence* is an analysis and interpretation of your data. Finally, the *justification* of the evidence is why your group thinks the evidence matters. The justification of the evidence is important because scientists can use different kinds of evidence to support their claims. Your group will create your initial argument on a whiteboard. Your whiteboard should include all the information shown in Figure L3.2 (p. 84).

LAB 3

Argumentation Session

The argumentation session allows all of the groups to share their arguments. One or two members of each group will stay at the lab station to share that group's argument, while the other members of the group go to the other lab stations to listen to and critique the other arguments. This is similar to what scientists do when they propose, support, evaluate, and refine new ideas during a poster session at a conference. If you are presenting your group's argument, your goal is to share your ideas and answer questions. You should also keep a record of the critiques and suggestions made by your classmates so you can use this feedback to make your initial argument stronger. You can keep track of specific critiques and suggestions for improvement that your classmates mention in the space below.

Critiques about our initial argument and suggestions for improvement:

Argument presentation on a whiteboard

The Guiding Question:	
Our Claim:	
Our Evidence:	Our Justification of the Evidence:

If you are critiquing your classmates' arguments, your goal is to look for mistakes in their arguments and offer suggestions for improvement so these mistakes can be fixed. You should look for ways to make your initial argument stronger by looking for things that the other groups did well. You can keep track of interesting ideas that you see and hear during the argumentation in the space below. You can also use this space to keep track of any questions that you will need to discuss with your team.

Interesting ideas from other groups or questions to take back to my group:

Once the argumentation session is complete, you will have a chance to meet with your group and revise your initial argument. Your group might need to gather more data or design a way to test one or more alternative claims as part of this process. Remember, your goal at this stage of the investigation is to develop the best argument possible.

Report

Once you have completed your research, you will need to prepare an investigation report that consists of three sections. Each section should provide an answer to the following questions:

1. What question were you trying to answer and why?

2. What did you do to answer your question and why?

3. What is your argument?

Your report should answer these questions in two pages or less. This report must be typed, and any diagrams, figures, or tables should be embedded into the document. Be sure to write in a persuasive style; you are trying to convince others that your claim is acceptable or valid!

Reference

Wilkins, J., B. J. Schoville, K. S. Brown, and M. Chazan. 2012. Evidence for early hafted hunting technology. *Science* 338 (6109): 942–946.

Checkout Questions

Lab 3. Projectile Motion: How Do Changes to the Launch Angle, the Initial Velocity, and the Mass of a Projectile Affect Its Hang Time?

1. Given the models you created during your investigation of $t(\mathbf{v}_0)$ and $t(\theta)$, choose two instances for $\mathbf{v}_0$ and θ for which the hang time of the projectile is 2 seconds.

2. Let the acceleration due to gravity be expressed as $\mathbf{a} = -\mathbf{g}$ and the initial velocity of the projectile in the $\mathbf{y}$ direction as $\mathbf{v}_0 = \mathbf{v}\sin\theta$. Given the equation for the change in $\mathbf{y}$ below,

$$\Delta\mathbf{y} = \mathbf{v}_0 t + \frac{1}{2}\mathbf{a}t^2$$

write an equation for hang time, t, as a function of $\mathbf{v}_0$, θ, and $\mathbf{g}$.

3. Current scientific knowledge and the perspectives of individual scientists influence inferences but not observations.

   a. I agree with this statement.

   b. I disagree with this statement.

   Explain your answer, using an example from your investigation of projectile motion.

4. Scientists use experiments to prove ideas.

    a. I agree with this statement.

    b. I disagree with this statement.

    Explain your answer, using an example from your investigation of projectile motion.

5. Why is it useful to identify patterns during an investigation? In your answer, be sure to include examples from at least two different investigations.

6. Why is identifying cause-and-effect relationships so important in science? In your answer, be sure to include examples from at least two different investigations.

Application Lab

LAB 4

Teacher Notes

Lab 4. The Coriolis Effect: How Do the Direction and Rate of Rotation of a Spinning Surface Affect the Path of an Object Moving Across That Surface?

Purpose

The purpose of this lab is for students to *apply* what they know about the core idea of forces and motion, part of the disciplinary core idea (DCI) of Motion and Stability: Forces and Interactions from the *NGSS,* to determine how the movement of a rotating platform affects the movement of a marble rolled across it as viewed from different frames of reference. (This lab is not intended to introduce the idea of rotational motion, which is a focus of other labs in the book.) This lab also gives students an opportunity to learn about the crosscutting concepts (CCs) of (a) Scale, Proportion, and Quantity; and (b) Systems and System Models from the *NGSS.* In addition, this lab can be used to help students understand two big ideas from AP Physics: (a) the interactions of an object with other objects can be described by forces and (b) interactions between systems can result in changes in those systems. As part of the explicit and reflective discussion, students will also learn about (a) the difference between data and evidence in science and (b) the role of imagination and creativity in science.

Underlying Physics Concepts

To explain the Coriolis effect, imagine that instead of rolling a ball across a rotating platform, a person were to throw the ball across the platform, from a point near the center to a point near the edge. Figure 4.1 shows a diagram of this as viewed from above the rotating platform. In Figure 4.1, the platform rotates counterclockwise around the center of the circle (incidentally, this is the same direction Earth rotates as viewed from above the North Pole). In Figure 4.1, the person throws the ball from point A to point B. Points A and B are radially aligned with the center of the platform. All points on the platform have the same angular velocity, but not all points have the same tangential velocity. Instead, the tangential velocity increases as one moves radially outward. Point B therefore has a larger tangential velocity

FIGURE 4.1 _____

Diagram of velocity vectors on a rotating platform

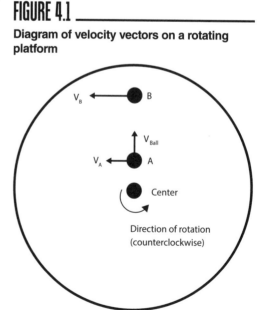

The Coriolis Effect

How Do the Direction and Rate of Rotation of a Spinning Surface Affect the Path of an Object Moving Across That Surface?

than point A. Finally, it is important to realize that, from the frame of reference of the platform, there is no sideways velocity component. That is, $\mathbf{v}_A$ and $\mathbf{v}_B$ are only perceived by reference frames off the platform.

Now it is possible to analyze the motion of the ball. First, we analyze the motion from the inertial reference frame looking down at the platform. Figure 4.2 shows the path the ball will take, due to the vector addition. Notice how this path is straight. This is due to the fact that no forces act on the ball once it is thrown (ignoring gravity). According to Newton's first law, if no external forces act on an object, it will move in a straight line with constant velocity.

When the ball reaches the edge of the platform, it will have missed the person at point B. With respect to the direction of rotation, the ball will be behind person B. This is because the tangential velocity with which the person at point B moved was greater than the tangential velocity with which the person at point A moved. The ball was thrown from point A, so its tangential velocity is equal to the tangential velocity at point A and is less than the tangential velocity of the person at point B. Figure 4.3 shows the path of the ball from the non-inertial reference frame; in other words, it illustrates the path the ball takes when viewed by a person standing on the platform.

From the reference frame of the platform, the ball only has a single velocity component, $\mathbf{V}_{Ball}$, in the radial direction from point A to B. We have already established that when the ball reaches the edge, it is behind the person at point B. The dotted line shows the path the ball takes when viewed from the platform.

Notice that the path of the ball is curved when viewed from the platform. The path is curved because of the difference in the tangential velocities at each point along the radial line from A to B (or along any radius). Figure 4.4 (p. 92) shows several points along the radial path from A to B, along with the tangential velocity vector (although this velocity vector is not noticeable from the reference frame of the platform). In order of least to greatest, the tangential velocities are A, C, D, and B. (Note: We define each of the points A–D as the point in space above the surface of the rotating surface. However, the tangential velocities are defined against the point on the surface, which

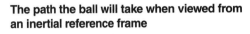

FIGURE 4.2

The path the ball will take when viewed from an inertial reference frame

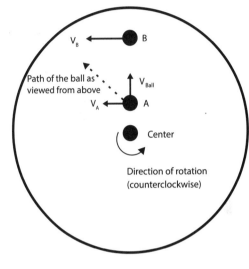

FIGURE 4.3

The path the ball takes when viewed from the non-inertial reference frame

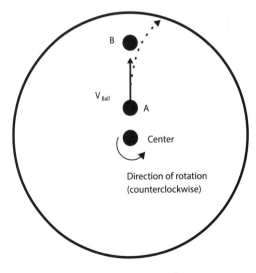

LAB 4

lies directly below points A–D. We use these definitions to keep the points through which the ball passes stationary relative to the moving disc, while recognizing that the points on the disc rotate). As the ball is thrown radially from point A to point B, it will pass through points C and D. The distance that the ball misses each point is a function of the difference in their tangential velocity. Equation 4.1 provides the mathematical relationship, where **d** is displacement of the ball (i.e., "the distance by which the ball misses"); **v** is velocity, and t is time (assuming the time is small compared with the period of rotation). In SI units, distance is measured in meters (m), velocity is measured in meters per second (m/s), and time is measured in seconds (s).

FIGURE 4.4

The tangential velocities for several points when viewed from an inertial reference frame

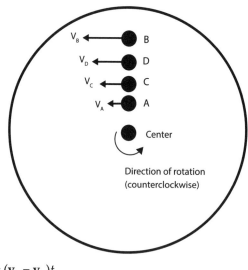

$$\text{(Equation 4.1)} \quad \mathbf{d} = (\mathbf{v_B} - \mathbf{v_A})t$$

As the ball moves outward, the difference in velocities increases between point A and each subsequent point. Thus, the distance by which the ball misses the person at each point also will increase as one moves radially outward, which makes the ball appear to take a curved path.

In this investigation, students investigate how the direction of rotation and the rate of rotation affect the movement of a marble rolled across the platform. The direction of rotation will affect the direction of curvature. When the platform rotates counterclockwise, the curve will be to the right of point B. When the direction of rotation is clockwise, the direction of curvature is to the left. The greater the rotational velocity, the larger the apparent curvature will be when viewed from the non-inertial reference frame.

Timeline

The instructional time needed to complete this lab investigation is 200–280 minutes. Appendix 3 (p. 531) provides options for implementing this lab investigation over several class periods. Option E (280 minutes) should be used if students are unfamiliar with scientific writing, because this option provides extra instructional time for scaffolding the writing process. You can scaffold the writing process by modeling, providing examples, and providing hints as students write each section of the report. Option E can also be used if you are introducing students to the video analysis programs. Option F (200 minutes) should be used if students are familiar with scientific writing and have developed the skills needed to write an investigation report on their own. In option F, students complete

How Do the Direction and Rate of Rotation of a Spinning Surface Affect the Path of an Object Moving Across That Surface?

stage 6 (writing the investigation report) and stage 8 (revising the investigation report) as homework.

Materials and Preparation

The materials needed to implement this investigation are listed in Table 4.1. The rotating "Coriolis" platform can be purchased from Ward's Science or Flinn Scientific. Ward's Science sells the platform as part of their "Coriolis Effect Lab Activity" (item 470218-770 for one lab group). Flinn Scientific sells the platform under the name "Coriolis Effect Apparatus" (item AP5113); we recommend using this product because it includes a reusable film that allows students to track the path of the rolling marble. The other equipment can be purchased from Flinn, Ward's Science, or any other science supply company. The video analysis software can be purchased from Vernier (Logger *Pro*) or PASCO Scientific (SPARKvue or Capstone). These companies also have apps that can be used on Apple- or Android-based tablets and cell phones. We recommend consulting with your school's information technology coordinator to determine the best option for your students.

TABLE 4.1 _____

Materials list for Lab 4

Item	Quantity
Safety glasses or goggles	1 per student
Video camera	1 per group
Computer or tablet with video analysis software	1 per group
Rotating Coriolis platform	1 per group
Marble	1 per group
Stopwatch	1 per student
Ruler	1 per student
Investigation Proposal A (optional)	1 per group
Whiteboard, 2' × 3'*	1 per group
Lab Handout	1 per student
Peer-review guide and teacher scoring rubric	1 per student
Checkout Questions	1 per student

* As an alternative, students can use computer and presentation software such as Microsoft PowerPoint or Apple Keynote to create their arguments.

Be sure to use a set routine for distributing and collecting the materials during the lab investigation. One option is to set up the materials for each group at each group's lab

station before class begins. This option works well when there is a dedicated section of the classroom for lab work and the materials are large and difficult to move (such as a dynamics track). A second option is to have all the materials on a table or cart at a central location. You can then assign a member of each group to be the "materials manager." This individual is responsible for collecting all the materials his or her group needs from the table or cart during class and for returning all the materials at the end of the class. This option works well when the materials are small and easy to move (such as stopwatches, metersticks, or marbles). It also makes it easy to inventory the materials at the end of the class before students leave for the day.

Safety Precautions

Remind students to follow all normal lab safety rules. In addition, tell students to take the following safety precautions:

1. Wear sanitized safety glasses or goggles during lab setup, hands-on activity, and takedown.

2. Keep fingers and toes out of the way of moving objects.

3. Do not throw marbles.

4. Pick up marbles or other materials on the floor immediately to avoid a trip, slip, or fall hazard.

5. Wash hands with soap and water after completing the lab.

Topics for the Explicit and Reflective Discussion

Reflecting on the Use of Core Ideas and Crosscutting Concepts During the Investigation

Teachers should begin the explicit and reflective discussion by asking students to discuss what they know about the core idea they used during the investigation. The following are some important concepts related to the core idea of forces and motion that students need to use to determine how the motion of a rotating platform affects the movement of a marble rolled across the platform as viewed from different frames of reference:

- Quantities such as position, displacement, distance, velocity, speed, and acceleration are used to describe the motion of an object.

- Displacement, velocity, and acceleration are vector quantities.

- *Displacement* is a change in position. *Velocity* is the rate of change of position over a period of time. *Acceleration* is the rate of change in velocity over a period of time.

- A reference frame is needed to determine the direction and the magnitude of the displacement, velocity, and acceleration of an object.

The Coriolis Effect

How Do the Direction and Rate of Rotation of a Spinning Surface Affect the Path of an Object Moving Across That Surface?

- An object at rest stays at rest and an object in motion stays in motion with the same speed and in the same direction unless acted on by an unbalanced force.

To help students reflect on what they know about forces and motion, we recommend showing them two or three images using presentation software that help illustrate these important ideas. You can then ask the students the following questions to encourage them to share how they are thinking about these important concepts:

1. What do we see going on in this image?

2. Does anyone have anything else to add?

3. What might be going on that we can't see?

4. What are some things that we are not sure about here?

You can then encourage students to think about how CCs played a role in their investigation. There are at least two CCs that students need to use to determine how the motion of a rotating platform affects the movement of a marble rolled across the platform as viewed from different frames of reference: (a) Scale, Proportion, and Quantity; and (b) Systems and System Models (see Appendix 2 [p. 527] for a brief description of these CCs). To help students reflect on what they know about these CCs, we recommend asking them the following questions:

1. Why is it important to think about issues of scale and quantity during an investigation?

2. What scales and quantities did you use during your investigation? Why was it useful?

3. Why do scientists often define a system and then develop a model of it as part of an investigation?

4. How did you use a model to understand the effect of non-inertial reference frames on motion? Why was that useful?

You can then encourage the students to think about how they used all these different concepts to help answer the guiding question and why it is important to use these ideas to help justify their evidence for their final arguments. Be sure to remind your students to explain why they included the evidence in their arguments and make the assumptions underlying their analysis and interpretation of the data explicit in order to provide an adequate justification of their evidence.

Reflecting on Ways to Design Better Investigations

It is important for students to reflect on the strengths and weaknesses of the investigation they designed during the explicit and reflective discussion. Students should therefore be encouraged to discuss ways to eliminate potential flaws, measurement errors, or sources of uncertainty in their investigations. To help students be more reflective about the design of their investigation and what they can do to make their investigations more rigorous in the future, you can ask them the following questions:

1. What were some of the strengths of the way you planned and carried out your investigation? In other words, what made it scientific?

2. What were some of the weaknesses of the way you planned and carried out your investigation? In other words, what made it less scientific?

3. What rules can we make, as a class, to ensure that our next investigation is more scientific?

Reflecting on the Nature of Scientific Knowledge and Scientific Inquiry

This investigation can be used to illustrate two important concepts related to the nature of scientific knowledge and the nature of scientific inquiry: (a) the difference between data and evidence in science and (b) the role of imagination and creativity in science (see Appendix 2 for a brief description of these two concepts). Be sure to review these concepts during and at the end of the explicit and reflective discussion. To help students think about these concepts in relation to what they did during the lab, you can ask them the following questions:

1. You had to talk about data and evidence during your investigation. Can you give me some examples of data and evidence from your investigation?

2. Can you work with your group to come up with a rule that you can use to decide if a piece of information is data or evidence? Be ready to share in a few minutes.

3. Some people think that there is no room for imagination or creativity in science. What do you think?

4. Can you work with your group to come up with different ways that you needed to use your imagination or be creative during this investigation? Be ready to share in a few minutes.

You can also use presentation software or other techniques to encourage your students to think about these concepts. You can show examples of information from the investigation that are either data or evidence and ask students to classify each example and explain their thinking. You can also show students an image of the following quote by E. O. Wilson from *Letters to a Young Scientist* (2013) and ask them what they think he meant by it:

> The ideal scientist thinks like a poet and only later works like a bookkeeper. Keep in mind that innovators in both literature and science are basically dreamers and storytellers. In the early stages of the creation of both literature and science,

The Coriolis Effect

How Do the Direction and Rate of Rotation of a Spinning Surface Affect the Path of an Object Moving Across That Surface?

everything in the mind is a story. There is an imagined ending, and usually an imagined beginning, and a selection of bits and pieces that might fit in between. In works of literature and science alike, any part can be changed, causing a ripple among the other parts, some of which are discarded and new ones added. (p. 74)

Be sure to remind your students that, to be proficient in science, it is important that they understand what counts as scientific knowledge and how that knowledge develops over time.

Hints for Implementing the Lab

- Allowing students to design their own procedures for collecting data gives students an opportunity to try, to fail, and to learn from their mistakes. However, you can scaffold students as they develop their procedure by having them fill out an investigation proposal. These proposals provide a way for you to offer students hints and suggestions without telling them how to do it. You can also check the proposals quickly during a class period. For this lab we suggest using Investigation Proposal A.

- Learn how to use the rotating Coriolis platform before the lab begins. It is important for you to know how to use the equipment so you can help students when technical issues arise.

- Allow the students to become familiar with the Coriolis platform as part of the tool talk before they begin to design their investigation. Giving them 5–10 minutes to examine the equipment will let them see what they can and cannot do with it.

- If the film is not tracking the path of the ball, heat the film and platform with a hairdryer for a few minutes.

- For the best results when using the Flinn equipment, let the marble roll down the platform before turning the platform. Once the marble begins to roll, rotate the platform.

- We suggest allowing students to familiarize themselves with the video analysis software before they finalize the procedure for the investigation, especially if they have not used such software previously. This gives students an opportunity to learn how to work with the software and to improve the quality of the video they take.

- Remind students to hold the video camera as still as possible. Any movement of the camera will introduce error into their analysis. If using actual camcorders, we recommend using a tripod to hold the camera steady. If students are using a camera on a cell phone or tablet, we recommend using a table to help steady the camera.

- Remind students to place a meterstick in the same field of view as the motion they are capturing with the video camera. Also, the meterstick should be approximately the same distance from the camera as the motion. Most video analysis software requires the user to define a scale in the video (this allows the software to establish distances and, subsequently, other variables dependent on distance and displacement).

- Be sure to allow students to go back and re-collect data at the end of the argumentation session. Students often realize that they made numerous mistakes when they were collecting data as a result of their discussions during the argumentation session. The students, as a result, will want a chance to re-collect data, and the re-collection of data should be encouraged when time allows. This also offers an opportunity to discuss what scientists do when they realize a mistake is made inside the lab.

Connections to Standards

Table 4.2 highlights how the investigation can be used to address learning objectives from AP Physics 1; learning objectives from AP Physics C: Mechanics, *Common Core State Standards*, in English language arts (*CCSS ELA*); and *Common Core State Standards*, Mathematics (*CCSS Mathematics*).

The Coriolis Effect

How Do the Direction and Rate of Rotation of a Spinning Surface Affect the Path of an Object Moving Across That Surface?

TABLE 4.2 _____

Lab 4 alignment with standards

***NGSS* performance expectations**	• None
AP Physics 1 learning objectives	• 3.A.1.1: The student is able to express the motion of an object using narrative, mathematical, and graphical representations. • 3.A.1.2: The student is able to design an experimental investigation of the motion of an object. • 3.A.1.3: The student is able to analyze experimental data describing the motion of an object and is able to express the results of the analysis using narrative, mathematical, and graphical representations. • 4.A.2.1: The student is able to make predictions about the motion of a system based on the fact that acceleration is equal to the change in velocity per unit time, and velocity is equal to the change in position per unit time.
AP Physics C: Mechanics learning objective	• I.D.3.b: Students should understand frames of reference.
Literacy connections (*CCSS ELA*)	• *Reading:* Key ideas and details, craft and structure, integration of knowledge and ideas • *Writing:* Text types and purposes, production and distribution of writing, research to build and present knowledge, range of writing • *Speaking and listening:* Comprehension and collaboration, presentation of knowledge and ideas
Mathematics connections (*CCSS Mathematics*)	• *Mathematical practices:* Reason abstractly and quantitatively, construct viable arguments and critique the reasoning of others, use appropriate tools strategically, attend to precision • *Number and quantity:* Reason quantitatively and use units to solve problems, represent and model with vector quantities, perform operations on vectors

Reference

Wilson, E. O. 2013. *Letters to a young scientist.* New York: Liveright Publishing.

LAB 4

Lab Handout

Lab 4. The Coriolis Effect: How Do the Direction and Rate of Rotation of a Spinning Surface Affect the Path of an Object Moving Across That Surface?

Introduction

When studying the motion of objects, one of the assumptions that we often make is that the ground underneath the object being studied is stationary. That is, the frame of reference being used to study the object is not moving. It turns out, however, that very few reference frames are perfectly stationary (or, more formally, inertial). For example, imagine that a person is standing on a flatbed train car. The train is moving on a track to the east with a constant velocity. Both the person and train, as a result, have the same velocity in the horizontal direction. Now imagine that the person standing on the flatbed train car were to drop a ball while the train was moving. The person who dropped the ball would see the ball fall straight down to the ground because that person has the same reference frame as the train. A second person that is watching the train go by, however,

FIGURE L4.1

A point on the solid circle travels a farther distance in one day than a point on the dashed circle. The tangential velocity is therefore greater on the solid circle than on the dashed circle.

would see something different. That person would see the same ball fall to the ground following a curved path. This is because the ball has both a vertical and horizontal component to its velocity as viewed from the reference frame of the person standing on the ground watching the train pass.

A French mathematician and physicist, Gaspard-Gustave de Coriolis, conducted numerous investigations in the 1800s to understand the movement of various bodies when the frame of reference was rotating. Initially, his studies were conducted against a rotating disc. Subsequent work by other scientists has applied his findings, known as the Coriolis effect, to the rotation of spheres, such as Earth. One of the interesting things that scientists have found is that while the rate of rotation is the same at all points on the globe, the tangential velocity is different depending on the latitude. Although every point makes one full rotation in one day, the radius of the disk traced by a point on the globe differs based on the latitude. Points near to the poles travel a smaller circle than points near the equator. Figure L4.1 illustrates this fact. What this means is that when an airplane travels with a north-south component, the place it takes off from will have a different east-west velocity than the place it intends to

The Coriolis Effect

How Do the Direction and Rate of Rotation of a Spinning Surface Affect the Path of an Object Moving Across That Surface?

land. It is therefore important to take into account the Earth's rotation when planning a flight path between two airports.

In this investigation you will have an opportunity to explore the motion of a marble as it moves across a rotating surface. Your goal will be to determine the path the marble will take when it is rolled in the y direction across a surface that is rotating in either a clockwise or counterclockwise direction. Much like the example of the person standing on a train that is moving, the rotation of the surface on which movement takes place can lead different observers to see different paths of motion based on their frame of reference. An observer on the rotating surface (formally, a non-inertial reference frame) will see the marble take a different path than a stationary person watching from outside the rotating surface (formally, an inertial reference frame). This is because, relative to each observer, the marble has different velocity vectors. To the stationary person, the marble has two velocity components, one in the y direction and one in the x direction that is due to the rotation of the surface. To the observer on the rotating platform, however, the marble only has a velocity component in the y direction.

Your Task

Use what you know about vector quantities, measurement and scale, and systems and system models to design several experiments to determine how the motion of a rotating platform affects the movement of a marble rolled across the platform as viewed from different frames of reference.

The guiding question of this investigation is, *How do the direction and rate of rotation of a spinning surface affect the path of an object moving across that surface?*

Materials

You may use any of the following materials during your investigation:

- Safety glasses or goggles (required)
- Video camera
- Computer or tablet with video analysis software
- Rotating "Coriolis" platform
- Marble
- Stopwatch
- Ruler

Safety Precautions

Follow all normal lab safety rules. In addition, take the following safety precautions:

1. Wear sanitized safety glasses or goggles during lab setup, hands-on activity, and takedown.

2. Keep fingers and toes out of the way of moving objects.

3. Do not throw marbles.

LAB 4

4. Pick up marbles or other materials on the floor immediately to avoid a trip, slip, or fall hazard.

5. Wash hands with soap and water after completing the lab.

Investigation Proposal Required? ☐ Yes ☐ No

Getting Started

To answer the guiding question, you will need to design and carry out at least two different experiments because a rotating platform can move in different directions (clockwise or counterclockwise) and at different rates. You will need to examine how both of these factors affect the movement of a marble rolled across the platform as viewed from different frames of reference. As you design your two experiments, you must decide what type of data you need to collect, how you will collect it, and how you will analyze it.

To determine *what type of data you need to collect,* think about the following questions:

- What are the boundaries and components of the system you are studying?
- How can you describe the components of the system quantitatively?
- What will be the independent variable and the dependent variable for each experiment?
- How will you quantify a change in the independent variable?
- Which variables are vector quantities, and which variables are scalar quantities?

To determine *how you will collect the data,* think about the following questions:

- How will you define the two reference frames?
- How will you vary the rotation rate of the platform?
- How will you measure the dependent variable?
- How will you measure the magnitude and direction of any vector quantities?
- What equipment will you need to take your measurements?
- How will you make sure that your data are of high quality (i.e., how will you reduce error)?
- How will you keep track of and organize the data you collect?

To determine *how you will analyze the data,* think about the following questions:

- What types of patterns could you look for in your data?
- How could you use mathematics to describe a relationship between variables?
- What type of calculations will you need to make?

The Coriolis Effect

How Do the Direction and Rate of Rotation of a Spinning Surface Affect the Path of an Object Moving Across That Surface?

- How will you determine whether the movement of a marble rolled across the platform as viewed from different frames of reference is the same or different?

- How will you model the motion of the marble as it moves across the rotating platform?

Connections to the Nature of Scientific Knowledge and Scientific Inquiry

As you work through your investigation, you may want to consider

- the difference between data and evidence in science, and

- the role of imagination and creativity in science.

Initial Argument

Once your group has finished collecting and analyzing your data, your group will need to develop an initial argument. Your argument needs to include a claim, evidence to support your claim, and a justification of the evidence. The *claim* is your group's answer to the guiding question. The *evidence* is an analysis and interpretation of your data. Finally, the *justification* of the evidence is why your group thinks the evidence matters. The justification of the evidence is important because scientists can use different kinds of evidence to support their claims. Your group will create your initial argument on a whiteboard. Your whiteboard should include all the information shown in Figure L4.2.

FIGURE L4.2

Argument presentation on a whiteboard

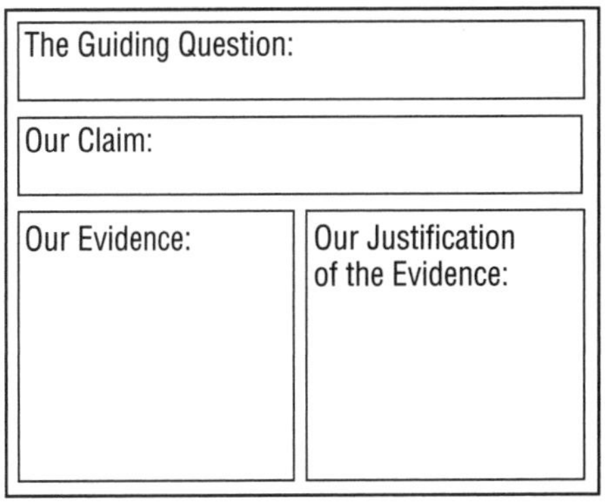

The Guiding Question:	
Our Claim:	
Our Evidence:	Our Justification of the Evidence:

Argumentation Session

The argumentation session allows all of the groups to share their arguments. One or two members of each group will stay at the lab station to share that group's argument, while the other members of the group go to the other lab stations to listen to and critique the other arguments. This is similar to what scientists do when they propose, support, evaluate, and refine new ideas during a poster session at a conference. If you are presenting your group's argument, your goal is to share your ideas and answer questions. You should also keep a record of the critiques and suggestions made by your classmates so you can use this feedback to make your initial argument stronger. You can keep track of specific critiques and suggestions for improvement that your classmates mention in the space below.

Critiques about our initial argument and suggestions for improvement:

If you are critiquing your classmates' arguments, your goal is to look for mistakes in their arguments and offer suggestions for improvement so these mistakes can be fixed. You should look for ways to make your initial argument stronger by looking for things that the other groups did well. You can keep track of interesting ideas that you see and hear during the argumentation in the space below. You can also use this space to keep track of any questions that you will need to discuss with your team.

Interesting ideas from other groups or questions to take back to my group:

Once the argumentation session is complete, you will have a chance to meet with your group and revise your initial argument. Your group might need to gather more data or design a way to test one or more alternative claims as part of this process. Remember, your goal at this stage of the investigation is to develop the best argument possible.

The Coriolis Effect

How Do the Direction and Rate of Rotation of a Spinning Surface Affect the Path of an Object Moving Across That Surface?

Report

Once you have completed your research, you will need to prepare an *investigation report* that consists of three sections. Each section should provide an answer to the following questions:

1. What question were you trying to answer and why?

2. What did you do to answer your question and why?

3. What is your argument?

Your report should answer these questions in two pages or less. This report must be typed, and any diagrams, figures, or tables should be embedded into the document. Be sure to write in a persuasive style; you are trying to convince others that your claim is acceptable or valid!

LAB 4

Checkout Questions

Lab 4. The Coriolis Effect: How Do the Direction and Rate of Rotation of a Spinning Surface Affect the Path of an Object Moving Across That Surface?

Use the figure on the right to answer questions 1 and 2.

1. When viewing the motion of the Earth from above the North Pole, the Earth appears to rotate counterclockwise. A plane is traveling from Miami, Florida, to Pittsburgh, Pennsylvania. Both cities are located at approximately 80° W longitude. If the plane takes off and continues in a straight line, how will its path appear to an observer in Pittsburgh?

   a. The path the plane takes will pass directly through Pittsburgh.

   b. The path the plane takes will pass to the east of Pittsburgh.

   c. The path the plane takes will pass to the west of Pittsburgh.

   How do you know?

Map of North America

2. A plane is traveling due west from Philadelphia to San Francisco. Both cities are on almost the same latitude. Will the plane be subject to the Coriolis effect?

   a. Yes

   b. No

The Coriolis Effect

How Do the Direction and Rate of Rotation of a Spinning Surface Affect the Path of an Object Moving Across That Surface?

Explain why or why not.

3. Scientists use their imagination to help them plan investigations and to analyze the results.

   a. I agree with this statement.

   b. I disagree with this statement.

   Explain your answer, using an example from your investigation of the Coriolis effect.

4. There is a difference between data and evidence in science.

   a. I agree with this statement.

   b. I disagree with this statement.

   Explain your answer, using an example from your investigation of the Coriolis effect.

5. Why is it important for scientists to think about issues related to quantity and scale as they plan or carry out an investigation? In your answer, be sure to include examples from at least two different investigations.

6. Why do scientists use models to understand and explain complex systems? In your answer, be sure to include examples from at least two different investigations.

SECTION 3
Forces and Motion

Dynamics

Introduction Labs

Teacher Notes

Lab 5. Force, Mass, and Acceleration: What Is the Mathematical Relationship Among the Net Force Exerted on an Object, the Object's Inertial Mass, and Its Acceleration?

Purpose

The purpose of this lab is to *introduce* students to the core idea of forces and motion, part of the disciplinary core idea (DCI) of Motion and Stability: Forces and Interactions from the *NGSS*, by having them determine the mathematical relationship among the net force acting on an object, its mass, and its acceleration. This lab also gives students an opportunity to learn about the crosscutting concepts (CCs) of (a) Cause and Effect: Mechanism and Explanation; and (b) Scale, Proportion, and Quantity from the *NGSS*. In addition, this lab can be used to help students understand three big ideas from AP Physics: (a) objects and systems have properties such as mass and charge, (b) the interactions of an object with other objects can be described by forces, and (c) interactions between systems can result in changes in those systems. As part of the explicit and reflective discussion, students will also learn about (a) how scientific knowledge changes over time and (b) the difference between laws and theories in science.

Underlying Physics Concepts

Isaac Newton revolutionized our understanding of forces and motion with the publication of *Philosophiae Naturalis Principia Mathematica* in 1687. Among the many important ideas in this book, Newton introduced three laws of motion. The first law states that, absent a net force, an object at rest will continue to be at rest (i.e., will not move) and an object in motion will continue to move in a straight line at constant velocity. There are several implications of Newton's first law. First, a net force is the sum of all the forces acting on the object with respect to their direction. Mathematically, this is shown in Equation 5.1, where ΣF is the symbol for the sum of the forces, F_1 is force 1, and F_n is the nth and last force acting on the object. In SI units, force is measured in newtons (N).

$$\textbf{(Equation 5.1)} \quad \Sigma F = F_1 + F_2 + F_3 \ldots + F_n$$

Although many non-zero forces can act on the object, it is possible that the addition of these forces (accounting for their direction) can lead to a net force equal to zero. When summing the forces, it is important to keep in mind the sign conventions for force, which allow us to mathematically account for direction as well as magnitude. Traditionally, vectors in the right, upward, north, or east direction are given a positive sign, whereas those in the left, downward, south, or west direction are given a negative sign.

Force, Mass, and Acceleration
*What Is the Mathematical Relationship Among the Net Force Exerted on an Object,
the Object's Inertial Mass, and Its Acceleration?*

Figure 5.1 shows two people pushing on a box in opposite directions. Person 1 is pushing to the right, so he or she is pushing with a positive force. Person 2 is pushing to the left, so he or she is pushing with a negative force. If the magnitude of each person's push is 20 N, then we can find the net force using Equation 5.1:

$$\Sigma F = F_1 + F_2 = 20\text{ N} + {-20}\text{ N} = 0\text{ N}$$

The net force on the box is zero, so the box will not change how it is moving. If the box was at rest before being pushed, then it will stay at rest. If, however, the box was moving before it was pushed, then the box will continue to move in the same direction with the same velocity.

FIGURE 5.1 _____

Two people pushing on a box with equal force but in opposite directions

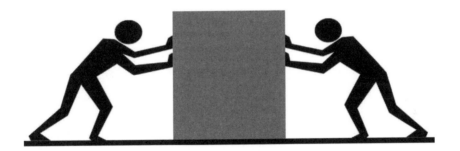

According to Newton's second law of motion, the acceleration of an object is directly proportional to the net force acting on it and inversely proportional to the mass of the object. This means that as the net force increases, the acceleration of the object will also increase. And, given the same net force acting on two objects, the object with the greater mass will have a smaller acceleration. Newton's second law is more commonly stated as follows: The net force on an object is equal to the mass of the object times the object's acceleration. Mathematically, this relationship is shown in Equation 5.2, where ΣF is the net force, m is the mass, and **a** is acceleration. In SI units, force is measured in newtons (N), mass in kilograms (kg); and acceleration in meters per second squared (m/s²).

(Equation 5.2) $\Sigma F = m\mathbf{a}$

Based on this equation, we also know that 1 N = 1 kg·m/s².

In this investigation, students will be gathering data about the acceleration of a cart and the force acting on the object. Some students may be familiar with running a regression to use mathematics to model a set of data with an equation. In this case, the regression will give them the function of a line $y = mx$, where m is the slope of the line. Other students may choose to graph the relationship between force and acceleration, resulting in a graph similar to the one in Figure 5.2. Note that in Figure 5.2, the data show acceleration as a function of force. If students treat force as the independent variable (as is the case in Figure 5.2), the slope of the line in the regression and in the graph will be equal to 1/*mass*. If, however, students treat force as the dependent variable, then the slope of the line will just be equal to the mass. In both cases, the students will find a linear relationship.

FIGURE 5.2

Acceleration as a function of force

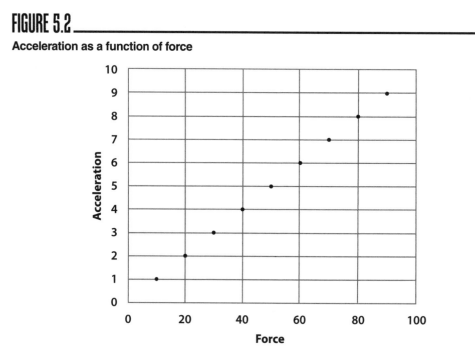

In this investigation, students are also instructed to conduct an experiment in which they vary the mass while leaving the net force acting on the object constant. In this case, the mathematical relationship is shown in Equation 5.3, and a graph of this relationship is shown in Figure 5.3.

(Equation 5.3) $\mathbf{a} = \Sigma \mathbf{F}/m$

FIGURE 5.3

Acceleration as a function of mass

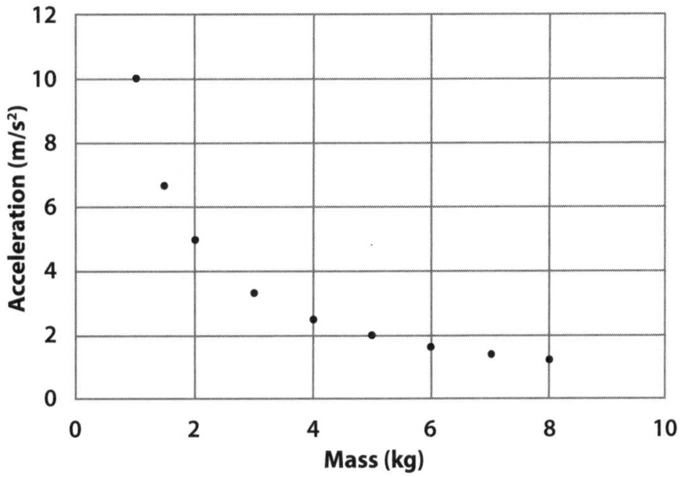

Timeline

The instructional time needed to complete this lab investigation is 200–280 minutes. Appendix 3 (p. 531) provides options for implementing this lab investigation over several class periods. Option A (280 minutes) should be used if students are unfamiliar with scientific writing, because this option provides extra instructional time for scaffolding the writing process. You can scaffold the writing process by modeling, providing examples, and providing hints as students write each section of the report. Option A can also be used if students are unfamiliar with any of the data collection and analysis tools. Option B (200 minutes) should be used if students are familiar with scientific writing and have developed the skills needed to write an investigation report on their own. In option B, students complete stage 6 (writing the investigation report) and stage 8 (revising the investigation report) as homework.

Materials and Preparation

The materials needed to implement this investigation are listed in Table 5.1 (p. 116). The equipment can be purchased from a science supply company such as Carolina, Flinn Scientific, or Ward's Science. Video analysis software can be purchased from Vernier (Logger *Pro*) or PASCO (SPARKvue or Capstone). These companies also have apps that can be used on Apple- or Android-based tablets and cell phones. We recommend consulting with your school's information technology coordinator to determine the best option for your students.

LAB 5

TABLE 5.1

Materials list for Lab 5

Item	Quantity
Consumables	
String or fishing line	1 meter per group
Tape	1 roll per group
Equipment and other materials	
Safety glasses or goggles	1 per student
Dynamics cart	1 per group
Dynamics track	1 per group
Hanging mass set	1 per group
Electronic or triple beam balance	1 per group
Pulley	2 per group
Stopwatch	1 per student
Ruler	1 per student
Meterstick	1 per group
Investigation Proposal C (optional)	1 per group
Whiteboard, 2' × 3'*	1 per group
Lab Handout	1 per student
Peer-review guide and teacher scoring rubric	1 per student
Checkout Questions	1 per student
Equipment for video analysis (optional)	
Video camera	1 per group
Computer or tablet with video analyis software	1 per group

* As an alternative, students can use computer and presentation software such as Microsoft PowerPoint or Apple Keynote to create their initial arguments.

Be sure to use a set routine for distributing and collecting the materials during the lab investigation. One option is to set up the materials for each group at each group's lab station before class begins. This option works well when there is a dedicated section of the classroom for lab work and the materials are large and difficult to move (such as a dynamics track). A second option is to have all the materials on a table or cart at a central location. You can then assign a member of each group to be the "materials manager." This individual is responsible for collecting all the materials his or her group needs from the table or cart during class and for returning all the materials at the end of the class. This option works well when the materials are small and easy to move (such as stopwatches, metersticks, or hanging masses). It also makes it easy to inventory the materials at the end of the class before students leave for the day.

Safety Precautions

Remind students to follow all normal lab safety rules. In addition, tell students to take the following safety precautions:

1. Wear sanitized safety glasses or goggles during lab setup, hands-on activity, and takedown.

2. Keep fingers and toes out of the way of moving objects.

3. Wash hands with soap and water after completing the lab.

Topics for the Explicit and Reflective Discussion

Reflecting on the Use of Core Ideas and Crosscutting Concepts During the Investigation

Teachers should begin the explicit and reflective discussion by asking students to discuss what they know about the core idea they used during the investigation. The following are some important concepts related to the core idea of forces and motion that students need to use to determine the mathematical relationship among the net force acting on an object, its mass, and its acceleration:

- Quantities such as position, displacement, distance, velocity, speed, and acceleration are used to describe the motion of an object.

- Displacement, velocity, and acceleration are vector quantities.

- *Displacement* is a change in position. *Velocity* is the rate of change of position over a period of time. *Acceleration* is the rate of change in velocity over a period of time.

- A reference frame is needed to determine the direction and the magnitude of the displacement, velocity, and acceleration of an object.

- Force is considered a vector quantity because a force has both magnitude and direction.

- The net force acting on an object is the vector sum of the individual forces acting on it.

- A non-zero net force acting on an object will cause an object to accelerate.

To help students reflect on what they know about forces and motion, we recommend showing them two or three images using presentation software that help illustrate these important ideas. You can then ask the students the following questions to encourage them to share how they are thinking about these important concepts:

1. What do we see going on in this image?

2. Does anyone have anything else to add?

3. What might be going on that we can't see?

4. What are some things that we are not sure about here?

You can then encourage students to think about how CCs played a role in their investigation. There are at least two CCs that students need to use to determine the mathematical relationship among the net force acting on an object, its mass, and its acceleration: (a) Cause and Effect: Mechanism and Explanation and (b) Scale, Proportion, and Quantity (see Appendix 2 [p. 527] for a brief description of these two CCs). To help students reflect on what they know about these CCs, we recommend asking them the following questions:

1. Why is it important to identify causal relationships in science?

2. What did you have to do during your investigation to determine the relationship between any two factors? Why was that useful to do?

3. Why is it important to think about issues related to scale, proportion, and quantity during an investigation?

4. How did you quantify changes in force, mass, and acceleration during your investigation? Why was this useful? What did it allow you to do?

You can then encourage the students to think about how they used all these different concepts to help answer the guiding question and why it is important to use these ideas to help justify their evidence for their final arguments. Be sure to remind your students to explain why they included the evidence in their arguments and make the assumptions underlying their analysis and interpretation of the data explicit in order to provide an adequate justification of their evidence.

Reflecting on Ways to Design Better Investigations

It is important for students to reflect on the strengths and weaknesses of the investigation they designed during the explicit and reflective discussion. Students should therefore be encouraged to discuss ways to eliminate potential flaws, measurement errors, or sources of uncertainty in their investigations. To help students be more reflective about the design of their investigation and what they can do to make their investigations more rigorous in the future, you can ask them the following questions:

1. What were some of the strengths of the way you planned and carried out your investigation? In other words, what made it scientific?

Force, Mass, and Acceleration
What Is the Mathematical Relationship Among the Net Force Exerted on an Object,
the Object's Inertial Mass, and Its Acceleration?

2. What were some of the weaknesses of the way you planned and carried out your investigation? In other words, what made it less scientific?

3. What rules can we make, as a class, to ensure that our next investigation is more scientific?

Reflecting on the Nature of Scientific Knowledge and Scientific Inquiry

This investigation can be used to illustrate two important concepts related to the nature of scientific knowledge and the nature of scientific inquiry: (a) how scientific knowledge changes over time and (b) the difference between laws and theories in science (see Appendix 2 for a brief description of these two concepts). Be sure to review these concepts during and at the end of the explicit and reflective discussion. To help students think about these concepts in relation to what they did during the lab, you can ask them the following questions:

1. Scientific knowledge can and does change over time. Can you tell me why it changes?

2. Can you work with your group to come up with some examples of how scientific knowledge has changed over time? Be ready to share in a few minutes.

3. Laws and theories are different in science. Why are Newton's first and second laws of motion considered laws and not theories?

4. Can you work with your group to come up with a rule that you can use to decide if something is a law or a theory? Be ready to share in a few minutes.

You can also use presentation software or other techniques to encourage your students to think about these concepts. You can show examples of how our thinking about motion has changed over time and ask students to discuss what they think led to those changes. You can also show images of either a law (such as $\mathbf{g} = GM/\mathbf{r}^2$) or a theory (such as *gravity is the curvature of four-dimensional space-time due to the presence of mass*) and ask students to indicate if they think it is a law or a theory and why.

Be sure to remind your students that, to be proficient in science, it is important that they understand what counts as scientific knowledge and how that knowledge develops over time.

Hints for Implementing the Lab

- Allowing students to design their own procedures for collecting data gives students an opportunity to try, to fail, and to learn from their mistakes. However, you can scaffold students as they develop their procedure by having them fill out an investigation proposal. These proposals provide a way for you to offer students hints and suggestions without telling them how to do it. You can also check the proposals quickly during a class period. For this lab we suggest using Investigation Proposal C.

- Have students fill out an investigation proposal for each experiment. The goal of the first experiment is to determine how changing the force acting on an object of a constant mass affects the acceleration of the object. The goal of the second experiment is determine how applying the same force to an object of different mass affects the acceleration of the object.

- Learn how to use the dynamics car and track before the lab begins. It is important for you to know how to use the equipment so you can help students when technical issues arise.

- Allow the students to become familiar with the dynamics car and track as part of the tool talk before they begin to design their investigation. Giving them 5–10 minutes to examine the equipment will let them see what they can and cannot do with it.

- Make sure that the mass of the hanging mass is much greater than the mass of the pulley. This way, the rotational inertia of the pulley will be minimized.

- We recommend using an experimental setup where students affix a string to the dynamics cart, run the string over a pulley, and attach the string to a mass. The students then let the mass fall to the floor, which exerts a constant force on the cart. They can then measure the acceleration of the cart. If students use this setup, we suggest telling them that the force the hanging mass exerts on the cart is equal to 10 times the mass itself. However, students may choose other experimental setups, which will provide a good opportunity for students to think about experimental design because they will need some way of controlling for the force exerted on the cart.

- Finally, as an additional option, you can have students only use a mass on one side of the cart. Thus, the net force acting on the cart will be from the single mass-and-pulley system.

- When students graph their results or run a regression, they may have the force as the independent variable. In other words, students may analyze the acceleration as a function of the force. If students go this route, they will get the following equation for Newton's second law: $\mathbf{a} = \dfrac{1}{m}\mathbf{F}$. This is a valid approach to mathematically describing Newton's second law.

- Be sure to allow students to go back and re-collect data at the end of the argumentation session. Students often realize that they made numerous mistakes when they were collecting data as a result of their discussions during the argumentation session. The students, as a result, will want a chance to re-collect data, and the re-collection of data should be encouraged when time allows.

This also offers an opportunity to discuss what scientists do when they realize a mistake is made inside the lab.

If students use video analysis

- We suggest allowing students to familiarize themselves with the video analysis software before they finalize the procedure for the investigation, especially if they have not used such software previously. This gives students an opportunity to learn how to work with the software and to improve the quality of the video they take.

- Remind students to hold the video camera as still as possible. Any movement of the camera will introduce error into their analysis. If using actual camcorders, we recommend using a tripod to hold the camera steady. If students are using a camera on a cell phone or tablet, we recommend using a table to help steady the camera.

- Remind students to place a meterstick in the same field of view as the motion they are capturing with the video camera. Also, the meterstick should be approximately the same distance from the camera as the motion. Most video analysis software requires the user to define a scale in the video (this allows the software to establish distances and, subsequently, other variables dependent on distance and displacement).

Connections to Standards

Table 5.2 (p. 122) highlights how the investigation can be used to address a specific performance expectation from the *NGSS;* learning objectives from AP Physics 1; learning objectives from AP Physics C: Mechanics, *Common Core State Standards*, in English language arts (*CCSS ELA*); and *Common Core State Standards*, Mathematics (*CCSS Mathematics*).

LAB 5

TABLE 5.2

Lab 5 alignment with standards

NGSS performance expectation	• HS-PS2-1: Analyze data to support the claim that Newton's second law of motion describes the mathematical relationship among the net force on a macroscopic object, its mass, and its acceleration.
AP Physics 1 learning objectives	• 3.A.1.1: The student is able to express the motion of an object using narrative, mathematical, and graphical representations. • 3.A.1.2: The student is able to design an experimental investigation of the motion of an object. • 3.A.1.3: The student is able to analyze experimental data describing the motion of an object and is able to express the results of the analysis using narrative, mathematical, and graphical representations. • 3.A.2.1: The student is able to represent forces in diagrams or mathematically using appropriately labeled vectors with magnitude, direction, and units during the analysis of a situation. • 3.B.1.1: The student is able to predict the motion of an object subject to forces exerted by several objects using an application of Newton's second law in a variety of physical situations with acceleration in one dimension. • 3.B.1.2: The student is able to design a plan to collect and analyze data for motion (static, constant, or accelerating) from force measurements and carry out an analysis to determine the relationship between the net force and the vector sum of the individual forces.
AP Physics C: Mechanics learning objectives	• I.B.2.a: Students should understand the relation between the force that acts on an object and the resulting change in the object's velocity. • I.B.2.b: Students should understand how Newton's second law applies to an object subject to forces such as gravity, the pull of strings, or contact forces. • I.B.2.c: Students should be able to analyze situations in which an object moves with specified acceleration under the influence of one or more forces so they can determine the magnitude and direction of the net force, or of one of the forces that makes up the net force, such as motion up or down with constant acceleration.
Literacy connections (CCSS ELA)	• *Reading:* Key ideas and details, craft and structure, integration of knowledge and ideas • *Writing:* Text types and purposes, production and distribution of writing, research to build and present knowledge, range of writing • *Speaking and listening:* Comprehension and collaboration, presentation of knowledge and ideas

Force, Mass, and Acceleration

What Is the Mathematical Relationship Among the Net Force Exerted on an Object, the Object's Inertial Mass, and Its Acceleration?

Mathematics connections (*CCSS Mathematics*)	• *Mathematical practices:* Make sense of problems and persevere in solving them, reason abstractly and quantitatively, construct viable arguments and critique the reasoning of others, model with mathematics, use appropriate tools strategically, attend to precision, look for and make use of structure, look for and express regularity in repeated reasoning • *Number and quantity:* Reason quantitatively and use units to solve problems, represent and model with vector quantities, perform operations on vectors • *Algebra:* Interpret the structure of expressions, write expressions in equivalent forms to solve problems, create equations that describe numbers or relationships, understand solving equations as a process of reasoning and explain the reasoning, solve equations and inequalities in one variable, represent and solve equations and inequalities graphically • *Functions:* Understand the concept of a function and use function notation; interpret functions that arise in applications in terms of the context; analyze functions using different representations; build a function that models a relationship between two quantities; construct and compare linear and exponential models and solve problems; interpret expressions for functions in terms of the situation they model • *Statistics and probability:* Summarize, represent, and interpret data on two categorical and quantitative variables; interpret linear models; understand and evaluate random processes underlying statistical experiments; make inferences and justify conclusions from sample surveys, experiments, and observational studies

Reference

Newton, I. 1687. *Philosophiae naturalis principia mathematica* [Mathematical principles of natural philosophy]. London: S. Pepys.

Lab Handout

Lab 5. Force, Mass, and Acceleration: What Is the Mathematical Relationship Among the Net Force Exerted on an Object, the Object's Inertial Mass, and Its Acceleration?

Introduction

Western scientific thought was dominated by Aristotle's views on physics for thousands of years. According to Aristotle, all objects in the universe were made of four elements: earth, water, fire, and air. Furthermore, all objects were believed to have both a *natural motion* and a *forced motion*. The natural motion of an element was toward the center of the universe (which, in Aristotle's view, was at the center of the Earth). Not all of the elements, however, had the same degree of natural motion. The elements of earth and water can move close to the center of the universe more easily than the elements of fire and air can. The natural motion of any object, as a result, depended on the relative amount of each element in it. For example, an object made up of half earth and half water would move closer to the center of the universe than an object made up of half earth and half air. The Earth exists, according to Aristotle, because objects made up of "earth" have come to rest as close as possible to the center of the universe. Forced motion, in contrast, was all the other types of motion that could not be described by natural motion. Later theorists elaborated on Aristotle's views and introduced the concept of *impetus*, the property of an object that created forced motion. When an object lost its impetus, it would then move with natural motion toward the center of the universe. In this view, when a ball is thrown in the air, the person throwing it "imparts impetus to the ball." When the ball runs out of impetus, its natural motion causes it to fall back down.

Beginning in the late Middle Ages, several scientists, including Galileo, Copernicus, and Newton, began to question Aristotle's explanation for how objects move. Galileo, for example, demonstrated that objects fall at the same rate independent of their mass, which was contrary to Aristotle's claims that heavy objects fall faster than lighter ones due to their natural motion. Galileo also observed moons orbiting Jupiter, a phenomenon that could not be explained using Aristotle's ideas. Copernicus later claimed that the Sun, and not the Earth, was at the center of the solar system. This claim, along with Galileo's observations of the moons of Jupiter, directly contradicted Aristotle's idea of natural motion because the Earth was no longer viewed as being located at the center of the universe. Isaac Newton delivered the final blow to Aristotle's explanation about how and why things move in his book *Philosophiae Naturalis Principia Mathematica* (commonly known as the *Principia*), which was published in 1687. In the *Principia*, Newton discarded natural and forced motion as an explanatory framework and introduced three basic laws of motion. *Newton's first law*, as it has become known, states that absent a net force acting on an object, an object at rest will

Force, Mass, and Acceleration
What Is the Mathematical Relationship Among the Net Force Exerted on an Object,
the Object's Inertial Mass, and Its Acceleration?

stay at rest and an object in motion will move in a straight line with constant velocity. This law also means that when a net force acts on an object, it will cause it to change how it is moving. In other words, a net force on an object will cause that object to accelerate. This law is so important that the unit for force is called a newton (N).

There are several important implications of Newton's first law. First, the idea of a net force is important, because Newton realized that multiple forces could act on an object at once. The net force, according to Newton, is just the sum of all the forces with respect to their direction. Newton showed that pushing an object to the right with a force of 10 N and pulling the object to the right with a force of 5 N is the same as if only one force of 15 N was moving the object to the right. Figure L5.1 shows another implication of Newton's first law. In this case, two people are each pushing on the box with a force of 25 N, but in opposite directions. According to Newton's first law, a push of 25 N to the left and a push of 25 N to the right results in a net force of zero (0 N). The final implication of Newton's first law is that the net force and the resulting change in motion must be in the same direction. For example, pushing an object with a force of 10 N to the right will cause it to accelerate to the right, and not upward toward the sky. Acceleration is equal to the rate of change of velocity with time, and velocity is equal to the rate of change of position with time.

FIGURE L5.1 _____

Two people pushing on a box with equal forces but in opposite directions

In the *Principia*, Newton also provided a mathematical relationship among the net force acting on an object, that object's mass, and the acceleration of the object. This mathematical relationship is now known as Newton's second law. This mathematical relationship was revolutionary at the time because it not only explained how objects move but also allowed for people to predict changes in an object's motion. In fact, physicists still use this mathematical relationship today to predict the motion of an object when a force acts on it and to determine the amount of force needed to move heavy objects.

LAB 5

Your Task

Use what you know about forces and motion, causal relationships, and the importance of scales, proportions, and quantity in science to design and carry out an investigation that will allow you to determine the mathematical relationship among the net force acting on an object, its mass, and its acceleration.

The guiding question of this investigation is, *What is the mathematical relationship among the net force exerted on an object, the object's inertial mass, and its acceleration?*

Materials

You may use any of the following materials during your investigation:

Consumables
- String or fishing line
- Tape

Equipment
- Safety glasses or goggles (required)
- Dynamics cart
- Dynamics track
- Set of masses
- Electronic or triple beam balance
- 2 Pulleys
- Stopwatch
- Ruler
- Meterstick

If you have access to the following equipment, you may also consider using a video camera and computer or tablet with video analysis software.

Safety Precautions

Follow all normal lab safety rules. In addition, take the following safety precautions:

1. Wear sanitized safety glasses or goggles during lab setup, hands-on activity, and takedown.

2. Keep fingers and toes out of the way of moving objects.

3. Wash hands with soap and water after completing the lab.

Investigation Proposal Required?   ☐ Yes   ☐ No

Getting Started

To answer the guiding question, you will need to design and carry out at least two different experiments. You will need to first determine how changing the force acting on an object of a constant mass affects the acceleration of the object. Then you will need to determine how applying the same force to an object of different mass affects the acceleration of the object. You can conduct these two experiments using a cart-and-track system; Figure L5.2 shows the basic setup of the system.

Force, Mass, and Acceleration

What Is the Mathematical Relationship Among the Net Force Exerted on an Object, the Object's Inertial Mass, and Its Acceleration?

FIGURE L5.2

Cart-and-track system that can be used to determine the mathematical relationship among the net force acting on an object, its mass, and its acceleration

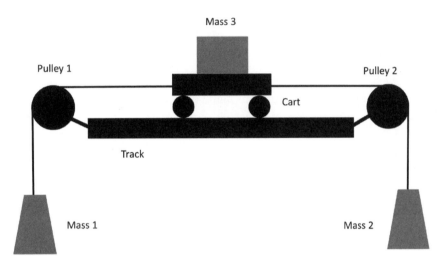

Before you can design and carry out these two experiments, you will need to decide what type of data you need to collect, how you will collect it, and how you will analyze it.

To determine *what type of data you need to collect,* think about the following questions:

- What are the boundaries and components of the system you are studying?
- How do the components of the system interact with each other?
- Which factors might control the rate of change in this system?
- How could you keep track of changes in this system quantitatively?
- What information will you need to be able to determine the acceleration of the cart?
- How will you measure the net force acting on the cart?
- What will be the independent variable and the dependent variable for each experiment?

To determine *how you will collect the data,* think about the following questions:

- How will you change the mass of the cart?
- How will you change the net force acting on the cart?
- What equipment will you need to collect the data?
- What conditions need to be satisfied to establish a cause-and-effect relationship?
- How will you measure the magnitude and direction of any vector quantities?

- How will you make sure that your data are of high quality (i.e., how will you reduce error)?
- How will you keep track of and organize the data you collect?

To determine *how you will analyze the data*, think about the following questions:

- How could you use mathematics to describe a relationship between variables?
- What type of calculations will you need to make?
- What types of graphs and equations signify a proportional relationship?
- What type of table or graph could you create to help make sense of your data?

Once you have carried out your experiments, your group will need to develop a mathematical function that you can use to predict the acceleration of an object based on its mass and the net force acting on it. The last step in this investigation will be to test your function. To accomplish this goal, you can use a different amount of mass on the cart and a different net force to determine if your function leads to accurate predictions about acceleration of the cart under different conditions. If you are able to use your function to make accurate predictions about the motion of the cart under different conditions, then you will be able to generate the evidence you need to convince others that the function you developed is valid.

Connections to the Nature of Scientific Knowledge and Scientific Inquiry

As you work through your investigation, you may want to consider

- how scientific knowledge changes over time, and
- the difference between laws and theories in science.

Initial Argument

Once your group has finished collecting and analyzing your data, your group will need to develop an initial argument. Your argument must include a claim, evidence to support your claim, and a justification of the evidence. The *claim* is your group's answer to the guiding question. The *evidence* is an analysis and interpretation of your data. Finally, the *justification* of the evidence is why your group thinks the evidence matters. The justification of the evidence is important because scientists can use different kinds of evidence to support their claims. Your group will create your initial argument on a whiteboard. Your whiteboard should include all the information shown in Figure L5.3.

Argumentation Session

The argumentation session allows all of the groups to share their arguments. One or two members of each group will stay at the lab station to share that group's argument, while the other members of the group go to the other lab stations to listen to and critique the

Force, Mass, and Acceleration

What Is the Mathematical Relationship Among the Net Force Exerted on an Object, the Object's Inertial Mass, and Its Acceleration?

other arguments. This is similar to what scientists do when they propose, support, evaluate, and refine new ideas during a poster session at a conference. If you are presenting your group's argument, your goal is to share your ideas and answer questions. You should also keep a record of the critiques and suggestions made by your classmates so you can use this feedback to make your initial argument stronger. You can keep track of specific critiques and suggestions for improvement that your classmates mention in the space below.

FIGURE L5.3

Argument presentation on a whiteboard

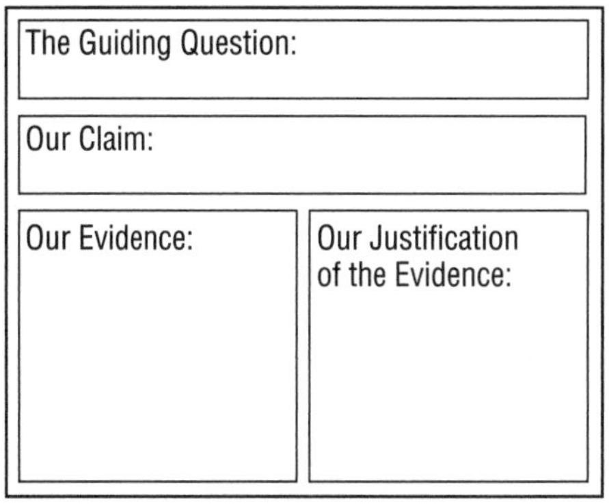

The Guiding Question:	
Our Claim:	
Our Evidence:	Our Justification of the Evidence:

Critiques about our initial argument and suggestions for improvement:

If you are critiquing your classmates' arguments, your goal is to look for mistakes in their arguments and offer suggestions for improvement so these mistakes can be fixed. You should look for ways to make your initial argument stronger by looking for things that the other groups did well. You can keep track of interesting ideas that you see and hear during the argumentation in the space below. You can also use this space to keep track of any questions that you will need to discuss with your team.

Interesting ideas from other groups or questions to take back to my group:

Once the argumentation session is complete, you will have a chance to meet with your group and revise your initial argument. Your group might need to gather more data or design a way to test one or more alternative claims as part of this process. Remember, your goal at this stage of the investigation is to develop the best argument possible.

Report

Once you have completed your research, you will need to prepare an investigation report that consists of three sections. Each section should provide an answer to the following questions:

1. What question were you trying to answer and why?

2. What did you do to answer your question and why?

3. What is your argument?

Your report should answer these questions in two pages or less. This report must be typed, and any diagrams, figures, or tables should be embedded into the document. Be sure to write in a persuasive style; you are trying to convince others that your claim is acceptable or valid!

Reference

Newton, I. 1687. *Philosophiae naturalis principia mathematica* [Mathematical principles of natural philosophy]. London: S. Pepys.

Force, Mass, and Acceleration
What Is the Mathematical Relationship Among the Net Force Exerted on an Object,
the Object's Inertial Mass, and Its Acceleration?

Checkout Questions

Lab 5. Force, Mass and Acceleration: What Is the Mathematical Relationship Among the Net Force Exerted on an Object, the Object's Inertial Mass, and Its Acceleration?

1. In mathematics, two variables are proportional if a change in one variable is always accompanied by a change in another variable. Which, if any, variables from your investigation are proportional? You may choose more than one answer.

   a. Force and mass

   b. Force and acceleration

   c. Mass and acceleration

   What evidence do you have to support your claim?

Use the following information to answer questions 2–4. Two people are playing a game of tug-of-war with the rope attached to a mass of 25 kg at the center. The person pulling to the left pulls with a force of 20 N. The person pulling to the right pulls with a force of 10 N.

2. Which direction will the 25 kg mass move?

   a. Left

   b. Right

   c. It will not move

   How do you know?

3. What will the velocity of the mass be after 1 second?

4. What will the velocity of the mass be after 2 seconds?

5. Two high school physics students are talking, and one says that an acceleration can cause a force, while the other says that a force causes an object to accelerate.

   a. I agree with the first person.
   b. I agree with the second person.

   Use the concept of a cause-and-effect relationship to explain your answer.

6. Force is directly proportional to both mass and acceleration.

   a. I agree with this statement.
   b. I disagree with this statement.

   Use the concept of a proportional relationship to explain your answer.

Force, Mass, and Acceleration
*What Is the Mathematical Relationship Among the Net Force Exerted on an Object,
the Object's Inertial Mass, and Its Acceleration?*

7. Aristotle's views about physics dominated scientific thought for thousands of years. Yet, ideas from Galileo, Newton, and others are now accepted scientific laws and theories. In other words, the accepted "scientific view" of the world changed. Explain what led others to discard Aristotle's views for Galileo's and Newton's, and why it is important for scientists to be open to differing ideas, using an example from your investigation about force, mass, and acceleration.

8. In science, there is a distinction between a law and a theory. What makes laws and theories different in science, and why does this distinction matter? Include an example from your investigation about force, mass, and acceleration in your answer.

LAB 6

Teacher Notes

Lab 6. Forces on a Pulley: How Does the Mass of the Counterweight Affect the Acceleration of a Pulley System?

Purpose

The purpose of this lab is to *introduce* students to the core idea of forces and motion, part of the disciplinary core idea (DCI) of Motion and Stability: Forces and Interactions from the *NGSS*, by having them explore the nature of the forces that act on two bodies connected to each other in a pulley system. This lab also gives students an opportunity to learn about the crosscutting concepts (CCs) of (a) Systems and System Models, (b) Structure and Function, and (c) Stability and Change from the *NGSS*. In addition, this lab can be used to help students understand two big ideas from AP Physics: (a) fields existing in space can be used to explain interactions and (b) the interactions of an object with other objects can be described by forces. As part of the explicit and reflective discussion, students will also learn about (a) the difference between observations and inferences in science and (b) how the culture of science, societal needs, and current events influence the work of scientists.

Underlying Physics Concepts

In mathematical form, Newton's second law is shown in Equation 6.1, where $\sum\mathbf{F}$ is the net force acting on an object, m is the mass of the object, and $\mathbf{a}$ is the acceleration of the object. In SI units, force is measured in units of newtons (N), mass is measured in kilograms (kg), and acceleration is measured in meters per second squared (m/s²).

FIGURE 6.1 _____

A basic pulley system

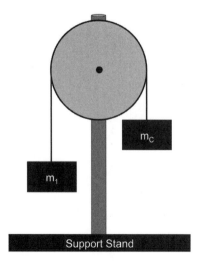

(Equation 6.1) $\sum\mathbf{F} = m\mathbf{a}$

We will use Newton's second law to analyze the movement of the pulley system, as shown in Figure 6.1. Also note that Figure 6.1 illustrates how to set up the equipment for this investigation. The mass on the left (m_1) remains constant, while the mass on the right (m_c) is the counterweight and varies throughout the experiment.

When studying systems of masses connected by a pulley, it is important to determine sign conventions to analyze different forces. Typically, any force that, when isolated, would result in the pulley turning clockwise is considered a force in the positive direction. A force that, when isolated, results in the pulley turning counterclockwise is considered a force in the negative direction. Figure 6.2 shows an example of this for m_1. In Figure 6.2a, the tension in the string, when isolated, would be in the direction associated with a clockwise rotation of the pulley. Thus, $\mathbf{T}_1$ is a positive force. In Figure 6.2b, the force of gravity acting on m_1, when isolated, would cause the pulley to rotate counterclockwise. Thus, $\mathbf{F}_{g1}$ is a negative force.

FIGURE 6.2

The forces acting on m_1: (a) A clockwise and therefore positive force acting on m_1; (b) a counterclockwise and therefore negative force acting on m_1

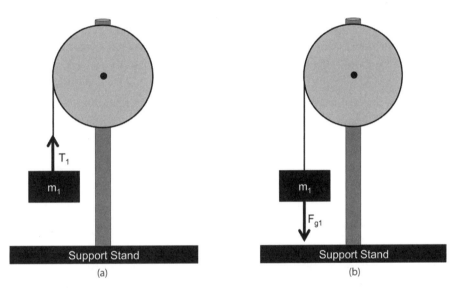

This sign convention is important when analyzing the nature of the forces that are acting on the various components of a pulley system. The sign convention is important because the force of gravity acting on m_1 and the force of gravity acting on m_c are both pointed down, but the force of gravity acting on them is acting on the pulley system in different directions. The force of gravity acting on m_1 acts on the pulley system in the negative direction because it causes the pulley to rotate counterclockwise, whereas the force of gravity acting on m_c acts on the pulley system in the positive direction because it causes the pulley to rotate clockwise.

Finally, it is important to distinguish between forces internal to the system and forces external to the system. Forces external to the system can cause the system as a whole to accelerate. The force of gravity is usually considered an external force.[1] Internal forces, on the other hand, are those forces that act on individual bodies within the system, but when the motion of the system as a whole is analyzed, internal forces can be ignored. As an example, we can see that the tension in the string acts on m_1 in the clockwise direction, so it is a positive force. At the same time, the tension in the string acts on the counterweight in

---

1. There are cases where gravity is an internal force. Gravity can be considered an internal force when the motion of two bodies is modeled as a system due to forces acting on the system as a whole. For example, in studying the motion of Earth and a person standing on Earth as they both move through the solar system, we assume that the gravitational attraction between Earth and the person is an internal force, while the gravitational attraction of the Sun on the Earth-person system is an external force.

the counterclockwise direction, so this is a negative force. Because we assume the tension in the string is constant, when analyzing the motion of the system as a whole the positive tension acting on m_1 is counteracted by the negative tension acting on m_c. These forces, as a result, do not influence the motion of the system. Also, as a rule of thumb, when two objects are connected by a string or rope, the tension in the string is an internal force and can be ignored when analyzing the entire system. However, when analyzing the motion of each mass individually, the tension in the string is no longer considered an internal force, because the system being analyzed is just the single mass.

When applying Newton's second law to our study of the pulley system (both m_1 and m_c), we must first identify the internal and external forces for our system. There are two external forces acting on the system and one internal force. The two external forces are the force of gravity acting on m_1 and m_c, respectively. The internal force is the tension in the string connecting the two masses. Equation 6.2 shows the application of Newton's second law to this system. In Equation 6.2, M_T represents the total mass of the system and is equal to $m_1 + m_c$.

$$\textbf{(Equation 6.2)} \quad \sum \mathbf{F} = \mathbf{F}_{g1} + \mathbf{F}_{gc} = M_T \mathbf{a}$$

Note that because $\mathbf{F}_{g1}$ results in a counterclockwise turn of the pulley, this is a negative force.

The guiding question for this lab asks how the counterweight affects the acceleration of the system. Figure 6.3 provides a graph of the mass of the counterweight versus the acceleration of the system. When there is no counterweight, the acceleration of the other mass will be equal to the acceleration due to gravity (-9.8 m/s^2). As the mass of the counterweight is increased, the acceleration of the system begins to decrease in magnitude. When the mass of the counterweight equals the mass of the other object (m_1), the acceleration of the system will equal zero. Finally, as the mass of the counterweight increases beyond the mass of the other object, the acceleration of the system will approach a positive 9.8 m/s^2.

Timeline

The instructional time needed to complete this lab investigation is 170–230 minutes. Appendix 3 (p. 531) provides options for implementing this lab investigation over several class periods. Option C (230 minutes) should be used if students are unfamiliar with scientific writing, because this option provides extra instructional time for scaffolding the writing process. You can scaffold the writing process by modeling, providing examples, and providing hints as students write each section of the report. Option C can also be used if you are introducing students to the video analysis programs. Option D (170 minutes) should be used if students are familiar with scientific writing and have developed the skills needed to write an investigation report on their own. In option D students complete

FIGURE 6.3 _____

Acceleration of the pulley system: graph of acceleration as a function of the mass of the counterweight

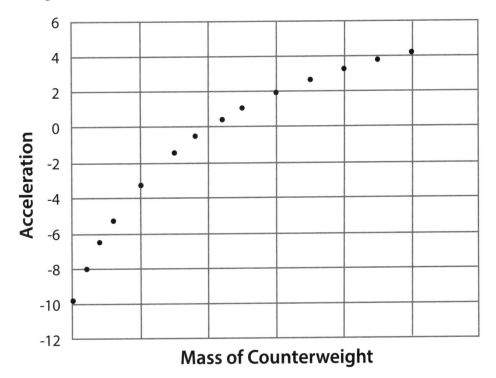

stage 6 (writing the investigation report) and stage 8 (revising the investigation report) as homework.

Materials and Preparation

The materials needed for this investigation are listed in Table 6.1 (p. 138). The consumables and equipment can be purchased from a science supply company such as Carolina, Flinn Scientific, PASCO, Vernier, or Ward's Science. Video analysis software can be purchased from Vernier (Logger *Pro*) or PASCO (SPARKvue or Capstone). These companies also have apps that can be used on Apple- or Android-based tablets and cell phones. We recommend consulting with your school's information technology coordinator to determine the best option for your students.

TABLE 6.1

Materials list for Lab 6

Item	Quantity
Safety glasses or goggles	1 per student
Ring stand	1 per group
Pulley	1 per group
Hanging mass set	1 per group
Electronic or triple beam balance	1 per group
String or fishing line	1 meter per group
Stopwatch	1 per student
Ruler	2–3 per group
Meterstick	1 per group
Investigation Proposal B (optional)	1 per group
Whiteboard, 2' × 3'*	1 per group
Lab Handout	1 per student
Peer-review guide and teacher scoring rubric	1 per student
Checkout Questions	1 per student
Equipment for video analysis (optional)	
Video camera	1 per group
Computer or tablet with video analysis software	1 per group

* As an alternative, students can use computer and presentation software such as Microsoft PowerPoint or Apple Keynote to create their arguments.

Be sure to use a set routine for distributing and collecting the materials during the lab investigation. One option is to set up the materials for each group at each group's lab station before class begins. This option works well when there is a dedicated section of the classroom for lab work and the materials are large and difficult to move (such as a dynamics track). A second option is to have all the materials on a table or cart at a central location. You can then assign a member of each group to be the "materials manager." This individual is responsible for collecting all the materials his or her group needs from the table or cart during class and for returning all the materials at the end of the class. This option works well when the materials are small and easy to move (such as stopwatches, metersticks, or hanging masses). It also makes it easy to inventory the materials at the end of the class before students leave for the day.

Safety Precautions

Remind students to follow all normal lab safety rules. In addition, tell students to take the following safety precautions:

1. Wear sanitized safety glasses or goggles during lab setup, hands-on activity, and takedown.

2. Keep fingers and toes out of the way of moving objects.

3. Wash hands with soap and water after completing the lab.

Topics for the Explicit and Reflective Discussion

Reflecting on the Use of Core Ideas and Crosscutting Concepts During the Investigation

Teachers should begin the explicit and reflective discussion by asking students to discuss what they know about the core idea they used during the investigation. The following are some important concepts related to the core idea of forces and motion that students need to use to explore the relationship between the mass of the counterweight and the acceleration of pulley system:

- Quantities such as position, displacement, distance, velocity, speed, and acceleration are used to describe the motion of an object.
- Displacement, velocity, and acceleration are vector quantities.
- *Displacement* is a change in position. *Velocity* is the rate of change of position over a period of time. *Acceleration* is the rate of change in velocity over a period of time.
- A reference frame is needed to determine the direction and the magnitude of the displacement, velocity, and acceleration of an object.
- Force is considered a vector quantity because a force has both magnitude and direction.
- The net force acting on an object is the vector sum of the individual forces acting on it.
- A *system* is an object or a collection of objects. An object is treated as if it has no internal structure.
- A non-zero net force acting on an object or system of objects will cause the object or system of objects to accelerate.

To help students reflect on what they know about forces and motion, we recommend showing them two or three images using presentation software that help illustrate these important ideas. You can then ask the students the following questions to encourage them to share how they are thinking about these important concepts:

1. What do we see going on in this image?

2. Does anyone have anything else to add?

3. What might be going on that we can't see?

4. What are some things that we are not sure about here?

You can then encourage students to think about how CCs played a role in their investigation. There are at least three CCs that students need to use to determine the relationship between the mass of the counterweight and the acceleration of pulley system: (a) Systems and System Models, (b) Structure and Function, and (c) Stability and Change (see Appendix 2 [p. 527] for a brief description of these CCs). To help students reflect on what they know about these CCs, we recommend asking them the following questions:

1. Why is it useful for scientists to define a system and then make a model of it during an investigation?

2. What are some examples of models that you used during your investigation? How were they useful? What were some of the limitations of your models?

3. The way an object is shaped or structured determines many of its properties and how it functions. Why is it useful to think about the relationship between structure and function during an investigation?

4. How did the structure of the pulley system affect the acceleration of the system as a whole?

5. Scientists are often interested in determining what makes a system stable or unstable and what controls rates of change in system. Why is this an important goal of science?

6. What did you have to do during your investigation to determine how the counterweight contributes to the stability of the pulley system and how it affects the acceleration of the primary mass? Why was that useful to do?

You can then encourage the students to think about how they used all these different concepts to help answer the guiding question and why it is important to use these ideas to help justify their evidence for their final arguments. Be sure to remind your students to explain why they included the evidence in their arguments and make the assumptions underlying their analysis and interpretation of the data explicit in order to provide an adequate justification of their evidence.

Reflecting on Ways to Design Better Investigations

It is important for students to reflect on the strengths and weaknesses of the investigation they designed during the explicit and reflective discussion. Students should therefore be encouraged to discuss ways to eliminate potential flaws, measurement errors, or sources of uncertainty in their investigations. To help students be more reflective about the design

of their investigation and what they can do to make their investigations more rigorous in the future, you can ask them the following questions:

1. What were some of the strengths of the way you planned and carried out your investigation? In other words, what made it scientific?

2. What were some of the weaknesses of the way you planned and carried out your investigation? In other words, what made it less scientific?

3. What rules can we make, as a class, to ensure that our next investigation is more scientific?

Reflecting on the Nature of Scientific Knowledge and Scientific Inquiry

This investigation can be used to illustrate two important concepts related to the nature of scientific knowledge and the nature of scientific inquiry: (a) the difference between observations and inferences in science and (b) how the culture of science, societal needs, and current events influence the work of scientists (see Appendix 2 for a brief description of these two concepts). Be sure to review these concepts during and at the end of the explicit and reflective discussion. To help students think about these concepts in relation to what they did during the lab, you can ask them the following questions:

1. You had to make observations and inferences during your investigation. Can you give some examples of these observations and inferences?

2. Can you work with your group to come up with a rule that you can use to determine if a piece of information is an observation or an inference? Be ready to share in a few minutes.

3. People view some types of research as being more important than other types of research because of cultural values and current events. Can you come up with some examples of how cultural values and current events have influenced the work of scientists?

4. Scientists share a set of values, norms, and commitments that shape what counts as knowing, how to represent or communicate information, and how to interact with other scientists. Can you work with your group to come up with a rule that you can use to decide if something is science or not science? Be ready to share in a few minutes.

You can also use presentation software or other techniques to encourage your students to think about these concepts. You can show examples of information from the investigation that are either observations or inferences and ask students to classify each example and explain their thinking. You can also show examples of research projects that were influenced by cultural values and current events and ask students to think about what was

LAB 6

going on at the time and why that research was viewed as being important for the greater good.

Be sure to remind your students that, to be proficient in science, it is important that they understand what counts as scientific knowledge and how that knowledge develops over time.

Hints for Implementing the Lab

- Allowing students to design their own procedures for collecting data gives students an opportunity to try, to fail, and to learn from their mistakes. However, you can scaffold students as they develop their procedure by having them fill out an investigation proposal. These proposals provide a way for you to offer students hints and suggestions without telling them how to do it. You can also check the proposals quickly during a class period. For this lab we suggest using Investigation Proposal B.

- Allow students to become familiar with the equipment as part of the tool talk before they begin to design their investigation. Giving them 5–10 minutes to examine the equipment will let them see what they can and cannot do with it. This also gives students a chance to begin thinking about what variables to test and how to control for other variables.

- Make sure that the mass of each hanging mass is much greater than the mass of the pulley. This way, the rotational inertia of the pulley will be minimized.

- If students use a range of masses for the counterweight that are close to the mass of the main weight in a two-mass system, the relationship will likely appear linear. For example, if students use 200 g masses for m_1, then mass values for the counterweight between 150 g and 250 g will appear to be linear. If students are providing a general answer to the guiding question, then the local linearity does not pose a problem. However, if you choose to have them create a mathematical model of the system, then they should use a broader range of values. One way to help them is to ask them what the acceleration would be if there were no counterweight at all. In this case, the acceleration would be the acceleration due to gravity (**g**), or 9.8 m/s². However, if students extrapolate their linear model to the case where there is no counterweight, they will find that it does not give a value equal to 9.8 m/s² (or −9.8m/s², if students choose a different direction to be the positive direction).

- Students may forget to establish the sign conventions during their initial data collection. We suggest that you let students proceed in this manner, because it provides an important learning opportunity regarding the importance of establishing such conventions.

- Be sure to allow students to go back and re-collect data at the end of the argumentation session. Students often realize that they made numerous mistakes when they were collecting data as a result of their discussions during the argumentation session. The students, as a result, will want a chance to re-collect data, and the re-collection of data should be encouraged when time allows. This also offers an opportunity to discuss what scientists do when they realize a mistake is made inside the lab.

If students use video analysis

- We suggest allowing students to familiarize themselves with the video analysis software before they finalize the procedure for the investigation, especially if they have not used such software previously. This gives students an opportunity to learn how to work with the software and to improve the quality of the video they take.

- Remind students to hold the video camera as still as possible. Any movement of the camera will introduce error into their analysis. If using actual camcorders, we recommend using a tripod to hold the camera steady. If students are using a camera on a cell phone or tablet, we recommend using a table to help steady the camera.

- Remind students to place a meterstick in the same field of view as the motion they are capturing with the video camera. Also, the meterstick should be approximately the same distance from the camera as the motion. Most video analysis software requires the user to define a scale in the video (this allows the software to establish distances and, subsequently, other variables dependent on distance and displacement).

Connections to Standards

Table 6.2 (p. 144) highlights how the investigation can be used to address specific performance expectations from the *NGSS*; learning objectives from AP Physics 1; learning objectives from AP Physics C: Mechanics, *Common Core State Standards*, in English language arts (*CCSS ELA*); and *Common Core State Standards*, Mathematics (*CCSS Mathematics*).

LAB 6

TABLE 6.2

Lab 6 alignment with standards

***NGSS* performance expectation**	• HS-PS2-1: Analyze data to support the claim that Newton's second law of motion describes the mathematical relationship among the net force on a macroscopic object, its mass, and its acceleration.
AP Physics 1 learning objectives	• 3.A.1.2: The student is able to design an experimental investigation of the motion of an object. • 3.A.1.3: The student is able to analyze experimental data describing the motion of an object and is able to express the results of the analysis using narrative, mathematical, and graphical representations. • 3.B.1.2: The student is able to design a plan to collect and analyze data for motion (static, constant, or accelerating) from force measurements and carry out an analysis to determine the relationship between the net force and the vector sum of the individual forces. • 3.B.1.3: The student is able to reexpress a free-body diagram representation into a mathematical representation and solve the mathematical representation for the acceleration of the object. • 4.A.2.1: The student is able to make predictions about the motion of a system based on the fact that acceleration is equal to the change in velocity per unit time, and velocity is equal to the change in position per unit time.
AP Physics C: Mechanics learning objectives	• I.B.2.b: Students should understand how Newton's second law applies to an object subject to forces such as gravity, the pull of strings, or contact forces. • I.B.2.c: Students should be able to analyze situations in which an object moves with specified acceleration under the influence of one or more forces so they can determine the magnitude and direction of the net force, or of one of the forces that makes up the net force, such as motion up or down with constant acceleration. • I.B.3.c: Students should know that the tension is constant in a light string that passes over a massless pulley and should be able to use this fact in analyzing the motion of a system of two objects joined by a string.
Literacy connections (*CCSS ELA*)	• *Reading:* Key ideas and details, craft and structure, integration of knowledge and ideas • *Writing:* Text types and purposes, production and distribution of writing, research to build and present knowledge, range of writing • *Speaking and listening:* Comprehension and collaboration, presentation of knowledge and ideas

Mathematics connections (*CCSS Mathematics*)	• *Mathematical practices:* Reason abstractly and quantitatively, construct viable arguments and critique the reasoning of others, use appropriate tools strategically, attend to precision • *Number and quantity:* Reason quantitatively and use units to solve problems, represent and model with vector quantities, perform operations on vectors • *Algebra:* Interpret the structure of expressions, understand solving equations as a process of reasoning and explain the reasoning, represent and solve equations and inequalities graphically • *Functions:* Understand the concept of a function and use function notation, interpret functions that arise in applications in terms of the context, interpret expressions for functions in terms of the situation they model • *Statistics and probability:* Summarize, represent, and interpret data on two categorical and quantitative variables; interpret linear models; understand and evaluate random processes underlying statistical experiments; make inferences and justify conclusions from sample surveys, experiments, and observational studies

LAB 6

Lab Handout

Lab 6. Forces on a Pulley: How Does the Mass of the Counterweight Affect the Acceleration of a Pulley System?

Introduction

There are many products that we now take for granted but were revolutionary when they were first invented. One of these products is the elevator. When elevators were first invented, there were numerous design problems related to moving the car up and down an elevator shaft. One of these problems was controlling the acceleration of the car as the mass of the car changed with the addition of people. In 1861, Elisha G. Otis patented a new design for an elevator; Figure L6.1 shows the patent drawing. To help moderate the acceleration of the car, Otis added a counterweight to the system. The counterweight is attached to the elevator via two pulleys at the top of the elevator shaft. This basic idea is still used in elevators today.

FIGURE L6.1

Drawings from Otis's 1861 patent application for an elevator

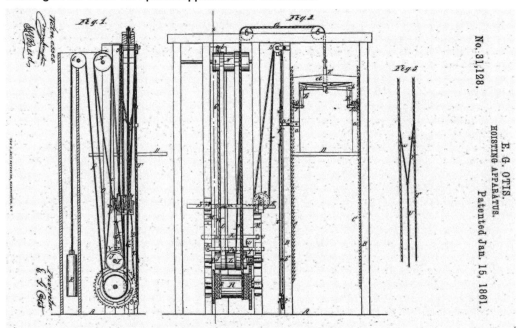

The basis for Otis's decision to add the counterweight to the system came from Newton's second law of motion, which states that the acceleration of a system is proportional to the net forces acting on the system. Acceleration is equal to the rate of change of velocity with time, and velocity is equal to the rate of change of position with time. A counterweight works because the force of gravity pulls both the elevator car down and the counterweight down. Because the counterweight is on a different side of the pulley from the elevator car, the force of gravity on the counterweight acts in a different direction (relative to the pulley) than the force of gravity acting on the elevator car. Thus, the total acceleration of the elevator car and counterweight, in accordance with Newton's second law of motion, is equal to the sum of the forces acting on each part of the system divided by the total mass of the system.

Otis's elevator design is an application of one class of simple machines called pulleys. The first scientist to study simple machines was Archimedes. Although he did some research on the behavior of pulleys, his main focus was on the behavior of a simple machine called the screw. Renaissance scientists took up the mantle from Archimedes and began to explore the behavior of other simple machines, including wheels, levers, and pulleys. However, it was not until the work of Isaac Newton that scientists were able to mathematically model the motion of simple machines. This ability to use equations to describe the forces acting on a simple machine allowed physicists and engineers to apply these equations to the design of new systems that incorporated much more complicated machines. Otis, for example, applied the mathematical insights of Newton's laws to study pulleys that could be used to improve on the design of an elevator.

Your Task

Use what you know about forces and motion, systems and system models, stability and change in systems, and the relationship between structure and function in designed systems to design and carry out an investigation to explore the relationship between the mass of the counterweight and the acceleration of the pulley system. You will then develop a conceptual model that can be used to explain the behavior of this simple machine. Once you have developed your model, you will need to test it to determine if it allows you to make accurate predictions about its behavior.

The guiding question of this investigation is, *How does the mass of the counterweight affect the acceleration of a pulley system?*

Materials

You may use any of the following materials during your investigation:

- Safety glasses or goggles (required)
- Ring stand
- Pulley
- Hanging mass set
- Electronic or triple beam balance
- String or fishing line
- Stopwatch
- Ruler
- Meterstick

If you have access to the following equipment, you may also consider using a video camera and a computer or tablet with video analysis software.

Safety Precautions

Follow all normal lab safety rules. In addition, take the following safety precautions:

1. Wear sanitized safety glasses or goggles during lab setup, hands-on activity, and takedown.

2. Keep your fingers and toes out of the way of moving objects.

3. Wash hands with soap and water after completing the lab.

Investigation Proposal Required?   ☐ Yes   ☐ No

Getting Started

The first step in developing your model is to determine how the mass of a counterweight affects the acceleration of a pulley system. One way to gather the information you need is to create a physical model of the system to see how it behaves. In this investigation, you can use a set of hanging masses and a pulley attached to a support stand to study the movement of the entire system. Figure L6.2 shows how you can set up a pulley system using the available materials. Before you can design your investigation, however, you must determine what type of data you need to collect, how you will collect it, and how you will analyze it.

To determine *what type of data you need to collect,* think about the following questions:

- What are the boundaries and components of the system you are studying?
- How might the structure of the pulley system determine its function?
- Which factor(s) might control the rate of change in this system?
- How could you keep track of changes in this system quantitatively?
- What forces are acting on the system?
- What forces act on each mass?
- How will you determine the acceleration of the system?
- What will be the independent variable and the dependent variable?

FIGURE L6.2

Equipment used to explore the movement of a pulley system

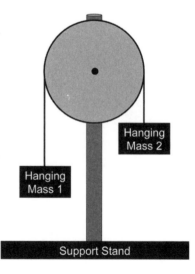

To determine *how you will collect the data,* think about the following questions:

- What other factors do you need to control for as you collect data?
- What measurement scale or scales should you use to collect data?
- What equipment will you need to take the measurements?
- For any vector quantities, which directions will be positive and which directions will be negative?
- How will you make sure that your data are of high quality (i.e., how will you reduce error)?
- How will you keep track of and organize the data you collect?

To determine *how you will analyze the data*, think about the following questions:

- How could you use mathematics to describe a change over time?
- How could you use mathematics to describe a relationship between variables?
- What type of calculations will you need to make?
- What will be the positive convention and what will be the negative convention?

Once you have determined how the mass of the counterweight affects the acceleration of a pulley system, your group will need to develop a conceptual model to explain it. Your model also must include information about all the forces acting on each component or structure found in the entire pulley system.

The last step in this investigation is to test your model. To accomplish this goal, you can use different hanging masses (ones that you did not use earlier) to determine if your model enables you to make accurate predictions about the acceleration of the entire pulley system. If you are able to use your model to make accurate predictions about the function of the system based on the structure of that system, then you will be able to generate the evidence you need to convince others that your model is valid.

Connections to the Nature of Scientific Knowledge and Scientific Inquiry

As you work through your investigation, you may want to consider

- the difference between observations and inferences in science, and
- how the culture of science, societal needs, and current events influence the work of scientists.

Initial Argument

Once your group has finished collecting and analyzing your data, your group will need to develop an initial argument. Your argument must include a claim, evidence to support your claim, and a justification of the evidence. The *claim* is your group's answer to the guiding question. The *evidence* is an analysis and interpretation of your data. Finally, the *justification* of the evidence is why your group thinks the evidence matters. The justification of the evidence is important because scientists can use different kinds of evidence to support their claims. Your group will create your initial argument on a whiteboard. Your whiteboard should include all the information shown in Figure L6.3.

Argumentation Session

The argumentation session allows all of the groups to share their arguments. One or two members of each group will stay at the lab station to share that group's argument, while the other members of the group go to the other lab stations to listen to and critique the other arguments. This is similar to what scientists do when they propose, support, evaluate, and refine new ideas during a poster session at a conference. If you are presenting your group's argument, your goal is to share your ideas and answer questions. You should also keep a record of the critiques and suggestions made by your classmates so you can use this feedback to make your initial argument stronger. You can keep track of specific critiques and suggestions for improvement that your classmates mention in the space below.

Critiques about our initial argument and suggestions for improvement:

FIGURE L6.3

Argument presentation on a whiteboard

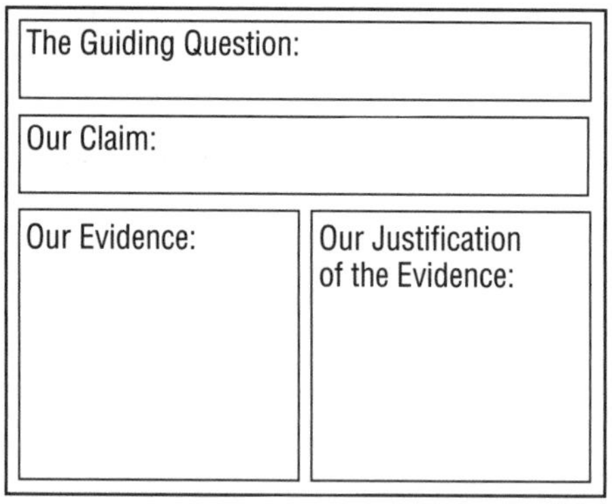

The Guiding Question:	
Our Claim:	
Our Evidence:	Our Justification of the Evidence:

If you are critiquing your classmates' arguments, your goal is to look for mistakes in their arguments and offer suggestions for improvement so these mistakes can be fixed. You should look for ways to make your initial argument stronger by looking for things that the other groups did well. You can keep track of interesting ideas that you see and hear during the argumentation in the space below. You can also use this space to keep track of any questions that you will need to discuss with your team.

Interesting ideas from other groups or questions to take back to my group:

Once the argumentation session is complete, you will have a chance to meet with your group and revise your initial argument. Your group might need to gather more data or design a way to test one or more alternative claims as part of this process. Remember, your goal at this stage of the investigation is to develop the best argument possible.

Report

Once you have completed your research, you will need to prepare an investigation report that consists of three sections. Each section should provide an answer to the following questions:

1. What question were you trying to answer and why?

2. What did you do to answer your question and why?

3. What is your argument?

Your report should answer these questions in two pages or less. This report must be typed, and any diagrams, figures, or tables should be embedded into the document. Be sure to write in a persuasive style; you are trying to convince others that your claim is acceptable or valid!

LAB 6

Lab 6. Forces on a Pulley: How Does the Mass of the Counterweight Affect the Acceleration of a Pulley System?

Use the information in the figure at right and Newton's laws of motion to answer questions 1 and 2. In the figure, assume that the 5 g mass is the counterweight.

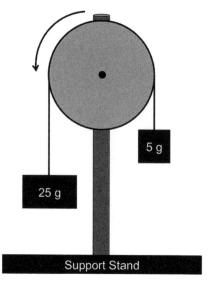

1. What is the tension in the string connecting the two masses?

    How do you know?

2. Is the tension in the string constant, or does it change depending on the mass of the counterweight? In answering this question, assume the string and the pulley are massless.

3. In physics, it is important to define the system under study in order to understand how objects move.

    a. I agree with this statement.

    b. I disagree with this statement.

Explain your answer, using an example from your investigation about a pulley system.

4. How a system is structured influences how that system functions.

    a. I agree with this statement.
    b. I disagree with this statement.

Explain your answer, using an example from your investigation about a pulley system.

5. In science, there is a distinction between observation and inference. Explain why this distinction is important, using an example from your investigation about a pulley system.

6. People view some research as being more important than other research in science because of the cultural values, societal needs, or current events. On another sheet of paper, explain why it is important to understand how the culture of science, societal needs, and current events influence the work of scientists, using an example from your investigation about a pulley system.

LAB 7

Teacher Notes

Lab 7. Forces on an Incline: What Is the Mathematical Relationship Between the Angle of Incline and the Acceleration of an Object Down the Incline?

Purpose

The purpose of this lab is to *introduce* students to the core idea of forces and motion, part of the disciplinary core idea (DCI) of Motion and Stability: Forces and Interactions from the *NGSS*, by having them explore the motion of an object moving down an incline plane with minimal friction. This lab also gives students an opportunity to learn about the cross-cutting concepts (CCs) of (a) Patterns and (b) Structure and Function from the *NGSS*. In addition, this lab can be used to help students understand three big ideas from AP Physics: (a) objects and systems have properties such as mass and charge, (b) fields existing in space can be used to explain interactions, and (c) the interactions of an object with other objects can be described by forces. As part of the explicit and reflective discussion, students will also learn about (a) the difference between observations and inferences in science and (b) how scientists use different methods to answer different types of questions.

Underlying Physics Concepts

To analyze the motion of an object on an incline plane, it helps to first draw a force diagram. Figure 7.1 is a general force diagram for an object on an incline. In this diagram, F_g is the force of gravity, $F_{g\parallel}$ is the component of the gravitational force that is parallel to the incline plane (and points down the plane), $F_{g\perp}$ is the component of the gravita-tional force that is perpendicular to the incline plane, F_N is the normal force acting on the object due to its resting on the plane, and F_{fr} is the frictional force. Finally, θ_1 is the angle of incline of the plane, and θ_2 is the angle between Fg and $Fg_\perp$.

Because the incline plane is also the hypotenuse to a right triangle, we can use some simple geometric and trigonometric relationships to determine the following:

FIGURE 7.1

Force diagram on an incline plane

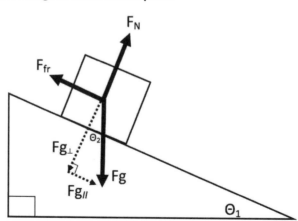

Forces on an Incline

What Is the Mathematical Relationship Between the Angle of Incline and the Acceleration of an Object Down the Incline?

1. The measure of θ_1 is equal to the measure of θ_2.

2. $\mathbf{F}_{g\parallel} = \mathbf{F}_g \sin\theta$

3. $\mathbf{F}_{g\perp} = \mathbf{F}_g \cos\theta$

Using Newton's second law—see Equation 7.1, where $\sum \mathbf{F}$ is the sum of the forces acting on an object, m is the mass of the object, and a is the acceleration—we can analyze the motion of the object on the incline. In SI units, force is measured in newtons (N); mass is measured in kilograms (kg); and acceleration is measured in meters per second squared (m/s^2).

$$\textbf{(Equation 7.1)} \quad \sum \mathbf{F} = m\mathbf{a}$$

It also helps to analyze the motion perpendicular to the incline and the motion parallel to the incline separately from each other. Starting with the perpendicular direction, we use our force diagram to identify those forces acting in the perpendicular direction, which are the normal force ($\mathbf{F}_N$) and the component of gravity in the perpendicular direction ($\mathbf{F}_{g\perp}$). We then set up the relationship shown in Equation 7.2:

$$\textbf{(Equation 7.2)} \quad \sum \mathbf{F} = \mathbf{F}_{g\perp} + \mathbf{F}_N$$

Because the object remains in contact with the surface of the incline, we know that there is not any motion in the perpendicular direction. Nor is there any net acceleration in the perpendicular direction. This means that $\sum \mathbf{F}_\perp = 0$ and that $\mathbf{F}_N = -\mathbf{F}_{g\perp}$.

Analyzing the motion in the parallel direction is a bit more complex. Drawing on Equation 7.1, we can set up the relationship shown in Equation 7.3:

$$\textbf{(Equation 7.3)} \quad \sum \mathbf{F} = \mathbf{F}_{g\parallel} + \mathbf{F}_{fr} = m\mathbf{a}_\parallel$$

Often in physics, we begin to analyze situations under the assumption that there is no frictional force acting on the object. If the frictional force is zero, then the acceleration down the incline is due to the parallel component of the force of gravity. Using Equation 7.3, and the known relationship that $\mathbf{F}_{g\parallel} = \mathbf{F}_g \sin\theta$, we get the following relationship: $m\mathbf{a}_\parallel = mg\sin\theta$. The mass on each side of the equation divides to 1, so the acceleration of an object down the incline (when friction is assumed to be zero) gives us Equation 7.4:

$$\textbf{(Equation 7.4)} \quad \mathbf{a}_\parallel = g\sin\theta$$

When the angle is 0°, sinθ has a value of 0; when the angle is 90°, sinθ has a value of 1. Thus, as the angle of incline increases from 0 to 90, the magnitude of the acceleration down the incline increases from a minimum value of 0 m/s² to 9.8 m/s².

When friction acts on the object, the motion is more complex. Equation 7.5 shows how to calculate the force of friction between two objects, where F_{fr} is the frictional force, F_N is the normal force, and μ is the coefficient of friction between the two objects.

$$\textbf{(Equation 7.5)} \quad \mathbf{F}_{fr} = \mu \mathbf{F}_N$$

Substituting the right side of Equation 7.5 and our geometrically derived relationships regarding the component forces of gravity into Equation 7.3, we get the following relationship, $m\mathbf{a} = mg\sin\theta - \mu mg\cos\theta$. The mass on each side of the equation divides to 1, so the acceleration of an object down an incline is given by Equation 7.6.

$$\textbf{(Equation 7.6)} \quad \mathbf{a}_{\|} = g\sin\theta - \mu g\cos\theta$$

Notice that the mass of the object does not affect the acceleration down the incline in either the case where we ignore friction or the case where we account for friction.

Finally, depending on the experimental setup (if students use a sum-of-the-forces approach, as opposed to using video analysis software), students may add an additional force to the force diagram. Figure 7.2 shows the experimental setup that we anticipate students will use if they do not have access to video analysis software. Figure 7.2 also shows the external forces acting on the cup and the object on the incline plane using dotted lines.

FIGURE 7.2

A possible experimental setup

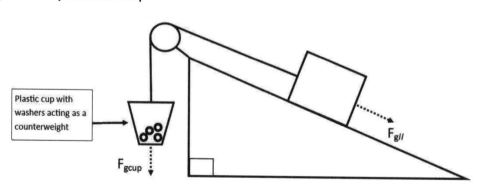

Plastic cup with washers acting as a counterweight

In this situation, students again start with the basic equation $\sum \mathbf{F} = m\mathbf{a}$. Furthermore, they can treat the object, cup, and string as one system that moves due to the external forces

Forces on an Incline

What Is the Mathematical Relationship Between the Angle of Incline and the Acceleration of an Object Down the Incline?

acting on the system as a whole. Finally, in Figure 7.2, we are ignoring friction. Thus, they set up the relationship shown in Equation 7.7:

$$\textbf{(Equation 7.7)} \quad M_T\mathbf{a} = \mathbf{F}_{gcup} + \mathbf{F}_{g/\!/}$$

It is important to realize that the mass value (M_T) on the left hand of the equation represents the total mass of the system—the mass of the cup, plus the mass of the washers, plus the mass of the object on the incline (the mass of the string is typically ignored). Furthermore, $\mathbf{F}_{gcup}$ is the force due to gravity on the cup and washers only, which means that when calculating $\mathbf{F}_{gcup}$ we only use the mass of the cup plus the mass of the washers. Likewise, $\mathbf{Fg}_{/\!/}$ is calculated only using the mass for the object on the incline plane. When the system is at rest, this means that the net acceleration of the system is zero. We can rearrange Equation 7.7 and substitute in for both $\mathbf{F}_{gcup}$ and $\mathbf{F}_{g/\!/}$ to get Equation 7.8:

$$\textbf{(Equation 7.8)} \quad m_{cup+washers}\mathbf{g} = m_{object}\mathbf{g}\sin\theta$$

As can be seen in Equation 7.8, when the angle of the incline is small, then the component force of gravity down the incline is small. This means that only a small number of washers are needed inside the cup to balance the force down the incline. As the angle of the incline increases, it will also require more washers in the cup to balance the component force down the incline.

Timeline

The instructional time needed to complete this lab investigation is 200–280 minutes. Appendix 3 (p. 531) provides options for implementing this lab investigation over several class periods. Option E (280 minutes) should be used if students are unfamiliar with scientific writing, because this option provides extra instructional time for scaffolding the writing process. You can scaffold the writing process by modeling, providing examples, and providing hints as students write each section of the report. Option E can also be used if you are introducing students to the video analysis programs. Option F (200 minutes) should be used if students are familiar with scientific writing and have developed the skills needed to write an investigation report on their own. In option F, students complete stage 6 (writing the investigation report) and stage 8 (revising the investigation report) as homework.

Materials and Preparation

The materials needed to implement this investigation are listed in Table 7.1 (p. 158). The equipment can be purchased from a science supply company such as Carolina, Flinn Scientific, PASCO, Vernier, or Ward's Science. Also, much of the equipment used in this lab, such as the toy cars and plastic cups, can be purchased from local stores such as Wal-Mart or Target. Video analysis software can be purchased from Vernier (Logger *Pro*) or PASCO (SPARKvue or Capstone). These companies also have apps that can be used on Apple- or Android-based tablets and cell phones. We recommend consulting with your school's information technology coordinator to determine the best option for your students.

TABLE 7.1

Materials list for Lab 7

Item	Quantity
Consumables	
Tape	As needed per group
String or fishing line	As needed per group
Equipment and other materials	
Safety glasses or goggles	1 per student
Plastic cup	1 per group
Small toy car or cart	1 per group
Electronic or triple beam balance	1 per group
Small masses (e.g., washers or pennies)	Several per group
Meterstick	2–3 per group
Protractor	2–3 per group
Ramp or board to roll car down	1 per group
Stopwatch	2–3 per group
Pulley	1 per group
Investigation Proposal A (optional)	1 per group
Whiteboard, 2' × 3'*	1 per group
Lab Handout	1 per student
Peer-review guide and teacher scoring rubric	1 per student
Checkout Questions	1 per student
Equipment for video analysis (optional)	
Video camera	1 per group
Computer or tablet with video analysis software	1 per group

* As an alternative, students can use computer and presentation software such as Microsoft PowerPoint or Apple Keynote to create their arguments.

Forces on an Incline

What Is the Mathematical Relationship Between the Angle of Incline and the Acceleration of an Object Down the Incline?

Be sure to use a set routine for distributing and collecting the materials during the lab investigation. One option is to set up the materials for each group at each group's lab station before class begins. This option works well when there is a dedicated section of the classroom for lab work and the materials are large and difficult to move (such as a dynamics track). A second option is to have all the materials on a table or cart at a central location. You can then assign a member of each group to be the "materials manager." This individual is responsible for collecting all the materials his or her group needs from the table or cart during class and for returning all the materials at the end of the class. This option works well when the materials are small and easy to move (such as stopwatches, metersticks, or hanging masses). It also makes it easy to inventory the materials at the end of the class before students leave for the day.

Safety Precautions

Remind students to follow all normal lab safety rules. In addition, tell the students to take the following safety precautions:

1. Wear sanitized safety glasses or goggles during lab setup, hands-on activity, and takedown.

2. Keep fingers and toes out of the way of moving objects.

3. Do not stand on tables or chairs.

4. Wash hands with soap and water after completing the lab activity.

Topics for the Explicit and Reflective Discussion

Reflecting on the Use of Core Ideas and Crosscutting Concepts During the Investigation

Teachers should begin the explicit and reflective discussion by asking students to discuss what they know about the core idea they used during the investigation. The following are some important concepts related to the core idea of forces and motion that students need to use to determine the mathematical relationship between the angle of incline and the acceleration of an object as it moves down the incline:

- Quantities such as position, displacement, distance, velocity, speed, and acceleration are used to describe the motion of an object.

- Displacement, velocity, and acceleration are vector quantities.

- *Displacement* is a change in position. *Velocity* is the rate of change of position over a period of time. *Acceleration* is the rate of change in velocity over a period of time.

- Force is considered a vector quantity because a force has both magnitude and direction.

- The net force acting on an object is the vector sum of the individual forces acting on it.

- A *system* is an object or a collection of objects. An object is treated as if it has no internal structure.

- Gravity is an attractive force between two objects that have mass.

- There may be forces acting on an object at rest.

- A non-zero net force acting on an object or system of objects will cause the object or system of objects to accelerate.

To help students reflect on what they know about forces and motion, we recommend showing them two or three images using presentation software that help illustrate these important ideas. You can then ask the students the following questions to encourage them to share how they are thinking about these important concepts:

1. What do we see going on in this image?

2. Does anyone have anything else to add?

3. What might be going on that we can't see?

4. What are some things that we are not sure about here?

You can then encourage students to think about how CCs played a role in their investigation. There are at least two CCs that students need to use to determine the mathematical relationship between the angle of incline and the acceleration of an object as it moves down the incline: (a) Patterns and (b) Structure and Function (see Appendix 2 [p. 527] for a brief description of these CCs). To help students reflect on what they know about these CCs, we recommend asking them the following questions:

1. Why is it important to look for patterns during an investigation?

2. What patterns did you identify during your investigation? What did the identification of these patterns allow you to do?

3. The way an object is shaped or structured determines many of its properties and how it functions. Why is it useful to think about the relationship between structure and function during an investigation?

4. How did the structure of the incline affect the acceleration of the toy car? How did the structure of the toy car affect its function?

You can then encourage the students to think about how they used all these different concepts to help answer the guiding question and why it is important to use these ideas to help justify their evidence for their final arguments. Be sure to remind your students to explain why they included the evidence in their arguments and make the assumptions

Forces on an Incline

What Is the Mathematical Relationship Between the Angle of Incline and the Acceleration of an Object Down the Incline?

underlying their analysis and interpretation of the data explicit in order to provide an adequate justification of their evidence.

Reflecting on Ways to Design Better Investigations

It is important for students to reflect on the strengths and weaknesses of the investigation they designed during the explicit and reflective discussion. Students should therefore be encouraged to discuss ways to eliminate potential flaws, measurement errors, or sources of uncertainty in their investigations. To help students be more reflective about the design of their investigation and what they can do to make their investigations more rigorous in the future, you can ask them the following questions:

1. What were some of the strengths of the way you planned and carried out your investigation? In other words, what made it scientific?

2. What were some of the weaknesses of the way you planned and carried out your investigation? In other words, what made it less scientific?

3. What rules can we make, as a class, to ensure that our next investigation is more scientific?

Reflecting on the Nature of Scientific Knowledge and Scientific Inquiry

This investigation can be used to illustrate two important concepts related to the nature of scientific knowledge and the nature of scientific inquiry: (a) the difference between observations and inferences in science and (b) how scientists use different methods to answer different types of questions (see Appendix 2 for a brief description of these two concepts). Be sure to review these concepts during and at the end of the explicit and reflective discussion. To help students think about these concepts in relation to what they did during the lab, you can ask them the following questions:

1. You had to make observations and inferences during your investigation. Can you give some examples of these observations and inferences?

2. Can you work with your group to come up with a rule that you can use to determine if a piece of information is an observation or an inference? Be ready to share in a few minutes.

3. There is no universal step-by-step scientific method that all scientists follow. Why do you think there is no universal scientific method?

4. Think about what you did during this investigation. How would you describe the method you used to determine the mathematical relationship between the angle of incline and the acceleration of an object as it moves down the incline? Why would you call it that?

LAB 7

You can also use presentation software or other techniques to encourage your students to think about these concepts. You can show examples of information from the investigation that are either observations or inferences and ask students to classify each example and explain their thinking. You can also show one or more images of "a universal scientific method" that misrepresent the nature of scientific inquiry (see, e.g., *https://commons.wikimedia.org/wiki/File:The_Scientific_Method_as_an_Ongoing_Process.svg*) and ask students why each image is *not* a good representation of what scientists do to develop scientific knowledge. You can ask students to suggest revisions to the image that would make it more consistent with the way scientists develop scientific knowledge.

Be sure to remind your students that, to be proficient in science, it is important that they understand what counts as scientific knowledge and how that knowledge develops over time.

Hints for Implementing the Lab

- Allowing students to design their own procedures for collecting data gives students an opportunity to try, to fail, and to learn from their mistakes. However, you can scaffold students as they develop their procedure by having them fill out an investigation proposal. These proposals provide a way for you to offer students hints and suggestions without telling them how to do it. You can also check the proposals quickly during a class period. For this lab we suggest using Investigation Proposal A.

- Allow the students to become familiar with the equipment and materials as part of the tool talk before they begin to design their investigation. Giving them 5–10 minutes to examine the equipment and materials will let students see what they can and cannot do with them.

- We suggest using a toy car or an object with plastic wheels as opposed to an object that slides down the incline, because using plastic wheels will minimize the effect due to friction. However, for an advanced class, you may choose to substitute an object that will experience a greater frictional force.

- We also suggest making sure that the masses to be placed in the cup (if students are not using video analysis software) are proportional to the mass of the car or other object on the incline. If you are using a toy car that has a mass of around 50 g, then students can use very small washers. If you are using a large car or a cart from a physics equipment kit, then you will want to use larger washers. For the counterweight, you can use objects other than washers.

- We also recommend using masses that are considerably larger than the mass of the pulley. This will minimize the effect of the rotational inertia of the pulley on the movement of the car down the incline.

Forces on an Incline

What Is the Mathematical Relationship Between the Angle of Incline and the Acceleration of an Object Down the Incline?

- Students may first graph the acceleration down the incline as a function of the angle of incline. As shown earlier in the Teacher Notes, the correct relationship is that the acceleration down the incline is a function of the *sine of the angle*. If students graph the acceleration as a function of the angle (and not $\sin\theta$), the mathematical formula they determine will have a high degree of error, depending on the type of regression that students try to run (e.g., linear regression or exponential regression). If students do this, there are two options that you can use to help students with this lab. First, you may provide a hint to the class that any graphs or regressions with the angle as a variable should use $\sin\theta$ as opposed to θ. A graph of the acceleration as a function of $\sin\theta$, on the other hand, will give a linear relationship (i.e., $y = mx + b$). Second, you may choose before beginning the lab to ask students for a qualitative relationship or a graphical relationship for the claim, as opposed to the mathematical relationship. You can then introduce the mathematical relationship during the explicit and reflective discussion or after the lab reports have been written.

If students use video analysis

- We suggest allowing students to familiarize themselves with the video analysis software before they finalize the procedure for the investigation, especially if they have not used such software previously. This gives students an opportunity to learn how to work with the software and to improve the quality of the video they take.

- Remind students to hold the video camera as still as possible. Any movement of the camera will introduce error into their analysis. If using actual camcorders, we recommend using a tripod to hold the camera steady. If students are using a camera on a cell phone or tablet, we recommend using a table to help steady the camera.

- Remind students to place a meterstick in the same field of view as the motion they are capturing with the video camera. Also, the meterstick should be approximately the same distance from the camera as the motion. Most video analysis software requires the user to define a scale in the video (this allows the software to establish distances and, subsequently, other variables dependent on distance and displacement).

Connections to Standards

Table 7.2 (p. 164) highlights how the investigation can be used to address specific performance expectations from the *NGSS;* learning objectives from AP Physics 1; learning objectives from AP Physics C: Mechanics, *Common Core State Standards,* in English language arts (*CCSS ELA*); and *Common Core State Standards,* Mathematics (*CCSS Mathematics*).

LAB 7

TABLE 7.2 _____

Lab 7 alignment with standards

NGSS performance expectation	• HS-PS2-1: Analyze data to support the claim that Newton's second law of motion describes the mathematical relationship among the net force on a macroscopic object, its mass, and its acceleration.
AP Physics learning objectives	• 1.C.1.1: The student is able to design an experiment for collecting data to determine the relationship between the net force exerted on an object, its inertial mass, and its acceleration. • 3.A.1.2: The student is able to design an experimental investigation of the motion of an object. • 3.A.1.3: The student is able to analyze experimental data describing the motion of an object and is able to express the results of the analysis using narrative, mathematical, and graphical representations. • 3.A.3.1: The student is able to analyze a scenario and make claims (develop arguments, justify assertions) about the forces exerted on an object by other objects for different types of forces or components of forces. • 3.B.1.2: The student is able to design a plan to collect and analyze data for motion (static, constant, or accelerating) from force measurements and carry out an analysis to determine the relationship between the net force and the vector sum of the individual forces. • 3.B.1.4: The student is able to predict the motion of an object subject to forces exerted by several objects using an application of Newton's second law in a variety of physical situations. • 4.A.3.1: The student is able to apply Newton's second law to systems to calculate the change in the center-of-mass velocity when an external force is exerted on the system. • 4.A.3.2: The student is able to use visual or mathematical representations of the forces between objects in a system to predict whether or not there will be a change in the center-of-mass velocity of that system.
AP Physics C: Mechanics learning objectives	• I.A.2.a(1): Determine components of a vector along two specified, mutually perpendicular axes. • I.B.2.b(2): Write down the vector equation that results from applying Newton's second law to the object, and take components of this equation along appropriate axes. • I.B.3.c: Students should know that the tension is constant in a light string that passes over a massless pulley and should be able to use this fact in analyzing the motion of a system of two objects joined by a string.

Forces on an Incline

What Is the Mathematical Relationship Between the Angle of Incline and the Acceleration of an Object Down the Incline?

Literacy connections (*CCSS ELA*)	• *Reading:* Key ideas and details, craft and structure, integration of knowledge and ideas • *Writing:* Text types and purposes, production and distribution of writing, research to build and present knowledge, range of writing • *Speaking and listening:* Comprehension and collaboration, presentation of knowledge and ideas
Mathematics connections (*CCSS Mathematics*)	• *Mathematical practices:* Make sense of problems and persevere in solving them, reason abstractly and quantitatively, construct viable arguments and critique the reasoning of others, model with mathematics, use appropriate tools strategically, attend to precision, look for and make use of structure, look for and express regularity in repeated reasoning • *Number and quantity:* Reason quantitatively and use units to solve problems, represent and model with vector quantities, perform operations on vectors • *Algebra:* Interpret the structure of expressions, write expressions in equivalent forms to solve problems, create equations that describe numbers or relationships, understand solving equations as a process of reasoning and explain the reasoning, solve equations and inequalities in one variable, represent and solve equations and inequalities graphically • *Functions:* Understand the concept of a function and use function notation, interpret functions that arise in applications in terms of the context, analyze functions using different representations, build a function that models a relationship between two quantities, construct and compare linear and exponential models and solve problems, interpret expressions for functions in terms of the situation they model • *Statistics and probability:* Summarize, represent, and interpret data on two categorical and quantitative variables; interpret linear models; understand and evaluate random processes underlying statistical experiments; make inferences and justify conclusions from sample surveys, experiments, and observational studies

LAB 7

Lab 7. Forces on an Incline: What Is the Mathematical Relationship Between the Angle of Incline and the Acceleration of an Object Down the Incline?

Introduction

Physicists have been studying the motion of objects on an incline for centuries because most of the world is not flat. This line of research is important because many people live on or near hills or mountains and must build homes, store equipment, and travel from one location to another on an incline. Engineering of large structures has required further study of how objects do or do not move down an incline. For example, when building parking garages, many engineers build garages so that cars can be parked on the incline of the garage between levels, and not just on the flat surface.

Scientists often use Newton's laws of motion to understand the relationship between the net force exerted on an object on an incline plane, its inertial mass, and its acceleration. Acceleration is equal to the rate of change of velocity with respect to time, and velocity is equal to the rate of change of position with respect to time. Gravity is one of the forces that act on an object on an incline plane. The force of gravity pulls the object toward the center of the Earth. This is important to keep in mind because it means that gravity does not act directly down the incline. Figure L7.1 shows how the force of gravity (**Fg**) acts on a box sitting on a ramp. It also shows how scientists measure the angle of incline. As can be seen in this figure, the angle of incline (θ) is measured relative to a flat, horizontal surface (as opposed to measuring the angle from the vertical). Therefore, when the measure of the angle is 0°, the surface is flat. When the angle of incline is 90°, it is a vertical surface and an object no longer moves down the surface, but instead moves in free fall.

FIGURE L7.1

A force diagram for an object on an incline plane

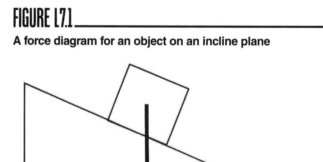

As mentioned above, it is important for us to understand how the angle of an incline plane affects the motion of an object on the incline for many different reasons. An engineer who is responsible for designing and building a parking garage, for example, needs to

Forces on an Incline

What Is the Mathematical Relationship Between the Angle of Incline and the Acceleration of an Object Down the Incline?

understand how the angle of incline in the garage will affect the motion of parked cars because if he or she makes the incline in the garage too steep, then the cars may roll or slide down the incline, even when they are in park. When this happens, property can be damaged and people can get hurt. In this investigation, you will have an opportunity to learn more about the relationship between the angle of an incline and the motion of object on that incline.

Your Task

Use what you know about forces and motion, the importance of looking for patterns, and the relationship between structure and function to design and carry out an investigation to examine how a toy car moves down the incline. The goal of this investigation is to come up with an equation that you can use to predict the acceleration of the toy car down as it moves down an incline.

The guiding question of this investigation is, *What is the mathematical relationship between the angle of incline and the acceleration of an object down the incline?*

Materials

You may use any of the following materials during your investigation:

Consumables	Equipment	
• Tape	• Safety glasses or	• Small masses
• String or fishing line	goggles (required)	• Meterstick
	• Plastic cup	• Protractor
	• Small toy car or cart	• Ramp or board
	• Electronic or triple	• Stopwatch
	beam balance	• Pulley

If you have access to the following equipment, you may also consider using a video camera and a computer or tablet with video analysis software.

Safety Precautions

Follow all normal lab safety rules. In addition, take the following safety precautions:

1. Wear sanitized safety glasses or goggles during lab setup, hands-on activity, and takedown.

2. Keep fingers and toes out of the way of moving objects.

3. Do not stand on tables and chairs.

4. Wash hands with soap and water after completing the lab.

LAB 7

Investigation Proposal Required? ☐ Yes   ☐ No

Getting Started

To answer the guiding question, you will need to design and carry out an experiment. To accomplish this task, you must determine what type of data you need to collect, how you will collect it, and how you will analyze it.

To determine *what type of data you need to collect*, think about the following questions:

- What are the boundaries and components of the system you are studying?
- How do the components of the system interact with each other?
- Which factors might control the rate of change in velocity of the toy car?
- How might a change in the structure of the incline affect how the moving object functions?
- How could you keep track of changes in this system quantitatively?
- What forces are acting on the toy car when it is on the incline?
- What information will you need to be able to determine the acceleration of the toy car?
- What will be the independent variable and the dependent variable for your experiment?

To determine *how you will collect the data*, think about the following questions:

- How will you change the independent variable during your experiment?
- What other factors will you need to hold constant during your experiment?
- How will you measure the magnitude and direction of any vector quantities?
- What equipment will you need to collect the data?
- How will you make sure that your data are of high quality (i.e., how will you reduce error)?
- How will you keep track of and organize the data you collect?

To determine *how you will analyze the data*, think about the following questions:

- What types of patterns might you look for as you analyze your data?
- How could you use mathematics to describe a relationship between variables?
- What type of calculations will you need to make?
- What types of graphs and equations signify a proportional relationship?
- What type of table or graph could you create to help make sense of your data?

Forces on an Incline

What Is the Mathematical Relationship Between the Angle of Incline and the Acceleration of an Object Down the Incline?

Connections to the Nature of Scientific Knowledge and Scientific Inquiry

As you work through your investigation, you may want to consider

- the difference between observations and inferences in science, and
- how scientists use different methods to answer different types of questions.

Initial Argument

Once your group has finished collecting and analyzing your data, your group will need to develop an initial argument. Your initial argument needs to include a claim, evidence to support your claim, and a justification of the evidence. The *claim* is your group's answer to the guiding question. The *evidence* is an analysis and interpretation of your data. Finally, the *justification* of the evidence is why your group thinks the evidence matters. The justification of the evidence is important because scientists can use different kinds of evidence to support their claims. Your group will create your initial argument on a whiteboard. Your whiteboard should include all the information shown in Figure L7.2.

FIGURE L7.2 _____

Argument presentation on a whiteboard

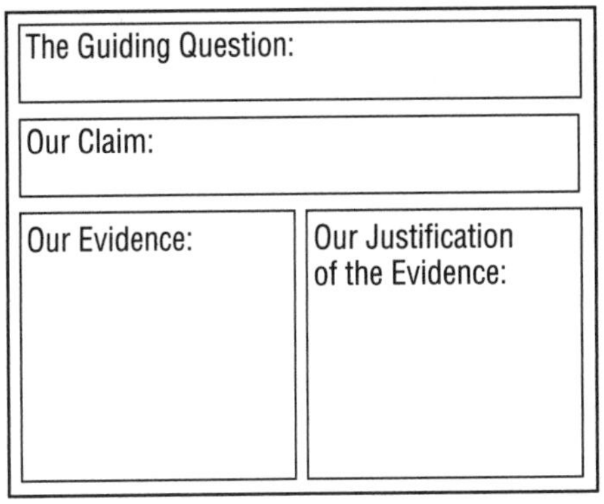

The Guiding Question:	
Our Claim:	
Our Evidence:	Our Justification of the Evidence:

Argumentation Session

The argumentation session allows all of the groups to share their arguments. One or two members of each group will stay at the lab station to share that group's argument, while the other members of the group go to the other lab stations to listen to and critique the other arguments. This is similar to what scientists do when they propose, support, evaluate, and refine new ideas during a poster session at a conference. If you are presenting your group's argument, your goal is to share your ideas and answer questions. You should also keep a record of the critiques and suggestions made by your classmates so you can use this feedback to make your initial argument stronger. You can keep track of specific critiques and suggestions for improvement that your classmates mention in the space below.

Critiques about our initial argument and suggestions for improvement:

If you are critiquing your classmates' arguments, your goal is to look for mistakes in their arguments and offer suggestions for improvement so these mistakes can be fixed. You should look for ways to make your initial argument stronger by looking for things that the other groups did well. You can keep track of interesting ideas that you see and hear during the argumentation in the space below. You can also use this space to keep track of any questions that you will need to discuss with your team.

Interesting ideas from other groups or questions to take back to my group:

Once the argumentation session is complete, you will have a chance to meet with your group and revise your initial argument. Your group might need to gather more data or design a way to test one or more alternative claims as part of this process. Remember, your goal at this stage of the investigation is to develop the best argument possible.

Report

Once you have completed your research, you will need to prepare an investigation report that consists of three sections. Each section should provide an answer to the following questions:

1. What question were you trying to answer and why?

2. What did you do to answer your question and why?

3. What is your argument?

Your report should answer these questions in two pages or less. This report must be typed, and any diagrams, figures, or tables should be embedded into the document. Be sure to write in a persuasive style; you are trying to convince others that your claim is acceptable or valid!

Forces on an Incline

What Is the Mathematical Relationship Between the Angle of Incline and the Acceleration of an Object Down the Incline?

Checkout Questions

Lab 7. Forces on an Incline: What Is the Mathematical Relationship Between the Angle of Incline and the Acceleration of an Object Down the Incline?

1. Draw a free-body diagram of the forces acting on a toy car accelerating down an incline.

2. In your investigation, you determined the relationship between the angle of incline and the acceleration of a toy car moving down the incline. Assuming there is no friction between the incline and the object that is placed on it, would your equation change if the mass of the object were increased?

   a. Yes

   b. No

   Explain your answer.

3. The wheels on the toy car reduced the impact of the force of friction on the motion of the toy car as it traveled down the incline. What would your data have looked like if the frictional force acting on the incline were larger? Use data from your investigation as part of your answer.

4. Scientists can make different inferences from the same observations.

    a. I agree with this statement.
    b. I disagree with this statement.

    Explain your answer, using an example from your investigation about objects moving down an incline.

5. There is a scientific method that all scientists must follow.

    a. I agree with this statement.
    b. I disagree with this statement.

    Explain your answer, using an example from your investigation about objects moving down an incline.

Forces on an Incline

What Is the Mathematical Relationship Between the Angle of Incline and the Acceleration of an Object Down the Incline?

6. Why is important to look for patterns in science? In your answer, be sure to include examples from at least two different investigations.

7. Why is important to think about the relationship between structure and function in nature during an investigation? In your answer, be sure to include examples from at least two different investigations.

Application Labs

LAB 8

Teacher Notes

Lab 8. Friction: Why Are Some Lubricants Better Than Others at Reducing the Coefficient of Friction Between Metal Plates?

Purpose

The purpose of this lab is for students to *apply* what they know about the core idea of forces and motion, part of the disciplinary core idea (DCI) of Motion and Stability: Forces and Interactions from the *NGSS*, to determine how different types of oils change the coefficient of kinetic friction between two metal plates and explain why some oils reduce the coefficient of friction more than other ones. This lab also gives students an opportunity to learn about the crosscutting concepts (CCs) of (a) Systems and System Models and (b) Structure and Function from the *NGSS*. In addition, this lab can be used to help students understand three big ideas from AP Physics: (a) objects and systems have properties such as mass and charge, (b) the interactions of an object with other objects can be described by forces and (c) interactions between systems can result in changes in those systems. As part of the explicit and reflective discussion, students will also learn about (a) how the culture of science, societal needs, and current events influence the work of scientists; and (b) how scientists use different methods to answer different types of questions.

Underlying Physics Concepts

The magnitude of the frictional force between an object and a surface is described by Equation 8.1, where $\mathbf{F}_{fr}$ is the frictional force, μ is the coefficient of friction, and $\mathbf{F}_N$ is the normal force of the surface acting on the object. In SI units, frictional force and normal force are measured in newtons (N). The coefficient of friction is the ratio of the frictional force to the normal force. As such, the coefficient of friction is unitless.

$$\text{(Equation 8.1)} \quad \mathbf{F}_{fr} = \mu\mathbf{F}_N$$

The coefficient of friction depends on the two surfaces in contact and is a measure of how much two surfaces grip each other. Typical values range from nearly zero (steel on ice) to one (rubber on dry asphalt). A frictionless interaction would correspond to μ equal to zero.

The coefficient of friction also depends on whether the two surfaces are in relative motion. *Static* friction is the friction force between two surfaces at rest with respect to each other, described by a coefficient of *static* friction, μ_s. When there is relative motion between two surfaces, *kinetic* friction acts between them. The coefficient of kinetic friction, μ_k, is almost always less than μ_s. An example of this is the reduced braking ability of a car sliding on the road surface—a problem that prompted the development of antilock brakes, which keep the tire surface in static contact with the road.

As seen in Equation 8.1, when one object is resting on top of another object or surface, the frictional force between the two surfaces is proportional to the normal force acting on the top object. Although the direction of the normal force is always perpendicular to the surface (in fact, that's what the word *normal* means in a geometric context), the magnitude of the normal force must be calculated using measurements of force and acceleration. The frictional force cannot be directly measured. Both coefficients (static and kinetic) will need to be determined by sketching a free-body diagram and applying Newton's second law to the object's motion. This is simplified for an object on a horizontal surface with no unbalanced external forces acting vertically, either at rest or moving at a constant velocity, while a known horizontally acting force is applied to it. The normal force and weight are equal and opposite, as are the frictional force and the pulling force. That is, if an object is moving at constant velocity in the horizontal direction, the frictional force is equal to the pulling or pushing force. Figure 8.1 illustrates a way that students can find the coefficient of friction between two plates using a force probe or spring scale. The frictional force is equal to the pulling force if the velocity of the top plate is constant.

FIGURE 8.1

Equipment used to measure the force required to pull a metal plate across another metal plate

If, however, the velocity is not constant, then students can apply Newton's second law to find the frictional force. The mathematical formula of Newton's second law is shown in Equation 8.2, where $\sum\mathbf{F}$ is the net force acting on an object; $\mathbf{F}_1$, $\mathbf{F}_2$, etc. are the individual forces acting on the object; $\mathbf{F}_n$ is the nth and final force acting on the object; m is the mass of the object; and $\mathbf{a}$ is the acceleration of the object. In SI units, mass is measured in kilograms (kg) and acceleration is measured in meters per second squared (m/s^2).

(Equation 8.2) $\sum\mathbf{F} = \mathbf{F}_1 + \mathbf{F}_2 \ldots + \mathbf{F}_n = m\mathbf{a}$

We can then use Equation 8.2 to find the frictional force acting on the top plate as we drag it across the bottom plate, as shown in Equation 8.3 (p. 178), where $\mathbf{F}_p$ is the pulling force and $\mathbf{F}_{fr}$ is the frictional force. Remember, friction always acts against motion, so if we define the positive direction as the pulling force, friction has a negative value.

$$\text{(Equation 8.3)} \quad \mathbf{F}_p + \mathbf{F}_{fr} = m\mathbf{a}$$

Lubrication is a technique used to reduce friction as two surfaces move relative to each another. A substance, called the lubricant, is placed between the two surfaces. The lubricant forms a protective film that separates the two surfaces. The most commonly used lubricants today are oils. A good lubrication oil has a high boiling point, a low freezing point, and high viscosity.

Timeline

The instructional time needed to complete this lab investigation is 200–280 minutes. Appendix 3 (p. 531) provides options for implementing this lab investigation over several class periods. Option E (280 minutes) should be used if students are unfamiliar with scientific writing, because this option provides extra instructional time for scaffolding the writing process. You can scaffold the writing process by modeling, providing examples, and providing hints as students write each section of the report. Option F (200 minutes) should be used if students are familiar with scientific writing and have developed the skills needed to write an investigation report on their own. In option F, students complete stage 6 (writing the investigation report) and stage 8 (revising the investigation report) as homework.

Materials and Preparation

The materials needed to implement this investigation are listed in Table 8.1. The equipment can be purchased from a science supply company such as Carolina, Flinn Scientific, PASCO, Vernier, or Ward's Science. Motor oil can be purchased at an auto supply store such as Advance Auto Parts, AutoZone, or NAPA. Canola oil, castor oil, corn oil, mineral oil, olive oil, and peanut oil can be purchased from a local grocery store. The metal plates can be purchased from a hardware store or from a science supply company. The type of metal is not important, as long as there are at least three different types of metal.

TABLE 8.1

Materials list for Lab 8

Item	Quantity
Consumables	
Motor oil (SAE 40)	5–10 ml per group
Motor oil (SAE 30)	5–10 ml per group
Motor oil (SAE 20)	5–10 ml per group
Motor oil (SAE 10)	5–10 ml per group
Canola oil	5–10 ml per group
Castor oil	5–10 ml per group
Corn oil	5–10 ml per group
Mineral oil	5–10 ml per group
Olive oil	5–10 ml per group
Peanut oil	5–10 ml per group
Wax paper	As needed
Tape	As needed
String	As needed
Equipment and other materials	
Safety glasses or goggles	1 per student
Chemical-resistant apron	1 per student
Gloves	1 per student
Electronic or triple beam balance	1 per group
Aluminum metal plate	2 per group
Brass metal plates	2 per group
Steel metal plates	2 per group
Spring scale or force probe	1 per group
Mass set	1 per group
Tray, 12" × 16" (or larger)	1 per group
Stopwatch	1 per group
Meterstick	1 per group
Investigation Proposal A (optional)	1 per group
Whiteboard, 2' × 3'*	1 per group
Lab Handout	1 per student
Peer-review guide and teacher scoring rubric	1 per student
Checkout Questions	1 per student

* As an alternative, students can use computer and presentation software such as Microsoft PowerPoint or Apple Keynote to create their arguments.

LAB 8

Be sure to use a set routine for distributing and collecting the materials during the lab investigation. One option is to set up the materials for each group at each group's lab station before class begins. This option works well when there is a dedicated section of the classroom for lab work and the materials are large and difficult to move (such as a dynamics track). A second option is to have all the materials on a table or cart at a central location. You can then assign a member of each group to be the "materials manager." This individual is responsible for collecting all the materials his or her group needs from the table or cart during class and for returning all the materials at the end of the class. This option works well when the materials are small and easy to move (such as stopwatches, metersticks, or hanging masses). It also makes it easy to inventory the materials at the end of the class before students leave for the day.

Safety Precautions and Laboratory Waste Disposal

Remind students to follow all normal lab safety rules. In addition, tell students to take the following safety precautions:

1. Follow safety precautions noted on safety data sheets for hazardous chemicals.

2. Wear sanitized indirectly vented chemical-splash goggles and chemical-resistant, nonlatex gloves and aprons during lab setup, hands-on activity, and takedown.

3. Never put consumables in their mouths.

4. Do not eat or drink any food items used in the lab activity.

5. Consult materials safety data sheets about the disposal of waste oil. Do not pour any chemicals down the lab sink.

6. Clean up any spilled liquid on the floor immediately to avoid a slip or fall hazard.

7. Wash hands with soap and water after completing the lab.

The motor oil should be collected and taken to a collection site for disposal; most automotive repair shops will dispose of used motor oil. We recommend following Flinn laboratory waste disposal method 18b to dispose of the mineral oil and method 24a for disposing of the canola oil, castor oil, corn oil, olive oil, and peanut oil used in this lab. Information about laboratory waste disposal methods is included in the Flinn catalog and reference manual; you can request a free copy at *www.flinnsci.com/flinn-freebies/request-a-catalog*.

Topics for the Explicit and Reflective Discussion
Reflecting on the Use of Core Ideas and Crosscutting Concepts During the Investigation

Teachers should begin the explicit and reflective discussion by asking students to discuss what they know about the core idea they used during the investigation. The following are some important concepts related to the core idea of forces and motion that students need to use to determine how different types of oils change the coefficient of kinetic friction between two metal plates and explain why some oils reduce the coefficient of friction more than other ones:

- Force is considered a vector quantity because a force has both magnitude and direction.

- The net force acting on an object is the vector sum of the individual forces acting on it.

- There may be forces acting on an object at rest.

- *Contact forces,* which include tension, friction, normal, and spring, result from the interaction of one object touching another object.

- The *coefficient of friction* is the ratio of frictional force to normal force.

- A *system* is an object or a collection of objects. An object is treated as if it has no internal structure.

To help students reflect on what they know about forces and motion, we recommend showing them two or three images using presentation software that help illustrate these important ideas. You can then ask the students the following questions to encourage them to share how they are thinking about these important concepts:

1. What do we see going on in this image?

2. Does anyone have anything else to add?

3. What might be going on that we can't see?

4. What are some things that we are not sure about here?

You can then encourage students to think about how CCs played a role in their investigation. There are at least two CCs that students need to use to determine how different types of oils change the coefficient of kinetic friction between two metal plates and explain why some oils reduce the coefficient of friction more than other ones: (a) Systems and System Models and (2) Structure and Function (see Appendix 2 [p. 527] for a brief description of these CCs). To help students reflect on what they know about these CCs, we recommend asking them the following questions:

1. Why is it useful to define a system and then make a model of it in science? What were the boundaries and components of the system you studied during this investigation?

2. What models did you use during the investigation? What were some of the limitations of these models?

3. The way an object is shaped or structured determines many of its properties and how it functions. Why is it useful to think about the relationship between structure and function during an investigation?

4. Why was it important to examine the structure of the oils in order to explain why some oils work better than others at reducing the coefficient of friction? Why is an understanding of the relationship between the structure and function of oil more useful than simply knowing which oil is the best lubricant?

You can then encourage the students to think about how they used all these different concepts to help answer the guiding question and why it is important to use these ideas to help justify their evidence for their final arguments. Be sure to remind your students to explain why they included the evidence in their arguments and make the assumptions underlying their analysis and interpretation of the data explicit in order to provide an adequate justification of their evidence.

Reflecting on Ways to Design Better Investigations

It is important for students to reflect on the strengths and weaknesses of the investigation they designed during the explicit and reflective discussion. Students should therefore be encouraged to discuss ways to eliminate potential flaws, measurement errors, or sources of uncertainty in their investigations. To help students be more reflective about the design of their investigation and what they can do to make their investigations more rigorous in the future, you can ask them the following questions:

1. What were some of the strengths of the way you planned and carried out your investigation? In other words, what made it scientific?

2. What were some of the weaknesses of the way you planned and carried out your investigation? In other words, what made it less scientific?

3. What rules can we make, as a class, to ensure that our next investigation is more scientific?

Reflecting on the Nature of Scientific Knowledge and Scientific Inquiry

This investigation can be used to illustrate two important concepts related to the nature of scientific knowledge and the nature of scientific inquiry: (a) how the culture of science, societal needs, and current events influence the work of scientists; and (b) how scientists use different methods to answer different types of questions (see Appendix 2 for a brief description of these two concepts). Be sure to review these concepts during and at the end of the explicit and reflective discussion. To help students think about these concepts in relation to what they did during the lab, you can ask them the following questions:

1. People view some types of research as being more important than other types of research because of cultural values and current events. Can you come up with some examples of how cultural values and current events have influenced the work of scientists?

2. Scientists share a set of values, norms, and commitments that shape what counts as knowing, how to represent or communicate information, and how to interact with other scientists. Can you work with your group to come up with a rule that you can use to decide if something is science or not science? Be ready to share in a few minutes.

3. There is no universal step-by-step scientific method that all scientists follow. Why do you think there is no universal scientific method?

4. Think about what you did during this investigation. How would you describe the method you used to determine how different types of oils change the coefficient of static and/or kinetic friction between two metal plates and explain why some oils reduce the coefficient of friction more than other one? Why would you call it that?

You can also use presentation software or other techniques to encourage your students to think about these concepts. You can show examples of research projects that were influenced by cultural values or current events and ask students to think about what was going on at the time and why that research was viewed as being important for the greater good. You can also show one or more images of "a universal scientific method" that misrepresent the nature of scientific inquiry (see, e.g., *https://commons.wikimedia.org/wiki/File:The_Scientific_Method_as_an_Ongoing_Process.svg*) and ask students why each image is *not* a good representation of what scientists do to develop scientific knowledge. You can ask students to suggest revisions to the image that would make it more consistent with the way scientists develop scientific knowledge.

Be sure to remind your students that, to be proficient in science, it is important that they understand what counts as scientific knowledge and how that knowledge develops over time.

Hints for Implementing the Lab

- Allowing students to design their own procedures for collecting data gives students an opportunity to try, to fail, and to learn from their mistakes. However, you can scaffold students as they develop their procedure by having them fill out an investigation proposal. These proposals provide a way for you to offer students hints and suggestions without telling them how to do it. You can also check the proposals quickly during a class period. For this lab we suggest using Investigation Proposal A.

- Learn how to measure the coefficient of friction before the lab begins. It is important for you to know how to use the equipment so you can help students when technical issues arise.

- Allow the students to become familiar with the equipment and materials as part of the tool talk before they begin to design their investigation. Giving them 5–10 minutes to examine the equipment and materials will let students see what they can and cannot do with them.

- Students only need to put a few drops of oil on the surface of one plate to see a reduction in the coefficient of kinetic friction. Students can pour a few milliliters of oil onto the top of one plate and then rub the oil across the plate to ensure the entire surface is covered.

- We recommend that you have a large normal force acting on the top plate. This can be achieved either by using a plate with a large mass or by placing additional mass onto the top plate. This will increase the frictional force acting on the top plate as it is pulled.

- Another way to measure the coefficient of static friction is to place the two plates together and then tilt them until the top one slides. The coefficient of static friction is equal to the tangent of the angle at which the top object slides. A similar method can be used to measure the coefficient of kinetic friction. To do that you give the top object a push as you increase the angle. When the top object keeps sliding with constant velocity, the tangent of that angle is equal to coefficient of kinetic friction.

- Students can place wax paper on their tray for each trial. This will allow them to keep the various oils from mixing together. This will also aid in preventing the oil from spilling.

- Be sure to allow students to go back and re-collect data at the end of the argumentation session. Students often realize that they made numerous mistakes when they were collecting data as a result of their discussions during the argumentation session. The students, as a result, will want a chance to re-collect data, and the re-collection of data should be encouraged when time allows. This also offers an opportunity to discuss what scientists do when they realize a mistake is made inside the lab.

Connections to Standards

Table 8.2 highlights how the investigation can be used to address specific performance expectations from the *NGSS;* learning objectives from AP Physics 1; learning objectives from AP Physics C: Mechanics, *Common Core State Standards,* in English language arts (*CCSS ELA*); and *Common Core State Standards,* Mathematics (*CCSS Mathematics*).

TABLE 8.2 _____

Lab 8 alignment with standards

***NGSS* performance expectation**	• HS-PS2-1: Analyze data to support the claim that Newton's second law of motion describes the mathematical relationship among the net force on a macroscopic object, its mass, and its acceleration.
AP Physics 1 learning objectives	• 3.C.4.2: The student is able to explain contact forces (tension, friction, normal, buoyant, spring) as arising from interatomic electric forces and that they therefore have certain directions.
AP Physics C: Mechanics learning objectives	• I.B.2.d(1): Write down the relationship between the normal and frictional forces on a surface. • I.B.2.d(2): Analyze situations in which an object moves along a rough inclined plane or horizontal surface. • I.B.2.d(3): Analyze under what circumstances an object will start to slip, or to calculate the magnitude of the force of static friction.
Literacy connections (*CCSS ELA*)	• *Reading:* Key ideas and details, craft and structure, integration of knowledge and ideas • *Writing:* Text types and purposes, production and distribution of writing, research to build and present knowledge, range of writing • *Speaking and listening:* Comprehension and collaboration, presentation of knowledge and ideas
Mathematics connections (*CCSS Mathematics*)	• *Mathematical practices:* Reason abstractly and quantitatively, construct viable arguments and critique the reasoning of others, use appropriate tools strategically, attend to precision • *Number and quantity:* Reason quantitatively and use units to solve problems, represent and model with vector quantities, perform operations on vectors • *Algebra:* Interpret the structure of expressions, create equations that describe numbers or relationships, understand solving equations as a process of reasoning and explain the reasoning, solve equations and inequalities in one variable, represent and solve equations and inequalities graphically • *Functions:* Understand the concept of a function and use function notation, interpret functions that arise in applications in terms of the context, interpret expressions for functions in terms of the situations they model • *Statistics and probability:* Summarize, represent, and interpret data on two categorical and quantitative variables; understand and evaluate random processes underlying statistical experiments; make inferences and justify conclusions from sample surveys, experiments, and observational studies

LAB 8

Lab Handout

Lab 8. Friction: Why Are Some Lubricants Better Than Others at Reducing the Coefficient of Friction Between Metal Plates?

Introduction

Friction plays an important role in many of our daily experiences. In many cases, friction between two surfaces is beneficial. Friction between your shoes and the ground allows you to walk; friction between your tires and the road allows your car to move forward; and friction keeps your cell phone from sliding off a table if you accidently bump into it. Frictional forces also produce heat. When you rub your hands together on a cold day, for example, friction produces heat and warms your hands. Sometimes, however, producing heat from friction is an undesirable outcome or unwanted by-product of a specific process. Take the internal combustion engine found in most cars as an example. An internal combustion engine has a set of pistons that are constantly moving when the engine is on. Each of these pistons is found inside a metal cylinder (see Figure L8.1). For the engine to work, each piston must be able to slide up and down inside a metal cylinder. The sliding motion, however, produces heat because of the friction that exists between the piston and the cylinder, which can cause parts of the engine to heat up and eventually reach a temperature where the different parts of the engine expand and deform or break.

The coefficient of friction is the ratio of the frictional force and the normal force that exists between two surfaces when they are in contact. There are two types of friction coefficients. The first type is called the *coefficient of static friction*. This measure is used when there is no relative motion between the two surfaces. For example, when a car is at rest we describe the friction that exists between the tires and the ground using the coefficient of static friction. The second type is called the *coefficient of kinetic friction*. This measure is used when there is relative motion between two surfaces. The coefficient of kinetic friction affects the amount of heat that is generated when objects that are in contact move past each other. In general, as the coefficient of kinetic friction increases, so does the amount of heat that is produced as the result of friction. Finally, it is important to keep in mind that a coefficient of friction, whether it is static or kinetic, only exists between materials that are in contact. For example, the coefficient of kinetic friction between iron and copper is 0.29 when these two materials are in contact, and the coefficient of kinetic friction between iron and zinc is 0.21 when these two materials are in contact. Iron does not have a coefficient of kinetic friction unless it is in contact with another material, such as copper or zinc.

As mentioned earlier, when a piston slides up and down in a cylinder, friction causes the engine to heat up quickly. Engineers therefore use *lubricants* to reduce the coefficient of kinetic friction between the piston and the cylinder. A lubricant is simply a substance that is put in between two materials to reduce the heat generated when the two surfaces move past each other. Oils are often used as lubricants. An oil can be defined as a viscous liquid that is composed of hydrogen, oxygen, and carbon atoms. Oils do not mix with water, and they feel slippery. *Viscosity* is a measure of how resistant a liquid is to flow. Honey, for example, has a higher viscosity than water, so honey flows slower than water. There are many different types of oils. Each one has a unique set of physical properties. Table L8.1 includes the physical properties and sources of 10 different types of oil. Some of these oils are better than others at reducing the coefficient of kinetic friction between metals because of their unique physical properties.

To be able to make new and better lubricants for engines, it is important to understand what makes oil such a good lubricant. Scientists and engineers, in other words, need to understand how the different physical properties of an oil affect or do not affect its ability to reduce the coefficient of kinetic friction between two materials. This type of research is

FIGURE L8.1

A cross-section of an internal combustion engine, showing three different pistons. Each piston is found inside a cylinder.

TABLE L8.1

Sources and physical properties of different types of oils

Name	Source	Physical properties			
		Density (g/cm³)	Viscosity (cP)	Melting point (°C)	Boiling point (°C)
Motor (SAE 40)	Petroleum	0.90	319	−12	315
Motor (SAE 30)	Petroleum	0.89	200	−30	300
Motor (SAE 20)	Petroleum	0.88	125	No data	280
Motor (SAE 10)	Petroleum	0.87	65	No data	260
Canola	Plants	0.91	57	−10	205
Castor	Plants	0.96	985	−18	313
Corn	Plants	0.90	81	−11	230
Mineral	Petroleum	0.87	44	−9	300
Olive	Plants	0.91	84	−6.0	300
Peanut	Plants	0.91	68	3	255

Note: cP = centipoise.

LAB 8

important because it is not enough to determine which type of lubricant works best as a lubricant (like a product reviewer would) if the goal of the scientist or engineer is to create a better product. Instead, scientists or engineers must understand why a lubricant works. It is also important for scientists and engineers to understand how different conditions, such as when there is a high coefficient of kinetic friction or a low coefficient of kinetic friction between two objects, affect how well a lubricant works. Your goal for this investigation is to explain why a lubricant, such as oil, is able to reduce the coefficient of kinetic friction between two surfaces.

Your Task

Use what you know about forces and motion, structure and function, and the role models of systems play in science to design and carry out an investigation to determine how different types of oils change the coefficient of kinetic friction between two metal plates. You will then use what you know about structure and function to develop a model that can be used to explain why some oils reduce the coefficient of friction more than other oils. Your model can be conceptual, mathematical, or graphical. To be valid or acceptable, your model must take into account the different physical properties of the oils. Once you have developed your model, you will need to test it to determine if it allows you to predict how the use of other oils (which you have not tested before) will change the coefficient of kinetic friction between two plates, using only the physical properties of these other oils as a guide.

The guiding question of this investigation is, *Why are some lubricants better than others at reducing the coefficient of friction between metal plates?*

Materials

You may use any of the following materials during your investigation:

Consumables	Equipment
• Motor oil (SAE 10)	• Safety glasses or goggles (required)
• Motor oil (SAE 20)	• Chemical-resistant apron (required)
• Motor oil (SAE 30)	• Gloves (required)
• Motor oil (SAE 40)	• Electronic or triple beam balance
• Canola oil	• Aluminum metal plates
• Castor oil	• Brass metal plates
• Corn Oil	• Steel metal plates
• Mineral oil	• Spring scale or force probe
• Olive oil	• Mass set
• Peanut oil	• Tray
• Wax paper	• Stopwatch
• Tape	• Meterstick
• String	

Why Are Some Lubricants Better Than Others at Reducing the Coefficient of Friction Between Metal Plates?

Safety Precautions

Follow all normal lab safety rules. Your teacher will explain relevant and important information about working with the chemicals associated with this investigation. In addition, take the following safety precautions:

1. Follow safety precautions noted on safety data sheets for hazardous chemicals.

2. Wear sanitized indirectly vented chemical-splash goggles and chemical-resistant, nonlatex gloves and aprons during lab setup, hands-on activity, and takedown.

3. Never put consumables in your mouth.

4. Do not eat or drink any food items used in the lab activity.

5. Consult with the teacher about disposal of waste oil. Do not pour any chemicals down the lab sink.

6. Clean up any spilled liquid on the floor immediately to avoid a slip or fall hazard.

7. Wash hands with soap and water after completing the lab.

Investigation Proposal Required?   ☐ Yes   ☐ No

Getting Started

The first step in developing your model is to determine how the addition of different types of oils changes the coefficient of kinetic friction between two metal plates. One way to gather the information you need to determine the coefficient of kinetic friction between two metal plates is to measure the amount of force it takes to pull a metal plate across another one. Figure L8.2 shows how you can measure this force.

FIGURE L8.2

Equipment used to measure the force required to pull a metal plate across another metal plate

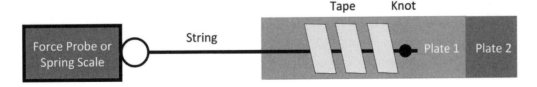

Before you can begin conducting your tests, however, you must decide what other type of data you will need to collect, how you will collect it, and how you will analyze it.

To determine *what type of data you need to collect,* think about the following questions:

- What are the boundaries and components of the system you are studying?

- How do the components of the system interact with each other?
- How could you keep track of changes in this system quantitatively?
- How might the structure of a lubricant impact its function in reducing the coefficient of friction?
- What information do you need to calculate the coefficient of kinetic friction?
- What are all the forces that are acting on each object?
- Are any of the pairs of forces balanced?

To determine *how you will collect the data,* think about the following questions:

- What comparisons will you need to make?
- What other factors will you need to control during your tests?
- How will you make sure that your data are of high quality (i.e., how will you reduce error)?
- How will you keep track of and organize the data you collect?

To determine *how you will analyze the data,* think about the following questions:

- What types of patterns could you look for in your data?
- What type of calculations will you need to make to determine a coefficient of kinetic friction?
- How could you use mathematics to describe a change in a coefficient of kinetic friction?
- How could you use mathematics to document a difference between types of lubricants?

Once you have determined how the addition of different types of oils changes the coefficient of kinetic friction between two metal plates, your group will need to develop your conceptual model. The model must be able to explain why some of the oils were better than others at reducing the coefficient of friction between metal plates. Your model also must include information about the physical properties of the oils. The physical properties of each oil are provided in Table L8.1. Your model should also include information about what you think is happening between the oils and the plates on the submicroscopic level.

The last step in this investigation is to test your model. To accomplish this goal, you can use different oils (ones that you did not test) to determine if your model enables you to make accurate predictions about how these oils will change the coefficient of kinetic friction between two metal plates. If you are able to use your model to make accurate predictions about how the oils function as a lubricant based on their structure, then you will be able to generate the evidence you need to convince others that your model is valid.

Connections to the Nature of Scientific Knowledge and Scientific Inquiry

As you work through your investigation, you may want to consider

- how the culture of science, societal needs, and current events influence the work of scientists; and
- how scientists use different methods to answer different types of questions.

Initial Argument

Once your group has finished collecting and analyzing your data, your group will need to develop an initial argument. Your initial argument needs to include a claim, evidence to support your claim, and a justification of the evidence. The *claim* is your group's answer to the guiding question. The *evidence* is an analysis and interpretation of your data. Finally, the *justification* of the evidence is why your group thinks the evidence matters. The justification of the evidence is important because scientists can use different kinds of evidence to support their claims. Your group will create your initial argument on a whiteboard. Your whiteboard should include all the information shown in Figure L8.3.

FIGURE L8.3

Argument presentation on a whiteboard

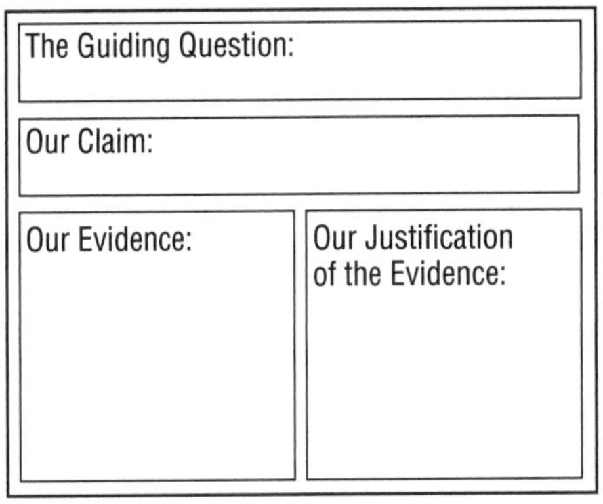

The Guiding Question:	
Our Claim:	
Our Evidence:	Our Justification of the Evidence:

Argumentation Session

The argumentation session allows all of the groups to share their arguments. One or two members of each group will stay at the lab station to share that group's argument, while the other members of the group go to the other lab stations to listen to and critique the other arguments. This is similar to what scientists do when they propose, support, evaluate, and refine new ideas during a poster session at a conference. If you are presenting your group's argument, your goal is to share your ideas and answer questions. You should also keep a record of the critiques and suggestions made by your classmates so you can use this feedback to make your initial argument stronger. You can keep track of specific critiques and suggestions for improvement that your classmates mention in the space below.

Critiques about our initial argument and suggestions for improvement:

If you are critiquing your classmates' arguments, your goal is to look for mistakes in their arguments and offer suggestions for improvement so these mistakes can be fixed. You should look for ways to make your initial argument stronger by looking for things that the other groups did well. You can keep track of interesting ideas that you see and hear during the argumentation in the space below. You can also use this space to keep track of any questions that you will need to discuss with your team.

Interesting ideas from other groups or questions to take back to my group:

Once the argumentation session is complete, you will have a chance to meet with your group and revise your initial argument. Your group might need to gather more data or design a way to test one or more alternative claims as part of this process. Remember, your goal at this stage of the investigation is to develop the best argument possible.

Report

Once you have completed your research, you will need to prepare an investigation report that consists of three sections. Each section should provide an answer to the following questions:

1. What question were you trying to answer and why?

2. What did you do to answer your question and why?

3. What is your argument?

Your report should answer these questions in two pages or less. This report must be typed, and any diagrams, figures, or tables should be embedded into the document. Be sure to write in a persuasive style; you are trying to convince others that your claim is acceptable or valid!

Checkout Questions

Lab 8. Friction: Why Are Some Lubricants Better Than Others at Reducing the Coefficient of Friction Between Metal Plates?

1. Doubling the coefficient of friction between two surfaces will cut the frictional force in half.

   a. I agree with this statement.
   b. I disagree with this statement.

   Explain your answer, using an example from your investigation about friction and lubricants.

2. Most pickup trucks have four tires, but some have six. How do you think having two additional tires affects the frictional force between a truck and the road?

   a. Adding two tires increases the frictional force between the truck and the road.
   b. Adding two tires decreases the frictional force between the truck and the road.
   c. Adding two tires has no effect on the frictional force between the truck and the road.

   Explain your answer.

3. Societal needs and current events can influence the research that scientists and engineers decide to do.

    a. I agree with this statement.

    b. I disagree with this statement.

Explain your answer, using an example from your investigation about friction and lubricants.

4. All scientists follow the same scientific method when doing research.

    a. I agree with this statement.

    b. I disagree with this statement.

Explain your answer, using an example from your investigation about friction and lubricants.

5. Why is it useful to identify a system under study and then make a model of it during an investigation? In your answer, be sure to include examples from at least two different investigations.

6. Why is it important to think about the relationship between structure and function when trying to develop an explanation for a natural phenomenon? In your answer, be sure to include examples from at least two different investigations.

LAB 9

Teacher Notes

Lab 9. Falling Objects and Air Resistance: How Does the Surface Area of a Parachute Affect the Force Due to Air Resistance as an Object Falls Toward the Ground?

Purpose

The purpose of this lab is for students to *apply* what they know about the core idea of forces and motion, part of the disciplinary core idea (DCI) of Motion and Stability: Forces and Interactions from the *NGSS*, to determine how the surface area of a parachute affects the force due to air resistance as an object falls toward the ground. This lab also gives students an opportunity to learn about the crosscutting concepts (CCs) of (a) Systems and System Models and (b) Structure and Function from the *NGSS*. In addition, this lab can be used to help students understand three big ideas from AP Physics: (a) fields existing in space can be used to explain interactions, (b) the interactions of an object with other objects can be described by forces, and (c) interactions between systems can result in changes in those systems. As part of the explicit and reflective discussion, students will also learn about (a) how the culture of science, societal needs, and current events influence the work of scientists; and (b) the role of imagination and creativity in science.

Underlying Physics Concepts

Newton's second law can be written in many ways, but the most common is shown in Equation 9.1, where $\sum F$ is the net force acting on an object, $\mathbf{a}$ is the acceleration of the object, and m is the mass of the object. In SI units, force is measured in newtons (N), mass in kilograms (kg), and acceleration in meters per second squared (m/s^2).

$$\text{(Equation 9.1)} \quad \sum\mathbf{F} = m\mathbf{a}$$

When ignoring the effect of air resistance on a falling object, we can easily solve the equation because the force of gravity is the only force acting on the object. This force will lead to an acceleration equal to -9.8 m/s^2. The equation, however, becomes more complex when we account for air resistance. When air resistance is included, we can rewrite Equation 9.1 as Equation 9.2, where $\mathbf{F}_g$ is the force of gravity and $\mathbf{F}_D$ is the force of air resistance, also referred to as the drag force, or the drag on the falling object.

$$\text{(Equation 9.2)} \quad \mathbf{F}_g + \mathbf{F}_D = m\mathbf{a}$$

$\mathbf{F}_g$ is easily calculated as the mass of the object times the acceleration due to gravity. The drag force, however, is a bit more complex. Equation 9.3 shows the drag force, where C_D is the coefficient of drag (related to the type of material used for the parachute) and is unitless; ρ (spelled rho and pronounced "row") is the density of air in kilograms per meter cubed (kg/m^3); $\mathbf{v}$ is velocity of the parachute in meters per second (m/s), and A is the surface area of the parachute in meters squared (m^2).

$$\textbf{(Equation 9.3)} \quad \mathbf{F}_D = C_D \rho \mathbf{v}^2 A/2$$

There are two important things to point to in this equation. First, and related most directly to this lab, is that the drag force on a parachute is directly proportional to the surface area of the parachute. The larger the surface area, the larger the drag force will be. This also means that if we find the value for the drag force and place it into Equation 9.2 that shows the sum of the forces acting on the falling object, then the relationship between the surface area of the parachute and the resulting acceleration is inversely proportional. As surface area increases, the net acceleration while falling will decrease.

The second important thing to point out with respect to Equation 9.3 is that the drag force is also a function of the velocity of the falling object. As the object's velocity increases, the drag force also increases and the resulting acceleration of the object decreases. When the object is at rest (i.e., when time [t] equals zero seconds, or when the object is released), there is no drag on the object due to the parachute. Only after the object begins to move does the drag force change the acceleration.

As the velocity increases, the drag on the parachute will continue to increase until the object reaches terminal velocity (this won't happen in this investigation, because students are not dropping from a great enough height). Terminal velocity occurs when the drag force is equal to the force of gravity. At this point, the sum of the forces equals zero, and the resulting acceleration will be zero. Furthermore, *this means that the acceleration is not constant when an object falls and we account for air resistance.* Instead, the acceleration is continuously changing until the object reaches terminal velocity, at which point the acceleration is zero.

That being said, students can find the average acceleration of their object during free fall. One way to do this is by using kinematics equations and solving for the acceleration of the object for each surface area tested. Students can do this using Equation 9.4, where $\mathbf{a}$ is acceleration, $\mathbf{y}$ is the vertical displacement, $\mathbf{v}_0$ is the initial velocity, and t is the time the object was falling. In SI units, displacement is measured in meters (m); velocity is measured in meters per second (m/s); and time is measured in seconds (s).

$$\textbf{(Equation 9.4)} \quad \mathbf{y} = \mathbf{v}_0 t + \tfrac{1}{2}\mathbf{a}t^2$$

If the object is released from rest, then $v_0 = 0$, and the average net acceleration over the displacement, **y,** can be calculated using Equation 9.5.

$$\textbf{(Equation 9.5)} \quad a = 2y/t^2$$

If students are using video analysis software, they can also use the software to find the function for acceleration as a function of time. If they are familiar with calculus, they can use the mean value theorem to find the average acceleration while the object was falling toward the ground.

Timeline

The instructional time needed to complete this lab investigation is 220–280 minutes. Appendix 3 (p. 531) provides options for implementing this lab investigation over several class periods. Option A (280 minutes) should be used if students are unfamiliar with scientific writing, because this option provides extra instructional time for scaffolding the writing process. You can scaffold the writing process by modeling, providing examples, and providing hints as students write each section of the report. Option A can also be used if you are introducing students to the video analysis programs. Option B (220 minutes) should be used if students are familiar with scientific writing and have developed the skills needed to write an investigation report on their own. In option B, students complete stage 6 (writing the investigation report) and stage 8 (revising the investigation report) as homework.

Materials and Preparation

The materials needed to implement this investigation are listed in Table 9.1. Most of the equipment can be purchased from a science supply company such as Carolina, Flinn Scientific, PASCO, Vernier, or Ward's Science. The washers and the plastic bags can be purchased from a general store, such as Wal-Mart or Target. Video analysis software can be purchased from Vernier (Logger *Pro*) or PASCO (SPARKvue or Capstone). These companies also have apps that can be used on Apple- or Android-based tablets and cell phones. We recommend consulting with your school's information technology coordinator to determine the best option for your students.

TABLE 9.1

Materials list for Lab 9

Item	Quantity
Consumables	
Large plastic trash bags	Several per group
Tape	1 roll per group
String or fishing line	1 roll per group
Equipment and other materials	
Safety glasses or goggles	1 per student
Electronic or triple beam balance	1 per group
Washers	Several per group
Stopwatch	1 per student
Ruler	1 per student
Meterstick	1 per group
Investigation Proposal C (optional)	1 per group
Whiteboard, 2' × 3'*	1 per group
Lab Handout	1 per student
Peer-review guide and teacher scoring rubric	1 per student
Checkout Questions	1 per student
Equipment for video analysis (optional)	
Video camera	1 per group
Computer or tablet with video analysis software	1 per group

* As an alternative, students can use computer and presentation software such as Microsoft PowerPoint or Apple Keynote to create their arguments.

Be sure to use a set routine for distributing and collecting the materials during the lab investigation. One option is to set up the materials for each group at each group's lab station before class begins. This option works well when there is a dedicated section of the classroom for lab work and the materials are large and difficult to move (such as a dynamics track). A second option is to have all the materials on a table or cart at a central location. You can then assign a member of each group to be the "materials manager." This individual is responsible for collecting all the materials his or her group needs from the table or cart during class and for returning all the materials at the end of the class. This option works well when the materials are small and easy to move (such as stopwatches,

metersticks, or hanging masses). It also makes it easy to inventory the materials at the end of the class before students leave for the day.

Safety Precautions

Remind students to follow all normal lab safety rules. In addition, tell the students to take the following safety precautions:

1. Wear sanitized safety glasses or goggles during lab setup, hands-on activity, and takedown.

2. Do not throw the washers or parachutes.

3. Do not stand on tables or chairs.

4. Wash hands with soap and water after completing the lab.

Topics for the Explicit and Reflective Discussion
Reflecting on the Use of Core Ideas and Crosscutting Concepts During the Investigation

Teachers should begin the explicit and reflective discussion by asking students to discuss what they know about the core idea they used during the investigation. The following are some important concepts related to the core idea of forces and motion that students need to use to determine how the surface area of a parachute affects the force due to air resistance as an object falls toward the ground:

- Gravity is an attractive force between two objects that have mass.

- Objects accelerate toward the center of Earth when in free fall because of the gravitational force between the object and Earth.

- *Displacement* is a change in position. *Velocity* is the rate of change of position over a period of time. *Acceleration* is the rate of change in velocity over a period of time.

- *Air resistance* is the result of an object moving through a layer of air and colliding with air molecules. The force of air resistance acting on a falling object is therefore dependent on the velocity and the cross-sectional surface area of the falling object.

- Force is considered a vector quantity because a force has both magnitude and direction.

- The net force acting on an object is the vector sum of the individual forces acting on it.

- The acceleration of an object interacting with other objects can be predicted by using the equation $\mathbf{a} = \sum \mathbf{F}/m$.

To help students reflect on what they know about forces and motion, we recommend showing them two or three images using presentation software that help illustrate these important ideas. You can then ask the students the following questions to encourage them to share how they are thinking about these important concepts:

1. What do we see going on in this image?

2. Does anyone have anything else to add?

3. What might be going on that we can't see?

4. What are some things that we are not sure about here?

You can then encourage students to think about how CCs played a role in their investigation. There are at least two CCs that students need to use to determine how the surface area of a parachute affects the force due to air resistance as an object falls toward the ground: (a) Systems and System Models and (b) Structure and Function (see Appendix 2 [p. 527] for a brief description of these CCs). To help students reflect on what they know about these CCs, we recommend asking them the following questions:

1. Why is it useful to define a system and then make a model of it in science? What were the boundaries and components of the system you studied during this investigation?

2. What models did you use during the investigation? What were some of the limitations of these models?

3. The way an object is shaped or structured determines many of its properties and how it functions. Why is it useful to think about the relationship between structure and function during an investigation?

4. Why was it important to examine the structure of a parachute in order to determine its ability to slow the acceleration of a falling object? Why is an understanding of the relationship between the structure and function of parachute more useful than simply knowing which parachute works the best?

You can then encourage the students to think about how they used all these different concepts to help answer the guiding question and why it is important to use these ideas to help justify their evidence for their final arguments. Be sure to remind your students to explain why they included the evidence in their arguments and make the assumptions underlying their analysis and interpretation of the data explicit in order to provide an adequate justification of their evidence.

Reflecting on Ways to Design Better Investigations

It is important for students to reflect on the strengths and weaknesses of the investigation they designed during the explicit and reflective discussion. Students should therefore be encouraged to discuss ways to eliminate potential flaws, measurement errors, or sources of uncertainty in their investigations. To help students be more reflective about the design

of their investigation and what they can do to make their investigations more rigorous in the future, you can ask them the following questions:

1. What were some of the strengths of the way you planned and carried out your investigation? In other words, what made it scientific?

2. What were some of the weaknesses of the way you planned and carried out your investigation? In other words, what made it less scientific?

3. What rules can we make, as a class, to ensure that our next investigation is more scientific?

Reflecting on the Nature of Scientific Knowledge and Scientific Inquiry

This investigation can be used to illustrate two important concepts related to the nature of scientific knowledge and the nature of scientific inquiry: (a) how the culture of science, societal needs, and current events influence the work of scientists; and (b) the role of imagination and creativity in science (see Appendix 2 for a brief description of these two concepts). Be sure to review these concepts during and at the end of the explicit and reflective discussion. To help students think about these concepts in relation to what they did during the lab, you can ask them the following questions:

1. People view some types of research as being more important than other types of research because of cultural values and current events. Can you come up with some examples of how cultural values and current events have influenced the work of scientists?

2. Scientists share a set of values, norms, and commitments that shape what counts as knowing, how to represent or communicate information, and how to interact with other scientists. Can you work with your group to come up with a rule that you can use to decide if something is science or not science? Be ready to share in a few minutes.

3. Some people think that there is no room for imagination or creativity in science. What do you think?

4. Can you work with your group to come up with different ways that you needed to use your imagination or be creative during this investigation? Be ready to share in a few minutes.

You can also use presentation software or other techniques to encourage your students to think about these concepts. You can show examples of research projects that were influenced by cultural values or current events and ask students to think about what was going on at the time and why that research was viewed as being important for the greater good.

You can also show students an image of the following quote by E. O. Wilson from *Letters to a Young Scientist* (2013) and ask them what they think he meant by it:

> The ideal scientist thinks like a poet and only later works like a bookkeeper. Keep in mind that innovators in both literature and science are basically dreamers and storytellers. In the early stages of the creation of both literature and science, everything in the mind is a story. There is an imagined ending, and usually an imagined beginning, and a selection of bits and pieces that might fit in between. In works of literature and science alike, any part can be changed, causing a ripple among the other parts, some of which are discarded and new ones added. (p. 74)

Be sure to remind your students that, to be proficient in science, it is important that they understand what counts as scientific knowledge and how that knowledge develops over time.

Hints for Implementing the Lab

- Allowing students to design their own procedures for collecting data gives students an opportunity to try, to fail, and to learn from their mistakes. However, you can scaffold students as they develop their procedure by having them fill out an investigation proposal. These proposals provide a way for you to offer students hints and suggestions without telling them how to do it. You can also check the proposals quickly during a class period. For this lab we suggest using Investigation Proposal C.

- Allow the students to become familiar with the equipment and materials as part of the tool talk before they begin to design their investigation. Giving them 5–10 minutes to examine the equipment and materials will let students see what they can and cannot do with them.

- If too much mass is added to the parachute, this may result in the parachute ripping or breaking. The limit on the mass to be added to the parachute is dependent on the type of plastic. The thicker the plastic, the more mass it can support. In general, a mass of 50 g–250 g will be sufficient.

- Students will sometimes create parachutes that are rectangles to make it easier to calculate the surface area. As a result, they may create parachutes that are too long and thin to work well. Figure 9.1 (p. 204) shows an example of how students might create their parachutes, but there are other ways that they can design them.

- The higher up the students can be when they drop their parachutes, the more pronounced the effect. We suggest having students drop the parachutes from the top of the bleachers at the football field or from a balcony two stories high.

FIGURE 9.1

A possible design of a parachute

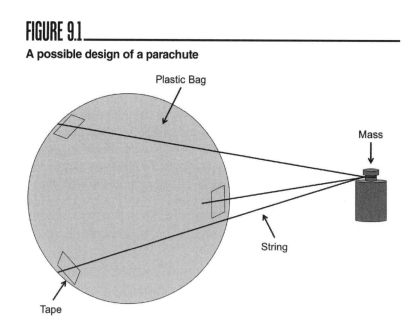

- Encourage students to derive an expression for the acceleration as a function of time for an object falling under the influence of drag forces as their answer to the guiding question.

- Be sure to allow students to go back and re-collect data at the end of the argumentation session. Students often realize that they made numerous mistakes when they were collecting data as a result of their discussions during the argumentation session. The students, as a result, will want a chance to re-collect data, and the re-collection of data should be encouraged when time allows. This also offers an opportunity to discuss what scientists do when they realize a mistake is made inside the lab.

If students use video analysis

- We suggest allowing students to familiarize themselves with the video analysis software before they finalize the procedure for the investigation, especially if they have not used such software previously. This gives students an opportunity to learn how to work with the software and to improve the quality of the video they take.

- Remind students to hold the video camera as still as possible. Any movement of the camera will introduce error into their analysis. If using actual camcorders, we recommend using a tripod to hold the camera steady. If students are using a camera on a cell phone or tablet, we recommend using a table to help steady the camera.

- Remind students to place a meterstick in the same field of view as the motion they are capturing with the video camera. Also, the meterstick should be approximately the same distance from the camera as the motion. Most video analysis software requires the user to define a scale in the video (this allows the software to establish distances and, subsequently, other variables dependent on distance and displacement).

Connections to Standards

Table 9.2 (p. 206) highlights how the investigation can be used to address specific performance expectations from the *NGSS;* learning objectives from AP Physics 1; learning objectives from AP Physics C: Mechanics, *Common Core State Standards,* in English language arts (*CCSS ELA*); and *Common Core State Standards,* Mathematics (*CCSS Mathematics*).

.

TABLE 9.2

Lab 9 alignment with standards

NGSS performance expectation	• HS-PS2-1: Analyze data to support the claim that Newton's second law of motion describes the mathematical relationship among the net force on a macroscopic object, its mass, and its acceleration.
AP Physics 1 learning objectives	• 3.A.1.2: The student is able to design an experimental investigation of the motion of an object. • 3.A.1.3: The student is able to analyze experimental data describing the motion of an object and is able to express the results of the analysis using narrative, mathematical, and graphical representations. • 3.B.1.2: The student is able to design a plan to collect and analyze data for motion (static, constant, or accelerating) from force measurements and carry out an analysis to determine the relationship between the net force and the vector sum of the individual forces.
AP Physics C: Mechanics learning objectives	• I.B.2.a: Students should understand the relation between the force that acts on an object and the resulting change in the object's velocity so they can: • I.B.2.a(1): Calculate, for an object moving in one dimension, the velocity change that results when a constant force **F** acts over a specified time interval. • I.B.2.a(2): Calculate, for an object moving in one dimension, the velocity change that results when a force **F**(t) acts over a specified time interval. • I.B.2.b: Students should understand how Newton's second law, applies to an object subject to forces such as gravity, the pull of strings, or contact forces. • I.B.2.c: Students should be able to analyze situations in which an object moves with specified acceleration under the influence of one or more forces so they can determine the magnitude and direction of the net force, or Öf one of the forces that makes up the net force, such as motion up or down with constant acceleration. • I.B.2.e: Students should understand the effect of drag forces on the motion of an object, so they can: • I.B.2.e(1): Find the terminal velocity of an object moving vertically under the influence of a retarding force dependent on velocity. • I.B.2.e(2): Describe qualitatively, with the aid of graphs, the acceleration, velocity, and displacement of such a particle when it is released from rest or is projected vertically with specified initial velocity. • I.B.2.e(3): Use Newton's second law to write a differential equation for the velocity of the object as a function of time. • I.B.2.e(4): Use the method of separation of variables to derive the equation for the velocity as a function of time from the differential equation that follows from Newton's second law. • I.B.2.e(5): Derive an expression for the acceleration as a function of time for an object falling under the influence of drag forces.

Literacy connections (*CCSS ELA*)	• *Reading:* Key ideas and details, craft and structure, integration of knowledge and ideas • *Writing:* Text types and purposes, production and distribution of writing, research to build and present knowledge, range of writing • *Speaking and listening:* Comprehension and collaboration, presentation of knowledge and ideas
Mathematics connections (*CCSS Mathematics*)	• *Mathematical practices:* Make sense of problems and persevere in solving them, reason abstractly and quantitatively, construct viable arguments and critique the reasoning of others, model with mathematics, use appropriate tools strategically, attend to precision, look for and make use of structure, look for and express regularity in repeated reasoning • *Number and quantity:* Reason quantitatively and use units to solve problems, represent and model with vector quantities, perform operations on vectors • *Algebra:* Interpret the structure of expressions, create equations that describe numbers or relationships, understand solving equations as a process of reasoning and explain the reasoning, solve equations and inequalities in one variable, represent and solve equations and inequalities graphically • *Functions:* Understand the concept of a function and use function notation, interpret functions that arise in applications in terms of the context, analyze functions using different representations, build a function that models a relationship between two quantities, construct and compare linear and exponential models and solve problems, interpret expressions for functions in terms of the situation they model • *Statistics and probability:* Summarize, represent, and interpret data on two categorical and quantitative variables; interpret linear models; understand and evaluate random processes underlying statistical experiments; make inferences and justify conclusions from sample surveys, experiments, and observational studies

Reference

Wilson, E. O. 2013. *Letters to a young scientist.* New York: Liveright Publishing.

Lab Handout

Lab 9. Falling Objects and Air Resistance: How Does the Surface Area of a Parachute Affect the Force Due to Air Resistance as an Object Falls Toward the Ground?

Introduction

When we solve motion problems in physics, we often neglect to take into account the effects of air resistance because, at slow speeds, they are relatively small compared with the force of gravity. Other times, we ignore air resistance when we perform calculations in order to simplify the problem. However, some devices like kites and parachutes are designed to use air resistance in order to function. In these cases, scientists need to account for the effect of air resistance on falling objects.

Besides being used for recreational purposes such as skydiving, parachutes play an important role in the humanitarian efforts of many governments. One of the first uses of parachutes to aid humanitarian efforts was the Berlin Airlift of 1948–1949 (*www. history.com/this-day-in-history/berlin-airlift-begins*). As tensions rose at the onset of the Cold War, the Soviet Union prevented any people or goods from entering West Berlin in Germany. In response, the United States and United Kingdom organized efforts to airdrop food, supplies, and coal (for power) into West Berlin. By the end of the Soviet blockade in 1949, over 200,000 flights had been made into and over Berlin.

FIGURE L9.1

An airdrop of food and medical supplies after a major earthquake in Haiti

The airdrop remains one of the more effective tools for bringing food and necessary supplies, such as medicine, to people that need it. Figure L9.1, for example, is a picture of the airdrop that took place in Haiti after the 2010 earthquake that nearly destroyed the city of Port-au-Prince.

Air resistance affects the net force acting on a falling object, although in some conditions the effect is negligible and/or not observable. Newton's second law states that the acceleration produced by a net force on an object is directly proportional to the magnitude of the net force, is in the same direction as the net force, and is inversely proportional to the mass of an object; or, in mathematical terms, acceleration equals net force divided by mass.

The acceleration of a falling object without air resistance is -9.8 m/s^2 because the net force acting on the falling object is equal to the force of gravity. However, when air resistance is present, then the net force on the object changes, because the force of air resistance counters the force of gravity.

An engineer needs to consider several different issues and work through a multistep design process in order to create a new parachute. The first step in the design process is to determine the performance specifications of the new parachute. This step requires the engineer to think about the minimum and maximum mass of any object that will be attached to the parachute and the maximum terminal velocity that the object will reach as it falls to the ground. Terminal velocity is the highest velocity attainable by an object as it falls through the air. Terminal velocity occurs when the drag force acting on the falling object is equal to the force of gravity. At this point, the sum of the forces acting on the object equals zero, and the resulting acceleration will be zero. A safe landing velocity for an object is usually between 2 and 5 m/s. The second step in the design process is to build a parachute with a specific surface area that will meet these important performance specifications. It is therefore important for engineers to understand how the surface area of a parachute affects the force of air resistance that acts on an object as it falls to the ground.

Your Task

Use what you know about forces and motion, structure and function, and models to design and carry out an investigation to determine how parachute surface area affects the force due to air resistance.

The guiding question of this investigation is, *How does the surface area of a parachute affect the force due to air resistance as an object falls toward the ground?*

Materials

You may use any of the following materials during your investigation:

Consumables
- Large trash bags
- Tape
- String or fishing line

Equipment
- Safety glasses or goggles (required)
- Electronic or triple beam balance
- Washers
- Stopwatch
- Ruler
- Meterstick

If you have access to the following equipment, you may also consider using a video camera and a computer or tablet with video analysis software.

LAB 9

Safety Precautions

Follow all normal lab safety rules. In addition, take the following safety precautions:

1. Wear sanitized safety goggles or glasses during lab setup, hands-on activity, and takedown.

2. Do not throw the washers or the parachutes.

3. Do not stand on tables or chairs.

4. Wash hands with soap and water after completing the lab.

Investigation Proposal Required?   ☐ Yes   ☐ No

Getting Started

To answer the guiding question, you will need to design and carry out an experiment. To accomplish this task, you must determine what type of data you need to collect, how you will collect it, and how you will analyze it.

To determine *what type of data you need to collect,* think about the following questions:

- What are the boundaries and components of the system you are studying?
- How do the components of the system interact with each other?
- How might the structure of a parachute relate to its function?
- How will you determine the surface area of a parachute?
- How will you measure the force of air resistance?
- What will be the independent variable and the dependent variable for your experiment?

To determine *how you will collect the data,* think about the following questions:

- What conditions need to be satisfied to establish a cause-and-effect relationship?
- What measurement scale or scales should you use to collect data?
- What equipment will you need to make the measurements?
- What other variables will you need to control during your experiment?
- Do you need to include a control group?
- How will you make sure that your data are of high quality (i.e., how will you reduce error)?
- How will you keep track of and organize the data you collect?

To determine *how you will analyze the data,* think about the following questions:

- What type of calculations will you need to make?

- What types of models can you use to help you analyze the motion of a parachute?

- How could you use mathematics to describe a relationship between variables?

- What types of patterns might you look for as you analyze your data?

- Are there any proportional relationships that you can identify?

- What type of table or graph could you create to help make sense of your data?

Connections to the Nature of Scientific Knowledge and Scientific Inquiry

As you work through your investigation, you may want to consider

- how the culture of science, societal needs, and current events influence the work of scientists; and

- the role of imagination and creativity in science.

Initial Argument

Once your group has finished collecting and analyzing your data, your group will need to develop an initial argument. Your initial argument needs to include a claim, evidence to support your claim, and a justification of the evidence. The *claim* is your group's answer to the guiding question. The *evidence* is an analysis and interpretation of your data. Finally, the justification of the evidence is why your group thinks the evidence matters. The *justification* of the evidence is important because scientists can use different kinds of evidence to support their claims. Your group will create your initial argument on a whiteboard. Your whiteboard should include all the information shown in Figure L9.2.

FIGURE L9.2

Argument presentation on a whiteboard

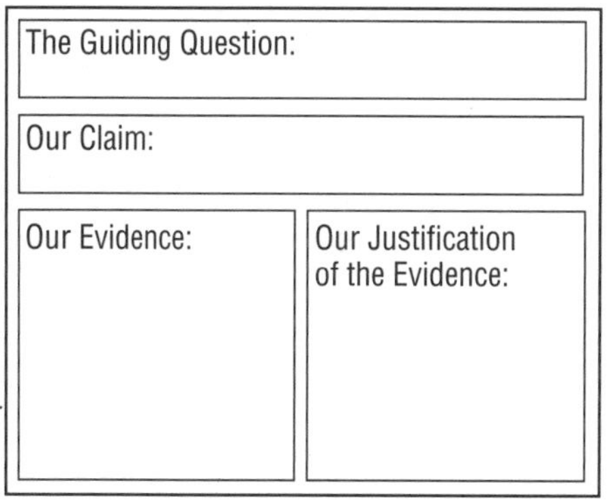

The Guiding Question:	
Our Claim:	
Our Evidence:	Our Justification of the Evidence:

Argumentation Session

The argumentation session allows all of the groups to share their arguments. One or two members of each group will stay at the lab station to share that group's argument, while the other members of the group go to the other lab stations to listen to and critique the other arguments. This is similar to what scientists do when they propose, support, evaluate, and refine new ideas during a poster session at a conference. If you are presenting your group's argument, your goal is to share your ideas and answer questions. You should also keep a record of the critiques and suggestions made by your classmates so you can use this feedback to make your initial argument stronger. You can keep track of specific critiques and suggestions for improvement that your classmates mention in the space below.

Critiques about our initial argument and suggestions for improvement:

If you are critiquing your classmates' arguments, your goal is to look for mistakes in their arguments and offer suggestions for improvement so these mistakes can be fixed. You should look for ways to make your initial argument stronger by looking for things that the other groups did well. You can keep track of interesting ideas that you see and hear during the argumentation in the space below. You can also use this space to keep track of any questions that you will need to discuss with your team.

Interesting ideas from other groups or questions to take back to my group:

Once the argumentation session is complete, you will have a chance to meet with your group and revise your initial argument. Your group might need to gather more data or design a way to test one or more alternative claims as part of this process. Remember, your goal at this stage of the investigation is to develop the best argument possible.

Report

Once you have completed your research, you will need to prepare an *investigation report* that consists of three sections. Each section should provide an answer to the following questions:

1. What question were you trying to answer and why?

2. What did you do to answer your question and why?

3. What is your argument?

Your report should answer these questions in two pages or less. This report must be typed, and any diagrams, figures, or tables should be embedded into the document. Be sure to write in a persuasive style; you are trying to convince others that your claim is acceptable or valid!

LAB 9

Checkout Questions

Lab 9. Falling Objects and Air Resistance: How Does the Surface Area of a Parachute Affect the Force Due to Air Resistance as an Object Falls Toward the Ground?

1. Is there a maximum force due to air resistance that can act on a parachute?

   a. Yes

   b. No

   How do you know?

   What does your answer suggest about the effect of increasing the size of the parachute?

2. The equation for the force of air resistance (more formally, the drag) on a parachute is $F_D = C_D \rho v^2 A/2$. In this equation, F_D is the drag force and v is the current velocity of the falling parachute and mass system. Is the drag force constant as a function of time?

   a. Yes

   b. No

   Justify your answer using the equation provided and/or data from your investigation.

3. Scientists share a set of values, norms, and commitments that shape what counts as knowing, how to represent or communicate information, and how to interact with other scientists.

   a. I agree with this statement.

   b. I disagree with this statement.

   Explain your answer, using an example from your investigation about air resistance and parachutes.

4. Scientists must use their imagination and creativity to figure out new ways to test ideas and collect or analyze data.

   a. I agree with this statement.

   b. I disagree with this statement.

   Explain your answer, using an example from your investigation about air resistance and parachutes.

5. Why is it useful to identify a system under study and then make a model of it during an investigation? In your answer, be sure to include examples from at least two different investigations.

6. Why is it important to think about the relationship between structure and function when trying to develop an explanation for a natural phenomenon? In your answer, be sure to include examples from at least two different investigations.

SECTION 4
Forces and Motion

Circular Motion and Rotation

Introduction Labs

LAB 10

Teacher Notes

Lab 10. Rotational Motion: How Do the Mass and the Distribution of Mass in an Object Affect Its Rotation?

Purpose

The purpose of this lab is to *introduce* students to the core idea of forces and motion, part of the disciplinary core idea (DCI) of Motion and Stability: Forces and Interactions from the *NGSS,* by giving them an opportunity to explore some factors that affect the rotation of an object. This lab also gives students an opportunity to learn about the crosscutting concepts (CCs) of (a) Patterns and (b) Structure and Function from the *NGSS.* In addition, this lab can be used to help students understand two big ideas from AP Physics: (a) the interactions of an object with other objects can be described by forces and (b) interactions between systems can result in changes in those systems. As part of the explicit and reflective discussion, students will also learn about (a) the difference between observations and inferences in science and (b) the role of imagination and creativity in science.

Underlying Physics Concepts

Rotational motion and linear motion are very similar, and the components of motion in each context are described using analogical quantities. For example, velocity in linear motion is described as angular velocity in rotational motion, and linear acceleration is described as angular acceleration in rotational motion. Angular descriptors are used in rotational contexts because any point on a rotating object is moving in a circular path around a fixed axis. Therefore any object along a line from the axis of rotation to the edge of the object moves through the same angle in a given amount of time.

Consider the rotating disc pictured in Figure 10.1; point P is some distance ($\mathbf{r}$) away from the axis of rotation and moving in a counterclockwise direction (we use $\mathbf{r}$ to denote the distance from the center because segment $\mathbf{r}$ is a radial distance). When the disc has rotated through angle theta (θ), point P has moved some arc-length ($\mathbf{l}$) from the horizontal (x). Because the disc is uniform, all points along line $\mathbf{r}$ have also moved through the same angle. However, $\mathbf{l}$ would be different for various points along the line. This example illustrates the need for defining the angular velocity of the disc rather than the linear velocity. Any point along line $\mathbf{r}$ would

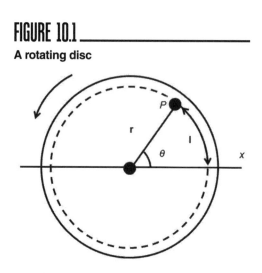

FIGURE 10.1

A rotating disc

have the same angular velocity, but because the circumference of the circular path grows when moving from the axis of rotation to the edge of the disc, the linear velocity is not consistent. Different points along line **r** will move through different arc-lengths **l** in the same amount of time even though all the points have moved through the same angle.

Whereas linear velocity is measured in meters per second (m/s), angular velocity is measured in radians per second (rad/s). These units indicate that linear velocity is proportional to linear displacement and angular velocity is proportional to angular displacement. There are 2π radians in a complete 360° circle; therefore, 1 radian is approximately equal to 57.3°. The average angular velocity denoted by $\overline{\omega}$ (we use the bar over the ω to indicate average) can be calculated using Equation 10.1, where $\Delta\theta$ is the angle the object has rotated, and Δt is amount of time for the rotation. In SI units, angular velocity is measured in radians per second (rad/s), the angle of rotation is measured in radians (rad), and time is measured in seconds (s).

$$\textbf{(Equation 10.1)} \quad \overline{\omega} = \frac{\Delta\theta}{\Delta t}$$

Much like calculating linear acceleration, angular acceleration is simply the ratio of the change in angular velocity with respect to the elapsed time. The average angular acceleration ($\bar{\alpha}$) of a rotating object can be determined using Equation 10.2, where $\Delta\omega$ is the change in angular velocity of the object during time Δt. In SI units, angular acceleration is measured in radians per second squared (rad/s^2).

$$\textbf{(Equation 10.2)} \quad \bar{\alpha} = \frac{\Delta\omega}{\Delta t}$$

In addition to similarities in describing linear motion and rotational motion, the influence of forces in linear and rotational contexts is also very similar. For linear contexts, Newton's second law of motion ($\Sigma\textbf{F} = m\textbf{a}$) describes the relationship between the net force acting on an object ($\Sigma\textbf{F}$), the object's mass (m), and the acceleration of the object (**a**). For rotational contexts, an analogous equation describes the relationship between torque (the force applied to a lever arm that results in rotation; the lever arm is simply the distance between the axis of rotation and the point where the force is applied), moment of inertia, and angular acceleration. This relationship is described in Equation 10.3, where $\Sigma\tau$ represents the sum of the torques acting on the object, I represents the moment of inertia, and α is the angular acceleration for the object. In SI units, torque is measured in Newton meters (N·m), moment of inertia is measured in kilogram-meters squared (kg·m^2), and angular acceleration is measured in radians per second squared (rad/s^2).

$$\textbf{(Equation 10.3)} \quad \Sigma\tau = I\alpha$$

In linear contexts the mass of an object can be assumed to be concentrated at the center of the object and therefore is treated as a point mass. For rotational motion, however, the mass cannot be assumed to be concentrated at the axis of rotation; its distribution must be taken into account. Figures 10.2 and 10.3 show two levers, with each lever holding a box of equal mass, m. For the scenario presented in Figure 10.2, a greater torque (turning force) is required to lift the lever and cause it to rotate because the mass is located far away from the fulcrum, which serves as the axis of rotation. In contrast, for the scenario in Figure 10.3, the mass is half the distance from the axis of rotation, so half the torque is required to rotate the lever through the same distance or angle.

FIGURE 10.2 _____

Lever with mass positioned far away from the axis of rotation

FIGURE 10.3 _____

Lever with mass positioned near the axis of rotation

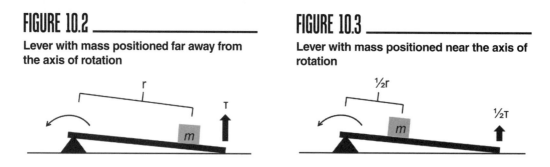

The examples in Figures 10.2 and 10.3 illustrate why the position or distribution of mass is important when investigating rotational motion. The lever in Figure 10.2 has high rotational inertia, which means that the lever has a high resistance to changes in its motion; in other words, it takes more effort or torque to cause it to rotate. In contrast, the lever in Figure 10.3 has a lower rotational inertia, meaning it will require less effort or torque to cause it to rotate. Imagine a lever with several boxes positioned side by side along the length of the lever. Each mass would have an impact on the rotational inertia of the entire lever, and rearranging the boxes (e.g., stacking the boxes instead of placing them side by side, creating two stacks of boxes side by side, or creating three stacks of boxes) would result in a variety of rotational inertia values for the system. When determining the rotational inertia of a system—whether it is a lever rotating or a rolling wheel—it is necessary to take into account how the distribution of mass contributes to the overall rotational inertia. The moment of inertia value, I, in Equation 10.3 is indeed the sum of the effects of all of the mass distributed within a rotating system.

Timeline

The instructional time needed to complete this lab investigation is 200–280 minutes. Appendix 3 (p. 531) provides options for implementing this lab investigation over several class periods. Option E (280 minutes) should be used if students are unfamiliar with scientific writing, because this option provides extra instructional time for scaffolding the writing process. You can scaffold the writing process by modeling, providing examples, and providing hints as students write each section of the report. Option E can also be used if you are introducing students to the video analysis programs. Option F (200 minutes) should be used if students are familiar with scientific writing and have developed the skills needed to write an investigation report on their own. In option F, students complete stage 6 (writing the investigation report) and stage 8 (revising the investigation report) as homework.

Materials and Preparation

The materials needed to implement this investigation are listed in Table 10.1 (p. 224). The consumables and equipment can be purchased from a science supply company such as Carolina, Flinn Scientific, or Ward's Science. Records can be purchased at used bookstores or online at Amazon. Video analysis software can be purchased from Vernier (Logger *Pro*) or PASCO (SPARKvue or Capstone). These companies also have apps that can be used on Apple- or Android-based tablets and cell phones. We recommend consulting with your school's information technology coordinator to determine the best option for your students.

For this investigation, students will collect data to determine how changes in mass and/ or changes in the distribution of mass affect the rotational motion of an object. The equipment setup shown in Figure 10.4 illustrates how two metersticks and a block (or stack of books) can be used to build a ramp structure that will serve as a track for a rolling record (i.e., a music record that predates tapes and CDs). The preferred center axle for the record can be made from a wooden dowel rod that has the same diameter as the central hole on the record. The axle should be long enough so that it spans the gap between the metersticks and can support the records. Students can change the total mass of the rotating system by adding additional records to the axle or by adding masses to the records. Simple slotted masses, coins, or washers can be taped directly to the sides of the records to increase the mass or placed in various arrangements (e.g., close to the outer perimeter or close to the center axle) to test different distributions of mass.

FIGURE 10.4

Sample equipment setup using metersticks, a block, a wood dowel, and several records as shown from the side (a) and from the top (b)

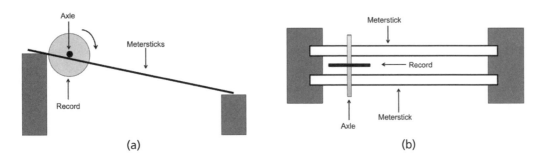

(a)   (b)

LAB 10

TABLE 10.1 _____

Materials list for Lab 10

Item	Quantity
Consumable	
Tape	1 roll per group
Equipment and other materials	
Safety glasses or goggles	1 per student
Records	6 per group
Center axle for records (wooden dowel rod)	1 per group
Stopwatch	1 per group
Slotted mass set, 30 coins, or 30 washers	1 set per group
Ramp (made from 2 metersticks and books or blocks)	1 per group
Electronic or triple beam balance	1 per group
Investigation Proposal C (optional)	1 per group
Whiteboard, 2' × 3'*	1 per group
Lab Handout	1 per student
Peer-review guide and teacher scoring rubric	1 per student
Checkout Questions	1 per student
Equipment for video analysis (optional)	
Video camera	1 per group
Computer or tablet with video analysis software	1 per group

* As an alternative, students can use computer and presentation software such as Microsoft PowerPoint or Apple Keynote to create their arguments.

Be sure to use a set routine for distributing and collecting the materials during the lab investigation. One option is to set up the materials for each group at each group's lab station before class begins. This option works well when there is a dedicated section of the classroom for lab work and the materials are large and difficult to move (such as a dynamics track). A second option is to have all the materials on a table or cart at a central location. You can then assign a member of each group to be the "materials manager." This individual is responsible for collecting all the materials his or her group needs from the table or cart during class and for returning all the materials at the end of the class. This option works well when the materials are small and easy to move (such as stopwatches,

metersticks, or hanging masses). It also makes it easy to inventory the materials at the end of the class before students leave for the day.

Safety Precautions

Remind students to follow all normal lab safety rules. In addition, tell students to take the following safety precautions:

1. Wear sanitized safety glasses or goggles during lab setup, hands-on activity, and takedown.

2. Keep fingers and toes out of the way of moving objects.

3. Wash hands with soap and water after completing the lab.

Topics for the Explicit and Reflective Discussion
Reflecting on the Use of Core Ideas and Crosscutting Concepts During the Investigation

Teachers should begin the explicit and reflective discussion by asking students to discuss what they know about the core idea they used during the investigation. The following are some important concepts related to the core idea of forces and motion that students need to use to determine how the mass and distribution of mass affect the rotation of an object:

- Quantities such as angular displacement, angular velocity, and angular acceleration are used to describe rotational motion about a fixed axis.

- *Angular displacement* of a body is the angle in radians (degrees, revolutions) through which a point or line has been rotated about a specified axis. *Angular velocity* is the rate of change of angular displacement. *Angular acceleration* is the rate of change in angular velocity.

- *Torque* is a measure of a force applied perpendicular to a lever arm multiplied by the distance from the point of rotation. Unbalanced torques cause an object to change its rotational motion.

- Torque, angular displacement, angular velocity, angular acceleration, and angular momentum are vector quantities. These quantities are characterized as being positive or negative depending on whether they give rise to or correspond to counterclockwise or clockwise rotation with respect to a specified axis.

To help students reflect on what they know about forces and motion, we recommend showing them two or three images using presentation software that help illustrate these important ideas. You can then ask the students the following questions to encourage them to share how they are thinking about these important concepts:

1. What do we see going on in this image?

2. Does anyone have anything else to add?

3. What might be going on that we can't see?

4. What are some things that we are not sure about here?

You can then encourage students to think about how CCs played a role in their investigation. There are at least two CCs that students need to determine how the mass and distribution of mass affect the rotation of an object: (a) Patterns and (b) Structure and Function (see Appendix 2 [p. 527] for a brief description of these CCs). To help students reflect on what they know about these CCs, we recommend asking them the following questions:

1. Why is it important to look for patterns during an investigation?

2. What patterns did you identify during your investigation? What did the identification of these patterns allow you to do?

3. The way an object is shaped or structured determines many of its properties and how it functions. Why is it useful to think about the relationship between structure and function during an investigation?

4. Why was it important to examine the structure of a rotating object while attempting to explain how fast or slow it rotates? Why is an understanding of the relationship between the structure and function of a rotating object more useful than simply knowing which object rotates faster than others?

You can then encourage the students to think about how they used all these different concepts to help answer the guiding question and why it is important to use these ideas to help justify their evidence for their final arguments. Be sure to remind your students to explain why they included the evidence in their arguments and make the assumptions underlying their analysis and interpretation of the data explicit in order to provide an adequate justification of their evidence.

Reflecting on Ways to Design Better Investigations

It is important for students to reflect on the strengths and weaknesses of the investigation they designed during the explicit and reflective discussion. Students should therefore be encouraged to discuss ways to eliminate potential flaws, measurement errors, or sources of uncertainty in their investigations. To help students be more reflective about the design of their investigation and what they can do to make their investigations more rigorous in the future, you can ask them the following questions:

1. What were some of the strengths of the way you planned and carried out your investigation? In other words, what made it scientific?

2. What were some of the weaknesses of the way you planned and carried out your investigation? In other words, what made it less scientific?

3. What rules can we make, as a class, to ensure that our next investigation is more scientific?

Reflecting on the Nature of Scientific Knowledge and Scientific Inquiry

This investigation can be used to illustrate two important concepts related to the nature of scientific knowledge and the nature of scientific inquiry: (a) the difference between observations and inferences in science and (b) the role of imagination and creativity in science (see Appendix 2 for a brief description of these two concepts). Be sure to review these concepts during and at the end of the explicit and reflective discussion. To help students think about these concepts in relation to what they did during the lab, you can ask them the following questions:

1. You had to make observations and inferences during your investigation. Can you give me some examples of these observations and inferences?

2. Can you work with your group to come up with a rule that you can use to decide if a piece of information is an observation or an inference? Be ready to share in a few minutes.

3. Some people think that there is no room for imagination or creativity in science. What do you think?

4. Can you work with your group to come up different ways that you needed to use your imagination or be creative during this investigation? Be ready to share in a few minutes.

You can also use presentation software or other techniques to encourage your students to think about these concepts. You can show examples of information from the investigation that are either observations or inferences and ask students to classify each example and explain their thinking. You can also show students an image of the following quote by E. O. Wilson from *Letters to a Young Scientist* (2013) and ask them what they think he meant by it:

> The ideal scientist thinks like a poet and only later works like a bookkeeper. Keep in mind that innovators in both literature and science are basically dreamers and storytellers. In the early stages of the creation of both literature and science, everything in the mind is a story. There is an imagined ending, and usually an imagined beginning, and a selection of bits and pieces that might fit in between. In works of literature and science alike, any part can be changed, causing a ripple among the other parts, some of which are discarded and new ones added. (p. 74)

Be sure to remind your students that, to be proficient in science, it is important that they understand what counts as scientific knowledge and how that knowledge develops over time.

LAB 10

Hints for Implementing the Lab

- Allowing students to design their own procedures for collecting data gives students an opportunity to try, to fail, and to learn from their mistakes. However, you can scaffold students as they develop their procedure by having them fill out an investigation proposal. These proposals provide a way for you to offer students hints and suggestions without telling them how to do it. You can also check the proposals quickly during a class period. For this lab we suggest using Investigation Proposal C.

- Data collection for this lab involves measuring the time it takes the record wheel and axle system to roll down the ramp. It is helpful for students to make distinct or bold marks on the surface of the record so they know how many revolutions have passed or when a certain point on the record has traveled the goal distance.

- Allow the students to become familiar with the record and ramp setup as part of the tool talk before they begin to design their investigation. Giving them 5–10 minutes to examine the equipment will let them see what they can and cannot do with it.

- We recommend that students create a ramp that is inclined at approximately 30° above the incline. This is the point at which the gravitational force will exert a large enough torque to overcome the rotational inertia. We recommend using the smallest incline possible that will still allow the record to roll down the ramp. The smaller the incline, the easier data collection is and the more pronounced the data will be.

- We recommend using a standard 33⅓ record (LP). This is the largest size of record and allows for more freedom in where students place additional masses on their records. If you want, you can have students mathematically model the relationship between the placement of mass and rotational acceleration by marking various radii on the surface of the record. This type of modeling is made easier with a larger record.

- As the wheel and axle roll down the ramp, the wheel (i.e., the record) will make contact with the table when the height of the metersticks is equal to the radius of the record. This provides students with a good opportunity to think about experimental design. When the record is rolling down the ramp, the only torque is due to the force of gravity. However, when the record makes contact with the table, gravity is no longer providing a torque. Instead, friction will provide a torque opposite the direction of rotation (which is why the record will eventually come to a stop). You can point this out ahead of time to students while they are designing their data collection procedure. Or, you can wait to see how students address this issue, if at all, and return to this point during the explicit and reflective discussion.

- Be sure to allow students to go back and re-collect data at the end of the argumentation session. Students often realize that they made numerous mistakes when they were collecting data as a result of their discussions during the argumentation session. The students, as a result, will want a chance to re-collect data, and the re-collection of data should be encouraged when time allows. This also offers an opportunity to discuss what scientists do when they realize a mistake is made inside the lab.

If students use video analysis

- We suggest allowing students to familiarize themselves with the video analysis software before they finalize the procedure for the investigation, especially if they have not used such software previously. This gives students an opportunity to learn how to work with the software and to improve the quality of the video they take.

- Remind students to hold the video camera as still as possible. Any movement of the camera will introduce error into their analysis. If using actual camcorders, we recommend using a tripod to hold the camera steady. If students are using a camera on a cell phone or tablet, we recommend using a table to help steady the camera.

- Remind students to place a meterstick in the same field of view as the motion they are capturing with the video camera. Also, the meterstick should be approximately the same distance from the camera as the motion. Most video analysis software requires the user to define a scale in the video (this allows the software to establish distances and, subsequently, other variables dependent on distance and displacement).

Connections to Standards

Table 10.2 (p. 230) highlights how the investigation can be used to address learning objectives from AP Physics 1; learning objectives from AP Physics C: Mechanics, *Common Core State Standards,* in English language arts (*CCSS ELA*); and *Common Core State Standards,* Mathematics (*CCSS Mathematics*).

LAB 10

TABLE 10.2

Lab 10 alignment with standards

***NGSS* performance expectations**	• None
AP Physics 1 learning objectives	• 4.D.1.1: The student is able to describe a representation and use it to analyze a situation in which several forces exerted on a rotating system of rigidly connected objects change the angular velocity and angular momentum of the system. • 4.D.1.2: The student is able to plan data collection strategies designed to establish that torque, angular velocity, angular acceleration, and angular momentum can be predicted accurately when the variables are treated as being clockwise or counterclockwise with respect to a well-defined axis of rotation, and refine the research question based on the examination of data.
AP Physics C: Mechanics learning objectives	• I.E.2.c(1): Determine by inspection which of a set of symmetrical objects of equal mass has the greatest rotational inertia. • I.E.2.c(2): Determine by what factor an object's rotational inertia changes if all its dimensions are increased by the same factor.
Literacy connections (*CCSS ELA*)	• *Reading:* Key ideas and details, craft and structure, integration of knowledge and ideas • *Writing:* Text types and purposes, production and distribution of writing, research to build and present knowledge, range of writing • *Speaking and listening:* Comprehension and collaboration, presentation of knowledge and ideas
Mathematics connections (*CCSS Mathematics*)	• *Mathematical practices:* Reason abstractly and quantitatively, construct viable arguments and critique the reasoning of others, use appropriate tools strategically, attend to precision • *Number and quantity:* Reason quantitatively and use units to solve problems, represent and model with vector quantities, perform operations on vectors • *Algebra:* Interpret the structure of expressions, understand solving equations as a process of reasoning and explain the reasoning, solve equations and inequalities in one variable, represent and solve equations and inequalities graphically • *Functions:* Interpret expressions for functions in terms of the situation they model • *Statistics and probability:* Understand and evaluate random processes underlying statistical experiments; make inferences and justify conclusions from sample surveys, experiments, and observational studies

Reference

Wilson, E. O. 2013. *Letters to a young scientist.* New York: Liveright Publishing.

Lab Handout

Lab 10. Rotational Motion: How Do the Mass and the Distribution of Mass in an Object Affect Its Rotation?

Introduction

The wheel and axle are arguably among the most important inventions of all time. When a wheel turns around an axle, every point on the wheel moves in a circle about an axis of rotation. The motion around an axis of rotation is called rotational motion. We can describe the motion of a rotating object just like we can describe the motion of objects moving in a straight line. There are many similarities between rotational motion and linear motion. Rotating objects also have inertia and are influenced by the net force acting on them just like an object undergoing linear motion. A force that causes an object to rotate is called a torque.

There are also several important differences between rotational motion and linear motion. One big difference is how the distribution of mass within a rotating object affects how fast that object rotates around an axis of rotation. Take ice-skaters, gymnasts, and high divers as an example. These athletes often need to spin during a routine or a dive. As they spin, they can position their bodies in a way that will make them rotate faster or slower, which means they can change their angular velocity. Angular velocity is measured in radians per second (rad/s) and describes how much of a circle is completed by a rotating object in 1 second. A complete circle is equal to 360° or

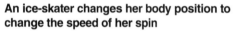

FIGURE L10.1

An ice-skater changes her body position to change the speed of her spin

2π radians (360° = 2π rad, therefore 1 radian ≈ 57.3°). An object with an angular velocity of 19 rad/s would complete approximately 3 rotations per second. Similar to linear motion, the mass of a rotating object influences how its velocity changes when a force acts on it. However, unlike linear motion, the angular velocity of a rotating object will also change when the position or distribution of the mass within that object is changed. Athletes, as a result, can change how far their body parts are away from the axis of rotation, which changes the distribution of their mass (their total mass stays constant, they do not gain or lose mass during the spin), in order to spin faster or slower during a routine or a dive (see Figure L10.1).

Understanding rotational motion is important in science and has implications for aspects of our everyday lives. One implication related to rotational motion is the behavior of automobile wheels. The early inventors of the wheel were not concerned with traveling at the speeds we do today. Understanding how torque, mass, angular velocity, and angular acceleration influence the behavior of a wheel is important for ensuring that vehicles operate efficiently and are safe to drive. Putting too large a wheel on a car will cause the engine to become overworked and use more gas than necessary; additionally, if a wheel is too large or too massive, the brakes on the vehicle may not be strong enough to stop the wheel's rotation.

Your Task

Use what you know about rotational motion, patterns, and structure and function to design and conduct an investigation to determine how the mass and distribution of mass affect the rotation of an object.

The guiding question of this investigation is, *How do mass and the distribution of mass in an object affect its rotation?*

Materials

You may use any of the following materials during your investigation:

Consumable	Equipment	
• Tape	• Safety glasses or goggles (required)	• Slotted masses, washers, or coins
	• Records	• Ramp (made from metersticks and a block)
	• Wood dowels	• Electronic or triple beam balance
	• Stopwatch	

If you have access to the following equipment, you may also consider using a video camera and a computer or tablet with video analysis software.

Safety Precautions

Follow all normal lab safety rules. In addition, take the following safety precautions:

1. Wear sanitized safety glasses or goggles during lab setup, hands-on activity, and takedown.

2. Keep fingers and toes out of the way of moving objects.

3. Wash hands with soap and water after completing the lab.

Investigation Proposal Required?   ☐ Yes   ☐ No

Getting Started

To answer the guiding question, you will need to design and carry out an experiment to determine how changes in mass and the distribution of mass affect the rotational motion of a record. The equipment setup shown in Figure L10.2 illustrates how you can use

FIGURE L10.2

Sample equipment setup using metersticks, a block, a wood dowel, and several records as shown from the side (a) and from the top (b)

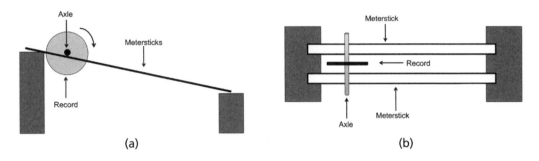

(a) (b)

metersticks and blocks (or a stack of books) to build a ramp for a rolling record. You can use a wooden dowel rod that has the same diameter as the center hole of a record for the axle (add tape to the rod if it is too thin). You can then change the total mass of the rotating system by adding additional records to the axle or by adding masses to the record. You can tape washers or coins to the sides of the records at different spots on the record (e.g., close to the outer perimeter or close to the center axle) to change the distribution of mass in the rotating system. Before you begin designing your experiment using this equipment, however, you must first determine what type of data you need to collect, how you will collect it, and how you will analyze it.

To determine *what type of data you need to collect,* think about the following questions:

- What are the boundaries and components of the system you are studying?
- How do the components of the system interact with each other?
- How might changes to the structure of what you are studying change how it functions?
- How could you keep track of changes in this system quantitatively?
- What variables do you need to compare?
- What is the outcome variable for your investigation?

To determine *how you will collect the data,* think about the following questions:

- What measurement scale or scales should you use to collect data?
- What other variables will you need to control during your investigation?
- How will you make sure that your data are of high quality (i.e., how will you reduce error)?
- How will you keep track of and organize the data you collect?

LAB 10

To determine *how you will analyze the data,* think about the following questions:

- What types of patterns might you look for as you analyze your data?
- What type of calculations will you need to make?
- How could you use mathematics to describe a relationship between variables?
- What type of table or graph could you create to help make sense of your data?

Connections to the Nature of Scientific Knowledge and Scientific Inquiry

As you work through your investigation, you may want to consider

- the difference between observations and inferences in science, and
- the role of imagination and creativity in science.

Initial Argument

Once your group has finished collecting and analyzing your data, your group will need to develop an initial argument. Your argument must include a claim, evidence to support your claim, and a justification of the evidence. The *claim* is your group's answer to the guiding question. The *evidence* is an analysis and interpretation of your data. Finally, the *justification* of the evidence is why your group thinks the evidence matters. The justification of the evidence is important because scientists can use different kinds of evidence to support their claims. Your group will create your initial argument on a whiteboard. Your whiteboard should include all the information shown in Figure L10.3.

FIGURE L10.3 _____

Argument presentation on a whiteboard

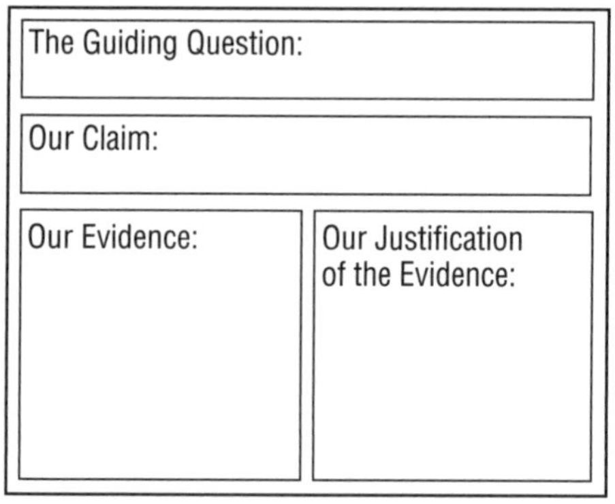

The Guiding Question:	
Our Claim:	
Our Evidence:	Our Justification of the Evidence:

Argumentation Session

The argumentation session allows all of the groups to share their arguments. One or two members of each group will stay at the lab station to share that group's argument, while the other members of the group go to the other lab stations to listen to and critique the other arguments. This is similar to what scientists do when they propose, support, evaluate, and refine new ideas during a poster session at a conference. If you are presenting your group's argument, your goal is to share your ideas and answer questions. You should also keep a record of the critiques and suggestions made by your classmates so you can use this feedback to make your initial argument stronger. You can keep track of specific critiques and suggestions for improvement that your classmates mention in the space below.

Critiques about our initial argument and suggestions for improvement:

If you are critiquing your classmates' arguments, your goal is to look for mistakes in their arguments and offer suggestions for improvement so these mistakes can be fixed. You should look for ways to make your initial argument stronger by looking for things that the other groups did well. You can keep track of interesting ideas that you see and hear during the argumentation in the space below. You can also use this space to keep track of any questions that you will need to discuss with your team.

Interesting ideas from other groups or questions to take back to my group:

LAB 10

Once the argumentation session is complete, you will have a chance to meet with your group and revise your initial argument. Your group might need to gather more data or design a way to test one or more alternative claims as part of this process. Remember, your goal at this stage of the investigation is to develop the best argument possible.

Report

Once you have completed your research, you will need to prepare an investigation report that consists of three sections. Each section should provide an answer to the following questions:

1. What question were you trying to answer and why?

2. What did you do to answer your question and why?

3. What is your argument?

Your report should answer these questions in two pages or less. This report must be typed, and any diagrams, figures, or tables should be embedded into the document. Be sure to write in a persuasive style; you are trying to convince others that your claim is acceptable or valid!

National Science Teachers Association

Checkout Questions

Lab 10. Rotational Motion: How Do the Mass and the Distribution of Mass in an Object Affect Its Rotation?

Pictured below are four objects (a solid disc, a hoop, a sphere, and a solid cylinder). The mass and radius for each object are also provided. Use this information to answer question 1.

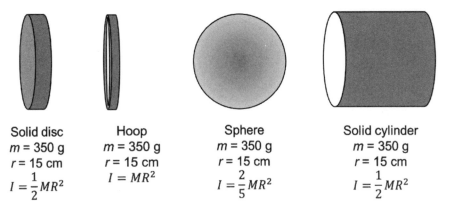

Solid disc
$m = 350$ g
$r = 15$ cm
$I = \frac{1}{2}MR^2$

Hoop
$m = 350$ g
$r = 15$ cm
$I = MR^2$

Sphere
$m = 350$ g
$r = 15$ cm
$I = \frac{2}{5}MR^2$

Solid cylinder
$m = 350$ g
$r = 15$ cm
$I = \frac{1}{2}MR^2$

1. If all four objects are released from the top of a ramp and allowed to roll to the bottom, what will be the finishing order for the objects?

   a. All objects will come to the bottom at the same time.

   b. The sphere will come to the bottom first, then the solid disc and solid cylinder at the same time, then the hoop.

   c. The hoop will come to the bottom first, then the solid disc and solid cylinder at the same time, then the sphere.

   d. The solid cylinder and the solid disc will come to the bottom first, then the sphere, then the hoop.

   Justify your answer using what you know about rotational motion.

2. Jeremy and Susan are playing on a merry-go-round. Susan says she wants to sit close to the center of the merry-go-round while Jeremy pushes the merry-go-round. Jeremy wants to sit closer to the outer edge. To get the merry-go-round to move at an angular velocity of one rotation per minute, who would need to turn the merry-go-round with a large force? Assume Jeremy and Susan each has a mass of 40 kg.

   a. Jeremy

   b. Susan

   c. The same force is needed.

Use what you know about rotational motion to justify your answer.

3. Scientific research requires imagination and creativity.

   a. I agree with this statement.

   b. I disagree with this statement.

Explain your answer, using an example from your investigation about the effect of a mass and the distribution of mass on a rotating object.

4. There is a difference between observations and inferences in science.

   a. I agree with this statement.

   b. I disagree with this statement.

   Explain your answer, using an example from your investigation about the effect of a mass and the distribution of mass on a rotating object.

5. Why is it important to identify patterns and the underlying causes for those patterns in science? Be sure to include examples from at least two different investigations in your answer.

6. Why is useful to think about the relationship between structure and function during an investigation? Be sure to include examples from at least two different investigations in your answer.

LAB 11

Teacher Notes

Lab 11. Circular Motion: How Does Changing the Angular Velocity of the Swinging Mass at the Top of a Whirligig and the Amount of Mass at the Bottom of a Whirligig Affect the Distance From the Top of the Tube to the Swinging Mass?

Purpose

The purpose of this lab is to *introduce* students to the core idea of forces and motion, part of the disciplinary core idea (DCI) of Motion and Stability: Forces and Interactions from the *NGSS*, by giving them an opportunity to explore the nature of circular motion. This lab also gives students an opportunity to learn about the crosscutting concepts (CCs) of (a) Systems and System Models and (b) Stability and Change from the *NGSS*. In addition, this lab can be used to help students understand two big ideas from AP Physics: (a) the interactions of an object with other objects can be described by forces and (b) interactions between systems can result in changes in those systems. As part of the explicit and reflective discussion, students will also learn about (a) the difference between laws and theories in science and (b) the difference between data and evidence in science.

Underlying Physics Concepts

A centripetal, or center-pointing, force sustains circular motion. In the case of planetary motion, the centripetal force is gravity.[1] A roller coaster does a loop using the track's normal force. For a ball on a string swung on a circular path, string tension (F_T) plays the role of the centripetal force (F_c) as shown in Equation 11.1. In SI units, force is measured in newtons (N).

$$\textbf{(Equation 11.1)} \quad F_T = F_c$$

The magnitude of a centripetal force required for circular motion depends on the object's mass (m), velocity (v), and the radius (r) of its circular path (or portion thereof), shown in Equation 11.2. In SI units, mass is measured in kilograms (kg), velocity is measured in meters per second (m/s), and radius is measured in meters (m).

$$\textbf{(Equation 11.2)} \quad F_c = m\,\frac{v^2}{r}$$

---

1 We recognize that planets move in elliptical, not circular, orbits. However, the eccentricity of Earth's orbit is 0.0167. *Eccentricity* is a measure of how much an object's orbit deviates from a perfect circle. The smaller the eccentricity, the smaller the deviation from a circle. An eccentricity of 0.0167 is small enough that, for most calculations in the realm of high school physics, it is possible to assume a circular orbit for Earth and the other planets in the solar system.

Circular Motion

How Does Changing the Angular Velocity of the Swinging Mass at the Top of a Whirligig and the Amount of Mass at the Bottom of a Whirligig Affect the Distance From the Top of the Tube to the Swinging Mass?

Mass and radius are directly measurable quantities, but velocity is not. If the speed, or the magnitude of the velocity, is assumed to be fairly constant, it can be determined by the quotient of distance (d) and time (t) as shown in Equation 11.3. In SI units, distance is measured in meters (m) and time is measured in seconds (s).

$$\text{(Equation 11.3)} \quad \mathbf{v} = \frac{d}{t}$$

For one complete circular revolution, the distance traveled is equal to the circumference. The time for one cycle of any repetitive motion is called the period, with symbol T (also measured in seconds). Often, we refer to repetitive or cyclical motion as "periodic." Thus, Equation 11.3 can be expanded into Equation 11.4.

$$\text{(Equation 11.4)} \quad \mathbf{v} = \frac{2\pi \mathbf{r}}{T}$$

In this lab, if the swinging mass at the end of a whirligig is swung quickly enough, the direction of the string tension is approximately in the horizontal plane (the plane perpendicular to the direction of the gravitational force). That allows a simpler mathematical treatment—that the entire magnitude of the string tension acts as the centripetal force. Otherwise, the angle between the string and the horizontal plane would need to be measured to determine the component of the string's tension in the horizontal direction.

The physics of the other end of the string is easier to analyze (i.e., the mass hanging vertically). Gravity exerts a downward force on the second mass ($\mathbf{F}_g$), equal to its mass (m) times the gravitational acceleration ($\mathbf{g}$), shown in Equation 11.5. In SI units, acceleration is measured in meters per second squared (m/s^2), and the acceleration due to gravity ($\mathbf{g}$) is a constant equal to -9.8 m/s^2.

$$\text{(Equation 11.5)} \quad \mathbf{F}_g = m\mathbf{g}$$

When the stopper is swung such that the string does not slide up or down, then the mass hanging from the string is at rest. By Newton's first and second laws, that implies balanced forces. Equation 11.6 relates the weight of the hanging mass ($\mathbf{F}_g$) to the string tension acting on the second hanging mass ($\mathbf{F}_T$).

$$\text{(Equation 11.6)} \quad \mathbf{F}_g = \mathbf{F}_T$$

Equation 11.1 established that the centripetal force is equal to the tension in the string. Assuming the tension in the string is constant, this implies that the tension in Equation 11.1 can be set equal to the tension in Equation 11.6. This gives us Equation 11.7.

$$\text{(Equation 11.7)} \quad \mathbf{F}_g = m\frac{\mathbf{v}^2}{\mathbf{r}}$$

With respect to the guiding question of this lab, students are to investigate (1) the relationship between angular velocity and the radius of the path that the swinging mass follows when it is stable and the hanging mass is constant, and (2) the relationship between the hanging mass and the radius of the path that the swinging mass follows when it is stable and the angular velocity is constant. As can be inferred from Equation 11.7, when the hanging mass is constant (i.e., when conducting experiments where the hanging mass is not changed), F_g will also be constant. Thus, an increase in the tangential or linear velocity will result in an increase in the radius of the swinging mass. And, if the tangential velocity is held constant, increasing the hanging mass on the end of the whirligig will result in a decrease in the swinging mass radius.

FIGURE 11.1

Force diagram of a whirligig

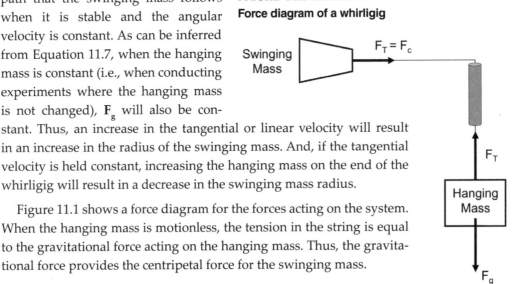

Figure 11.1 shows a force diagram for the forces acting on the system. When the hanging mass is motionless, the tension in the string is equal to the gravitational force acting on the hanging mass. Thus, the gravitational force provides the centripetal force for the swinging mass.

Timeline

The instructional time needed to complete this lab investigation is 170–230 minutes. Appendix 3 (p. 531) provides options for implementing this lab investigation over several class periods. Option C (230 minutes) should be used if students are unfamiliar with scientific writing, because this option provides extra instructional time for scaffolding the writing process. You can scaffold the writing process by modeling, providing examples, and providing hints as students write each section of the report. Option C can also be used if you are introducing students to the video analysis programs. Option D (170 minutes) should be used if students are familiar with scientific writing and have developed the skills needed to write an investigation report on their own. In option D, students complete stage 6 (writing the investigation report) and stage 8 (revising the investigation report) as homework.

Materials and Preparation

The materials needed to implement this investigation are listed in Table 11.1. The equipment can be purchased from a science supply company such as Carolina, Flinn Scientific, PASCO, Vernier, or Ward's Science. The use of video analysis is recommended for this lab. Video analysis software can be purchased from Vernier (Logger *Pro*) or PASCO (SPARKvue or Capstone). These companies also have apps that can be used on Apple- or Android-based tablets and cell phones. We recommend consulting with your school's information technology coordinator to determine the best option for your students.

Circular Motion

How Does Changing the Angular Velocity of the Swinging Mass at the Top of a Whirligig and the Amount of Mass at the Bottom of a Whirligig Affect the Distance From the Top of the Tube to the Swinging Mass?

TABLE 11.1

Materials list for Lab 11

Item	Quantity
Safety glasses or goggles	1 per student
Electronic or triple beam balance	1 per group
Variety of hooked masses, 10–100 g	1 set per group
String or nylon cord, ~1.5 m	1 per group
Rubber stopper or other pliable object, ~25–70 g	1 per group
PVC or metal tube, 10–25 cm long	1 per group
Stopwatch	1 per student
Investigation Proposal A (optional)	1 per group
Whiteboard, 2' × 3"	1 per group
Lab Handout	1 per student
Peer-review guide and teacher scoring rubric	1 per student
Checkout Questions	1 per student
Equipment for video analysis (optional)	
Video camera	1 per group
Computer or tablet with video analysis software	1 per group

* As an alternative, students can use computer and presentation software such as Microsoft PowerPoint or Apple Keynote to create their arguments.

Be sure to use a set routine for distributing and collecting the materials during the lab investigation. One option is to set up the materials for each group at each group's lab station before class begins. This option works well when there is a dedicated section of the classroom for lab work and the materials are large and difficult to move (such as a dynamics track). A second option is to have all the materials on a table or cart at a central location. You can then assign a member of each group to be the "materials manager." This individual is responsible for collecting all the materials his or her group needs from the table or cart during class and for returning all the materials at the end of the class. This option works well when the materials are small and easy to move (such as stopwatches, metersticks, or hanging masses). It also makes it easy to inventory the materials at the end of the class before students leave for the day.

LAB 11

Safety Precautions

Remind students to follow all normal lab safety rules. In addition, tell students to take the following safety precautions:

1. Wear sanitized safety glasses or goggles during lab setup, hands-on activity, and takedown.

2. Keep fingers and toes out of the way of moving objects.

3. Be aware of others around them when swinging the whirligig.

4. Wash hands with soap and water after completing the lab.

Topics for the Explicit and Reflective Discussion

Reflecting on the Use of Core Ideas and Crosscutting Concepts During the Investigation

Teachers should begin the explicit and reflective discussion by asking students to discuss what they know about the core idea they used during the investigation. The following are some important concepts related to the core idea of forces and motion that students need to use to (a) explain what keeps the swinging mass at the end of a whirligig moving in a stable circular path and (b) predict the radius of the circle that the swinging mass follows when it is stable:

- A change in the motion of an object can be used to detect the presence of a force.
- Force is considered a vector quantity because a force has both magnitude and direction.
- The net force acting on an object is the vector sum of the individual forces acting on it.
- A net force exerted on an object perpendicular to the direction of the displacement of the object can change the direction of the motion of the object without changing the kinetic energy of the object.

To help students reflect on what they know about forces and motion, we recommend showing them two or three images using presentation software that help illustrate these important ideas. You can then ask the students the following questions to encourage them to share how they are thinking about these important concepts:

1. What do we see going on in this image?

2. Does anyone have anything else to add?

3. What might be going on that we can't see?

4. What are some things that we are not sure about here?

Circular Motion

How Does Changing the Angular Velocity of the Swinging Mass at the Top of a Whirligig and the Amount of Mass at the Bottom of a Whirligig Affect the Distance From the Top of the Tube to the Swinging Mass?

You can then encourage students to think about how CCs played a role in their investigation. There are at least two CCs that students need to (a) explain what keeps the swinging mass at the end of a whirligig moving in a stable circular path and (b) predict the radius of the circle that the swinging mass follows when it is stable: (a) Systems and System Models and (b) Stability and Change (see Appendix 2 [p. 527] for a brief description of these CCs). To help students reflect on what they know about these CCs, we recommend asking them the following questions:

1. Why is it useful for scientists to define a system and then make a model of it during an investigation?

2. What are some examples of models that you used during your investigation? How were they useful? What were some of the limitations of your models?

3. Why is it important to think about what controls or affects the rate of change in system?

4. Which factor(s) might have controlled the rate of change in the velocity (i.e., acceleration) of the swinging mass? What did exploring these factors allow you to do?

You can then encourage the students to think about how they used all these different concepts to help answer the guiding question and why it is important to use these ideas to help justify their evidence for their final arguments. Be sure to remind your students to explain why they included the evidence in their arguments and make the assumptions underlying their analysis and interpretation of the data explicit in order to provide an adequate justification of their evidence.

Reflecting on Ways to Design Better Investigations
It is important for students to reflect on the strengths and weaknesses of the investigation they designed during the explicit and reflective discussion. Students should therefore be encouraged to discuss ways to eliminate potential flaws, measurement errors, or sources of uncertainty in their investigations. To help students be more reflective about the design of their investigation and what they can do to make their investigations more rigorous in the future, you can ask them the following questions:

1. What were some of the strengths of the way you planned and carried out your investigation? In other words, what made it scientific?

2. What were some of the weaknesses of the way you planned and carried out your investigation? In other words, what made it less scientific?

3. What rules can we make, as a class, to ensure that our next investigation is more scientific?

LAB 11

Reflecting on the Nature of Scientific Knowledge and Scientific Inquiry

This investigation can be used to illustrate two important concepts related to the nature of scientific knowledge and the nature of scientific inquiry: (a) the difference between laws and theories in science and (b) the difference between data and evidence in science (see Appendix 2 for a brief description of these two concepts). Be sure to review these concepts during and at the end of the explicit and reflective discussion. To help students think about these concepts in relation to what they did during the lab, you can ask them the following questions:

1. In this investigation, you developed a model that relates the components of the whirligig system to the circular motion of the mass at the end of the whirligig. Is your model an example of theory or a law? Why?

2. Can you work with your group to come up with a rule that you can use to decide if something is a theory or a law? Be ready to share in a few minutes.

3. You had to talk about data and evidence during your investigation. Can you give me some examples of data and evidence from your investigation?

4. Can you work with your group to come up with a rule that you can use to decide if a piece of information is data or evidence? Be ready to share in a few minutes.

You can also use presentation software or other techniques to encourage your students to think about these concepts. You can show examples of either a law (such as $\mathbf{g} = GM/\mathbf{r}^2$) or a theory (such as *gravity is the curvature of four-dimensional space-time due to the presence of mass*) and ask students to indicate if they think it is a law or a theory and why. You can also show examples of information from the investigation that are either data or evidence and ask students to classify each example and explain their thinking.

Be sure to remind your students that, to be proficient in science, it is important that they understand what counts as scientific knowledge and how that knowledge develops over time.

Hints for Implementing the Lab

- Allowing students to design their own procedures for collecting data gives students an opportunity to try, to fail, and to learn from their mistakes. However, you can scaffold students as they develop their procedure by having them fill out an investigation proposal. These proposals provide a way for you to offer students hints and suggestions without telling them how to do it. You can also check the proposals quickly during a class period. For this lab we suggest using Investigation Proposal A.

- Learn how to use the whirligig before the lab begins. It is important for you to know how to use the equipment so you can help students when technical issues arise.

- Allow the students to become familiar with the whirligig as part of the tool talk before they begin to design their investigation. Giving them 5–10 minutes to examine the equipment will let them see what they can and cannot do with it.

- We recommend giving students a wide berth for swinging the whirligig. You can ask the physical education teacher if you can use the gym for data collection. If the gym or another open space is unavailable, you can collect data outside. However, the wind may affect the quality of the data students collect.

- Be sure to allow students to go back and re-collect data at the end of the argumentation session. Students often realize that they made numerous mistakes when they were collecting data as a result of their discussions during the argumentation session. The students, as a result, will want a chance to re-collect data, and the re-collection of data should be encouraged when time allows. This also offers an opportunity to discuss what scientists do when they realize a mistake is made inside the lab.

If students use video analysis

- We recommend the use of video analysis in this lab, because uniform circular motion for an extended period of time may be difficult to achieve. The use of video analysis will allow students to isolate shorter periods of time where, over the small time interval, the motion is uniform. This will aid them in the development of their conceptual model.

- We suggest allowing students to familiarize themselves with the video analysis software before they finalize the procedure for the investigation, especially if they have not used such software previously. This gives students an opportunity to learn how to work with the software and to improve the quality of the video they take.

- Remind students to hold the video camera as still as possible. Any movement of the camera will introduce error into their analysis. If using actual camcorders, we recommend using a tripod to hold the camera steady. If students are using a camera on a cell phone or tablet, we recommend using a table to help steady the camera.

- Remind students to place a meterstick in the same field of view as the motion they are capturing with the video camera. Also, the meterstick should be approximately the same distance from the camera as the motion. Most video analysis software requires the user to define a scale in the video (this allows the software to establish distances and, subsequently, other variables dependent on distance and displacement).

Connections to Standards

Table 11.2 highlights how the investigation can be used to address learning objectives from AP Physics 1; learning objectives from AP Physics C: Mechanics, *Common Core State Standards,* in English language arts (*CCSS ELA*); and *Common Core State Standards,* Mathematics (*CCSS Mathematics*).

TABLE 11.2 _____

Lab 11 alignment with standards

NGSS performance expectations	• None
AP Physics 1 learning objectives	• 3.A.1.3: The student is able to analyze experimental data describing the motion of an object and is able to express the results of the analysis using narrative, mathematical, and graphical representations. • 3.A.3.1: The student is able to analyze a scenario and make claims (develop arguments, justify assertions) about the forces exerted on an object by other objects for different types of forces or components of forces. • 3.B.1.2: The student is able to design a plan to collect and analyze data for motion (static, constant, or accelerating) from force measurements and carry out an analysis to determine the relationship between the net force and the vector sum of the individual forces. • 3.B.2.1: The student is able to create and use free-body diagrams to analyze physical situations to solve problems with motion qualitatively and quantitatively.
AP Physics C: Mechanics learning objectives	• I.E.1.a: Relate the radius of the circle and the speed or rate of revolution of the particle to the magnitude of the centripetal acceleration. • I.E.1.b: Describe the direction of the particle's velocity and acceleration at any instant during the motion. • I.E.1.c: Determine the components of the velocity and acceleration vectors at any instant, and sketch or identify graphs of these quantities. • I.E.1.d: Analyze situations in which an object moves with specified acceleration under the influence of one or more forces so they can determine the magnitude and direction of the net force, or of one of the forces that makes up the net force, in situations such as the following: • I.E.1.d(1): Motion in a horizontal circle (e.g., mass on a rotating merry-go-round, or car rounding a banked curve).

Circular Motion

How Does Changing the Angular Velocity of the Swinging Mass at the Top of a Whirligig and the Amount of Mass at the Bottom of a Whirligig Affect the Distance From the Top of the Tube to the Swinging Mass?

Literacy connections (CCSS ELA)	• *Reading:* Key ideas and details, craft and structure, integration of knowledge and ideas • *Writing:* Text types and purposes, production and distribution of writing, research to build and present knowledge, range of writing • *Speaking and listening:* Comprehension and collaboration, presentation of knowledge and ideas
Mathematics connections (CCSS Mathematics)	• *Mathematical practices:* Reason abstractly and quantitatively, construct viable arguments and critique the reasoning of others, use appropriate tools strategically, attend to precision • *Number and quantity:* Reason quantitatively and use units to solve problems, represent and model with vector quantities, perform operations on vectors • *Algebra:* Interpret the structure of expressions, understand solving equations as a process of reasoning and explain the reasoning, solve equations and inequalities in one variable • *Functions:* Understand the concept of a function and use function notation, interpret functions that arise in applications in terms of the context, interpret expressions for functions in terms of the situation they model • *Statistics and probability:* Summarize, represent, and interpret data on two categorical and quantitative variables; understand and evaluate random processes underlying statistical experiments; make inferences and justify conclusions from sample surveys, experiments, and observational studies

LAB 11

Lab Handout

Lab 11. Circular Motion: How Does Changing the Angular Velocity of the Swinging Mass at the Top of a Whirligig and the Amount of Mass at the Bottom of a Whirligig Affect the Distance From the Top of the Tube to the Swinging Mass?

Introduction

Many things move in a circular path. For example, protons in the Large Hadron Collider, a car making a turn, and people on swing rides at amusement parks (see Figure L11.1) all move in a circular path. The orbital motion of many moons and planets is nearly circular as well. Even the stars in the spiral arms of the Milky Way galaxy, as they make their way around what may be a supermassive black hole at the galactic center, undergo circular motion.

Newton's first law of motion indicates that, without an unbalanced force acting on an object, an object moves in a straight line at constant speed. When we see an object traveling in a circular path, we must therefore assume that there is an unbalanced force acting on it. The unbalanced force is called a *centripetal* force. A force pulling toward the center of an object's circular path is what keeps it from moving in a straight line (*centripetal* comes from Latin, meaning "toward the center"). A centripetal force is whatever is causing the motion to be circular. For a planet, the centripetal force is gravity. For

FIGURE L11.1

The people on an a swing ride at an amusement park move in a circular path over time

a turning car, friction between the tires and the road is the centripetal force. The string tension in the swing ride shown in Figure L11.1 is the force that keeps the people moving in a circle.

We use the term *uniform circular motion* to describe the motion of an object when it follows the path of a circle with a constant speed. As it moves, it travels around the perimeter of the circle. The distance of one complete cycle around the perimeter of a circle is equal to

How Does Changing the Angular Velocity of the Swinging Mass at the Top of a Whirligig and the Amount of Mass at the Bottom of a Whirligig Affect the Distance From the Top of the Tube to the Swinging Mass?

the circumference of the circle. The average speed of an object in uniform circular motion can therefore be calculated using the following equation:

$$Average\ speed = \frac{2\pi \mathbf{r}}{T}$$

Velocity, unlike speed, has both a magnitude and a direction. The magnitude of the velocity vector is the instantaneous speed of the object. The velocity vector of an object in uniform circular motion is constant in magnitude but changing in direction. Since the object is constantly changing direction as it travels around the perimeter of a circle, it is also constantly accelerating. It is constantly accelerating because the direction of the velocity vector is constantly changing. This description of an object in uniform circular motion is also consistent with Newton's first and second laws. Because there is an unbalanced force acting on an object undergoing uniform circular motion, the object must also be accelerating.

In this investigation you will have an opportunity to learn more about the nature of uniform circular motion by attempting to explain how a whirligig works. A whirligig is a toy that has two masses attached to each other by a string. One mass, which is called the swinging mass, can be made to move in a stable circular path. Your goal is to develop a conceptual model of the whirligig that will allow you to explain what keeps the swinging mass moving in a stable circular path and predict the radius of the circle that it follows.

Your Task

Use what you know about uniform circular motion, centripetal force, vector and scalar quantities, and stability and change in systems to develop a conceptual model that will allow you to (a) explain what keeps the swinging mass at the end of a whirligig moving in a stable circular path and (b) predict the radius of the circle that the swinging mass follows when it is stable. To develop this conceptual model, you will also need to design and carry out an investigation to determine (a) the relationship between the radius of the circle that the swinging mass follows and its angular velocity and (b) the relationship between the radius of the circle that the swinging mass follows and the centripetal force caused by the mass at the bottom of the whirligig. You will then need to test it to determine if it is valid and acceptable.

The guiding question of this investigation is, *How does changing the angular velocity of the swinging mass at the top of a whirligig and the amount of mass at the bottom of a whirligig affect the distance from the top of the tube to the swinging mass?*

LAB 11

Materials

You may use any of the following materials during your investigation:

- Safety glasses or goggles (required)
- Electronic or triple beam balance
- Hooked masses
- String or cord
- Rubber stopper
- PVC or metal tube
- Stopwatch

If you have access to the following equipment, you may also consider using a video camera and a computer or tablet with video analysis software.

Safety Precautions

Follow all normal lab safety rules. In addition, take the following safety precautions:

1. Wear sanitized safety glasses or goggles during lab setup, hands-on activity, and takedown.

2. Keep fingers and toes out of the way of moving objects.

3. Be aware of others around you when swinging the whirligig.

4. Wash hands with soap and water after completing the lab.

Investigation Proposal Required?   ☐ Yes   ☐ No

Getting Started

The first step in developing your conceptual model is to build a whirligig. You can make a whirligig using a rubber stopper, some string, a PVC or metal tube, and a hooked mass. Figure L11.2 shows how to assemble a whirligig with these materials. Once assembled, you can make the swinging mass (the rubber stopper in Figure L11.2) at the end of the whirligig move in a stable circle by changing the amount of mass you hang from the bottom of the whirligig or changing the angular velocity of the swinging mass.

You will then design and carry out two experiments using your whirligig. In the first experiment you will need to determine how changing the angular velocity of swinging mass affects the radius of the circle that the swinging mass follows when it is stable. You

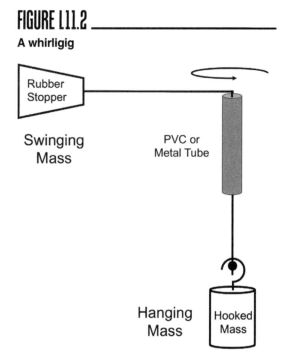

FIGURE L11.2
A whirligig

Rubber Stopper

Swinging Mass

PVC or Metal Tube

Hanging Mass

Hooked Mass

Circular Motion

How Does Changing the Angular Velocity of the Swinging Mass at the Top of a Whirligig and the Amount of Mass at the Bottom of a Whirligig Affect the Distance From the Top of the Tube to the Swinging Mass?

will then need to conduct a second experiment to determine how changing the centripetal force caused by the mass at the bottom of the whirligig affects the radius of the circle that the swinging mass follows when it is stable. Before you can design these two experiments, however, you must determine what type of data you need to collect, how you will collect it, and how you will analyze it.

To determine *what type of data you need to collect*, think about the following questions:

- What are the boundaries and the components of the system under study?
- How can you describe the components of the system quantitatively?
- When is this phenomenon stable, and under which conditions does it change?
- Which factor(s) might control the rate of change in this system?
- How could you keep track of changes in this system quantitatively?
- How will you determine the radius of the stopper's orbit?
- How will you determine the stopper's velocity?
- What will be the independent variable and the dependent variable for each experiment?

To determine *how you will collect the data*, think about the following questions:

- Which variables are scalar quantities, and which variables are vector quantities?
- What other factors will you need to account for or control during each experiment?
- What measurement scale or scales should you use to collect data?
- What equipment will you need to collect the data?
- How will you make sure that your data are of high quality (i.e., how will you reduce error)?
- How will you keep track of and organize the data you collect?

To determine *how you will analyze the data*, think about the following questions:

- What types of patterns might you look for as you analyze your data?
- Are there any proportional relationships that you can identify?
- How could you use mathematics to describe a relationship between variables?
- What type of calculations will you need to make?
- What type of table or graph could you create to help make sense of your data?

Once you have determined (a) the relationship between the radius of the circle that the swinging mass follows and its angular velocity and (b) the relationship between the radius of the circle that the swinging mass follows and the centripetal force caused by the mass at the bottom of the whirligig, your group will need to develop your conceptual model. The model must be able to explain what keeps the swinging mass moving in a stable circular path and allow you to predict the radius of the circle that it follows. To be considered complete, your model must include information about the forces acting on the swinging mass. You should therefore consider including free-body diagrams in your model.

The last step in this investigation is to test your model. To accomplish this goal, you can add different hanging masses to the end of the whirligig (amounts that you did not test) to determine if your model of the motion of the swinging mass enables you to accurately predict the radius of the circle that the swinging mass follows when it is stable. If you are able to use your model to make accurate predictions about the radius of the circle that the swinging mass follows when it is stable, then you will be able to generate the evidence you need to convince others that your model is valid or acceptable.

Connections to the Nature of Scientific Knowledge and Scientific Inquiry

As you work through your investigation, you may want to consider

- the difference between laws and theories in science, and
- the difference between data and evidence.

Initial Argument

Once your group has finished collecting and analyzing your data, your group will need to develop an initial argument. Your argument must include a claim, evidence to support your claim, and a justification of the evidence. The *claim* is your group's answer to the guiding question. The *evidence* is an analysis and interpretation of your data. Finally, the *justification* of the evidence is why your group thinks the evidence matters. The justification of the evidence is important because scientists can use different kinds of evidence to support their claims. Your group will create your initial argument on a whiteboard. Your whiteboard should include all the information shown in Figure L11.3.

FIGURE L11.3 _____

Argument presentation on a whiteboard

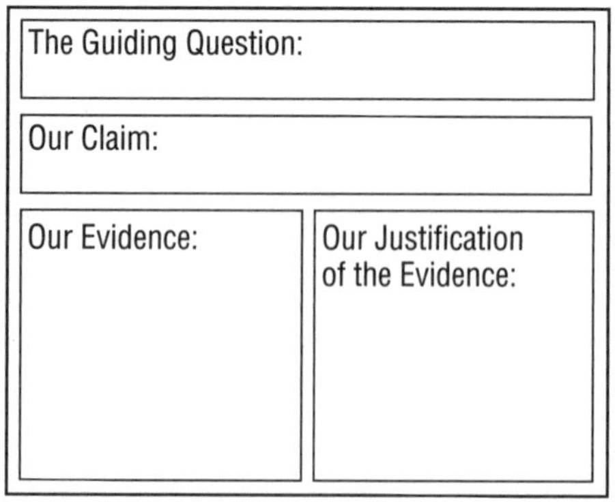

The Guiding Question:	
Our Claim:	
Our Evidence:	Our Justification of the Evidence:

Argumentation Session

The argumentation session allows all of the groups to share their arguments. One or two members of each group will stay at the lab station to share that group's argument, while the other members of the group go to the other lab stations to listen to and critique the

Circular Motion

How Does Changing the Angular Velocity of the Swinging Mass at the Top of a Whirligig and the Amount of Mass at the Bottom of a Whirligig Affect the Distance From the Top of the Tube to the Swinging Mass?

other arguments. This is similar to what scientists do when they propose, support, evaluate, and refine new ideas during a poster session at a conference. If you are presenting your group's argument, your goal is to share your ideas and answer questions. You should also keep a record of the critiques and suggestions made by your classmates so you can use this feedback to make your initial argument stronger. You can keep track of specific critiques and suggestions for improvement that your classmates mention in the space below.

Critiques about our initial argument and suggestions for improvement:

If you are critiquing your classmates' arguments, your goal is to look for mistakes in their arguments and offer suggestions for improvement so these mistakes can be fixed. You should look for ways to make your initial argument stronger by looking for things that the other groups did well. You can keep track of interesting ideas that you see and hear during the argumentation in the space below. You can also use this space to keep track of any questions that you will need to discuss with your team.

Interesting ideas from other groups or questions to take back to my group:

LAB 11

Once the argumentation session is complete, you will have a chance to meet with your group and revise your initial argument. Your group might need to gather more data or design a way to test one or more alternative claims as part of this process. Remember, your goal at this stage of the investigation is to develop the best argument possible.

Report

Once you have completed your research, you will need to prepare an investigation report that consists of three sections. Each section should provide an answer to the following questions:

1. What question were you trying to answer and why?

2. What did you do to answer your question and why?

3. What is your argument?

Your report should answer these questions in two pages or less. This report must be typed, and any diagrams, figures, or tables should be embedded into the document. Be sure to write in a persuasive style; you are trying to convince others that your claim is acceptable or valid!

Circular Motion

*How Does Changing the Angular Velocity of the Swinging Mass at the Top of a Whirligig and the Amount of Mass
at the Bottom of a Whirligig Affect the Distance From the Top of the Tube to the Swinging Mass?*

Checkout Questions

Lab 11. Circular Motion: How Does Changing the Angular Velocity of the Swinging Mass at the Top of a Whirligig and the Amount of Mass at the Bottom of a Whirligig Affect the Distance From the Top of the Tube to the Swinging Mass?

1. Sketch a free-body diagram for a stopper swung around on a whirligig.

2. Sketch a free-body diagram for the hanging mass on a whirligig.

3. Sketch a free-body diagram for a car on a flat road making a turn.

4. Sketch a free-body diagram for Jupiter orbiting the Sun.

5. How would a whirligig need to be swung to support a heavier hanging mass?

   a. Faster

   b. Slower

   Explain your reasoning.

6. Theories explain a phenomenon and laws describe the behavior of a phenomenon.

   a. I agree with this statement.

   b. I disagree with this statement.

   Explain your answer, using an example from your investigation about the whirligig.

Circular Motion

How Does Changing the Angular Velocity of the Swinging Mass at the Top of a Whirligig and the Amount of Mass at the Bottom of a Whirligig Affect the Distance From the Top of the Tube to the Swinging Mass?

7. Evidence is data collected during an investigation.

    a. I agree with this statement.

    b. I disagree with this statement.

    Explain your answer, using an example from your investigation about the whirligig.

8. Why is important to identify factors that contribute to the stability of a system? Be sure to include examples from at least two different investigations in your answer.

9. Why is it useful to create models of a system during an investigation? Be sure to include examples from at least two different investigations in your answer.

Application Lab

LAB 12

Teacher Notes

Lab 12. Torque and Rotation: How Can Someone Predict the Amount of Force Needed to Open a Bottle Cap?

Purpose

The purpose of this lab is for students to *apply* what they know about the core idea of forces and motion, part of the disciplinary core idea (DCI) of Motion and Stability: Forces and Interactions from the *NGSS*, to predict the amount of force needed to open a bottle cap. This lab also gives students an opportunity to learn about the crosscutting concepts (CCs) of (a) Scale, Proportion, and Quantity; and (b) Systems and System Models from the *NGSS*. In addition, this lab can be used to help students understand two big ideas from AP Physics: (a) the interactions of an object with other objects can be described by forces and (b) interactions between systems can result in changes in those systems. As part of the explicit and reflective discussion, students will also learn about (a) the difference between data and evidence in science and (b) the role of imagination and creativity in science.

Underlying Physics Concepts

Linear motion and rotational motion have many similarities, including analogous components such as linear velocity and angular velocity or linear acceleration and angular acceleration. In addition to the similarities in describing linear motion and rotational motion, the influence of forces in linear and rotational contexts is also very similar. For linear contexts, Newton's second law of motion ($\Sigma\mathbf{F} = m\mathbf{a}$) describes the relationship between the net force acting on an object ($\Sigma\mathbf{F}$), the object's mass (m), and the acceleration of the object ($\mathbf{a}$). For rotational contexts, an analogous equation describes the relationship between torque (the force applied to a lever arm that results in rotation), moment of inertia, and angular acceleration. This relationship is described in Equation 12.1, where $\Sigma\tau$ represents the sum of the torques acting on the object, I represents the moment of inertia, and α is the angular acceleration for the object. In SI units, torque is measured in Newton meters (N·m), moment of inertia is measured in kilogram-meters squared (kg·m^2), and angular acceleration is measured in radians per second squared (rad/s^2).

$$\text{(Equation 12.1)} \quad \Sigma\tau = I\alpha$$

In linear contexts the net force acting on an object is a simple push or pull that acts directly on the object and causes a change in motion in any direction, which is determined by the different force components acting on the object. For rotational contexts the net torque causes the object to rotate about an axis of rotation. Because the object is rotating about this axis of rotation, determining the net torque requires more than adding up all the pushes and pulls; not all forces acting on a rotating object have the same influence on how much that object rotates or on changes in the object's angular acceleration. Figure 12.1 shows how changes in the location of a force have a different influence on the torque.

Consider the overhead view of a door shown in Figure 12.1. If a force ($\mathbf{F}_1$) is applied to the doorknob at a radius of $\mathbf{r}_1$, it will cause a larger amount of rotation than if a second, equal force ($\mathbf{F}_2$) is applied at radius $\mathbf{r}_2$. This example shows how the lever arm influences the torque on an object. The lever arm is simply the distance between the axis of rotation and the point where the force is applied. When a force is applied near the doorknob, there is a large lever arm ($\mathbf{r}_1$). Alternatively, the lever arm is relatively short when the force is applied closer to the hinge ($\mathbf{r}_2$). Therefore, torque is proportional to the magnitude of the force being applied and the lever arm or radius. Equation 12.2 can be used to calculate the torque (τ) on an object, where $\mathbf{r}$ represents the length of the lever arm, $\mathbf{F}$ represents the force, and $\sin\theta$ represents the angle between the force and the lever arm. In SI units, the length of the lever arm is measured in meters (m); force is measured in newtons (N), and the angle is measured in radians (rad).

FIGURE 12.1

Overhead view of a door with a force applied at the doorknob and midway between the doorknob and the hinge

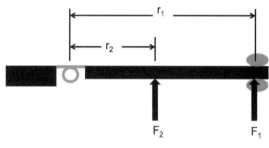

$$\textbf{(Equation 12.2)} \quad \tau = \mathbf{r}\mathbf{F}\sin\theta$$

The additional term $\sin\theta$ appears in Equation 12.2 because the force must be perpendicular to the radius or lever arm. When the lever arm is perpendicular to the radius (as shown in Figure 12.1 and a valid assumption for this lab), then the value of $\sin\theta$ is 1 because the angle between the force and the lever arm is 90°. If the applied force is not fully perpendicular to the radius, then $\mathbf{F}\sin\theta$ corresponds to the component of the force perpendicular to the radius.

Finally, for a disc or cylinder, the value of $\mathbf{r}$ in Equation 12.2 corresponds to the distance from the point of rotation (usually, but not always, the center of the disc or cylinder) of the object to the point where the force is being applied. Usually, the force is being applied at the edge of the disc, so that the value of $\mathbf{r}$ is the radius. However, it is possible to apply a torquing force at another point on the disc. For example, it is possible to spin a record by placing your finger on the surface of the record and pushing it, thereby causing the record to spin. In this case, the value of $\mathbf{r}$ in Equation 12.2 would correspond to the distance from the center of the record to the point on the surface of the record where your finger provides the push.

In this lab, students will also change the location of the force by using a meterstick attached to the bottle cap to vary the distance from the axis of rotation at which the force is applied. Thus, they will vary the torque at a point outside the radius of the disc (i.e., the surface of the bottle cap). If we assume (1) that the torque needed to open a given bottle cap is constant and (2) that the force is applied perpendicular to the radius of the bottle cap, then the relationship between the applied force and the location of the force is proportional. We can rewrite Equation 12.2 as Equation 12.3. As the radius of the applied force increases, the force required to open the bottle decreases.

$$(\text{Equation } 12.3) \quad F = \tau/r$$

Timeline

The instructional time needed to complete this lab investigation is 170–230 minutes. Appendix 3 (p. 531) provides options for implementing this lab investigation over several class periods. Option C (230 minutes) should be used if students are unfamiliar with scientific writing, because this option provides extra instructional time for scaffolding the writing process. You can scaffold the writing process by modeling, providing examples, and providing hints as students write each section of the report. Option D (170 minutes) should be used if students are familiar with scientific writing and have the developed the skills needed to write an investigation report on their own. In option D, students complete stage 6 (writing the investigation report) and stage 8 (revising the investigation report) as homework.

Materials and Preparation

The materials needed to implement this investigation are listed in Table 12.1. The consumables and equipment can be purchased from a science supply company such as Carolina, Flinn Scientific, PASCO, Vernier, or Ward's Science.

TABLE 12.1

Materials list for Lab 12

Item	Quantity
Consumables	
String	50 cm per group
Tape	1 roll per group
Equipment and other materials	
Safety glasses or goggles	1 per group
Plastic bottles with twist-style caps	1 per group

Meterstick	1 per group
Spring scales, 5 N–20 N (or force sensor with interface)	3 scales per group
Hanging mass set	1 per group
Table clamp	2 per group
Investigation Proposal A (optional)	1 per group
Whiteboard, 2' × 3'*	1 per group
Lab Handout	1 per student
Peer-review guide and teacher scoring rubric	1 per student
Checkout Questions	1 per student

* As an alternative, students can use computer and presentation software such as Microsoft PowerPoint or Apple Keynote to create their arguments.

Students will need to determine the torque required to unscrew a twist-style cap from a plastic bottle during this investigation. One potential way to set up the equipment for this lab is shown in Figure 12.2. In this setup, a meterstick is attached to the cap of a plastic bottle using two screws and secured with nuts on the underside of the cap (alternatively, you can use a smaller ruler instead of a meterstick). The meterstick acts as a lever arm that can support hanging weights at various distances to generate the torque needed to unscrew the cap. Alternatively, students can attach a spring scale or force sensor to the meterstick and measure the force required to generate the appropriate amount of torque to unscrew the cap. The bottle pictured in Figure 12.2 is a typical half-gallon juice bottle. Similar everyday bottles can be used for this investigation, or plastic bottles with caps can be purchased from the vendors listed earlier in this section.

FIGURE 12.2

Suggested equipment setup using a meterstick attached to a twist-style cap

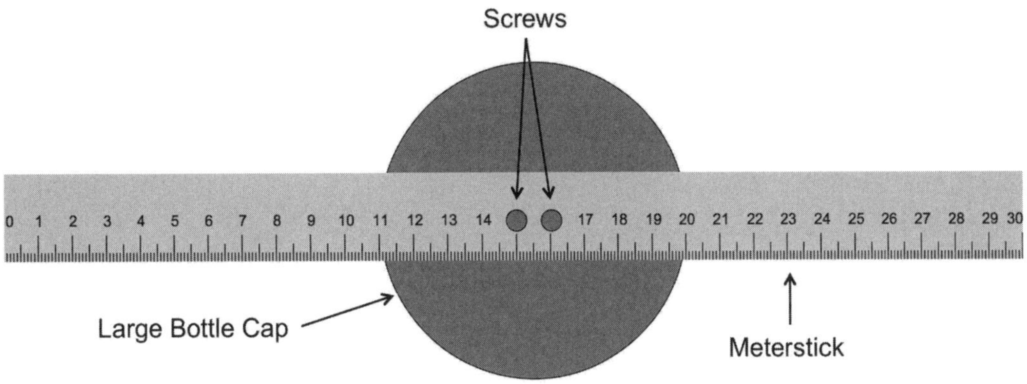

LAB 12

Be sure to use a set routine for distributing and collecting the materials during the lab investigation. One option is to set up the materials for each group at each group's lab station before class begins. This option works well when there is a dedicated section of the classroom for lab work and the materials are large and difficult to move (such as a dynamics track). A second option is to have all the materials on a table or cart at a central location. You can then assign a member of each group to be the "materials manager." This individual is responsible for collecting all the materials his or her group needs from the table or cart during class and for returning all the materials at the end of the class. This option works well when the materials are small and easy to move (such as stopwatches, metersticks, or hanging masses). It also makes it easy to inventory the materials at the end of the class before students leave for the day.

Safety Precautions

Remind students to follow all normal lab safety rules. In addition, tell students to take the following safety precautions:

1. Wear sanitized safety glasses or goggles during lab setup, hands-on activity, and takedown.

2. Wash hands with soap and water after completing the lab.

Topics for the Explicit and Reflective Discussion
Reflecting on the Use of Core Ideas and Crosscutting Concepts During the Investigation

Teachers should begin the explicit and reflective discussion by asking students to discuss what they know about the core idea they used during the investigation. The following are some important concepts related to the core idea of forces and motion that students need to use to predict the amount of force needed to open a bottle cap:

- A change in the motion of an object can be used to detect the presence of a force.
- Force is considered a vector quantity because a force has both magnitude and direction.
- Quantities such as angular displacement, angular velocity, and angular acceleration are used to describe rotational motion about a fixed axis.
- *Torque* is a measure of force applied perpendicular to a lever arm multiplied by the distance from the point of rotation.
- *Torque* is a vector quantity because torque has both magnitude and direction.
- The presence of a net torque along any fixed axis will cause a rigid system or an object to change its rotational motion about that axis.

To help students reflect on what they know about forces and motion, we recommend showing them two or three images using presentation software that help illustrate these important ideas. You can then ask the students the following questions to encourage them to share how they are thinking about these important concepts:

1. What do we see going on in this image?

2. Does anyone have anything else to add?

3. What might be going on that we can't see?

4. What are some things that we are not sure about here?

You can then encourage students to think about how CCs played a role in their investigation. There are at least two CCs that students need to predict the amount of force needed to open a bottle cap: (a) Scale, Proportion, and Quantity; and (b) Systems and System Models (see Appendix 2 [p. 527] for a brief description of these CCs). To help students reflect on what they know about these CCs, we recommend asking them the following questions:

1. Why is it important to think about issues of scale and quantity during an investigation?

2. What scales and quantities did you use during your investigation? Why was it useful?

3. Why do scientists often define a system and then develop a model of it as part of an investigation?

4. How did you use a model during your investigation to describe the relationship between force, location, and torque? Why was that useful?

You can then encourage the students to think about how they used all these different concepts to help answer the guiding question and why it is important to use these ideas to help justify their evidence for their final arguments. Be sure to remind your students to explain why they included the evidence in their arguments and make the assumptions underlying their analysis and interpretation of the data explicit in order to provide an adequate justification of their evidence.

Reflecting on Ways to Design Better Investigations

It is important for students to reflect on the strengths and weaknesses of the investigation they designed during the explicit and reflective discussion. Students should therefore be encouraged to discuss ways to eliminate potential flaws, measurement errors, or sources of uncertainty in their investigations. To help students be more reflective about the design of their investigation and what they can do to make their investigations more rigorous in the future, you can ask them the following questions:

1. What were some of the strengths of the way you planned and carried out your investigation? In other words, what made it scientific?

2. What were some of the weaknesses of the way you planned and carried out your investigation? In other words, what made it less scientific?

3. What rules can we make, as a class, to ensure that our next investigation is more scientific?

Reflecting on the Nature of Scientific Knowledge and Scientific Inquiry

This investigation can be used to illustrate two important concepts related to the nature of scientific knowledge and the nature of scientific inquiry: (a) the difference between data and evidence in science and (b) the role of imagination and creativity in science (see Appendix 2 for a brief description of these two concepts). Be sure to review these concepts during and at the end of the explicit and reflective discussion. To help students think about these concepts in relation to what they did during the lab, you can ask them the following questions:

1. You had to talk about data and evidence during your investigation. Can you give me some examples of data and evidence from your investigation?

2. Can you work with your group to come up with a rule that you can use to decide if a piece of information is data or evidence? Be ready to share in a few minutes.

3. Some people think that there is no room for imagination or creativity in science. What do you think?

4. Can you work with your group to come up with different ways that you needed to use your imagination or be creative during this investigation? Be ready to share in a few minutes.

You can also use presentation software or other techniques to encourage your students to think about these concepts. You can show examples of information from the investigation that are either data or evidence and ask students to classify each example and explain their thinking. You can also show students an image of the following quote by E. O. Wilson from *Letters to a Young Scientist* (2013) and ask them what they think he meant by it:

> The ideal scientist thinks like a poet and only later works like a bookkeeper. Keep in mind that innovators in both literature and science are basically dreamers and storytellers. In the early stages of the creation of both literature and science, everything in the mind is a story. There is an imagined ending, and usually an imagined beginning, and a selection of bits and pieces that might fit in between. In works of literature and science alike, any part can be changed, causing a ripple among the other parts, some of which are discarded and new ones added. (p. 74)

Be sure to remind your students that, to be proficient in science, it is important that they understand what counts as scientific knowledge and how that knowledge develops over time.

Hints for Implementing the Lab

- Allowing students to design their own procedures for collecting data gives students an opportunity to try, to fail, and to learn from their mistakes. However, you can scaffold students as they develop their procedure by having them fill out an investigation proposal. These proposals provide a way for you to offer students hints and suggestions without telling them how to do it. You can also check the proposals quickly during a class period. For this lab we suggest using Investigation Proposal A.

- Allow the students to become familiar with the equipment and materials as part of the tool talk before they begin to design their investigation. Giving them 5–10 minutes to examine the equipment and materials allows them to see what they can and cannot do with them.

- Students will need to come up with a method for controlling how tight the bottle cap is tightened between trials. We suggest that they use a controlled amount of turns of the bottle cap. This is an opportunity to help students identify possible sources of error and how experimental design can overcome said sources.

- When making the lid setups for this investigation, it is important to make sure that the meterstick is securely attached to the lid and that it does not slip when weights are added. Using screws as described in the "Materials and Preparation" section works well if the cap is large enough. Some caps may only accommodate one screw and may need superglue to ensure that the meterstick does not rotate independently of the cap.

- If you want to add more variables and increase the complexity of the lab, you may want to give students an opportunity to work with different-size bottle caps. In doing this, students will mathematically derive an inversely proportional relationship between force and radius for all bottle cap sizes. However, the constant of proportionality will change, because the torque required to open the bottle cap is different for different bottles.

- Be sure to allow students to go back and re-collect data at the end of the argumentation session. Students often realize that they made numerous mistakes when they were collecting data as a result of their discussions during the argumentation session. The students, as a result, will want a chance to re-collect data, and the re-collection of data should be encouraged when time allows. This also offers an opportunity to discuss what scientists do when they realize a mistake is made inside the lab.

LAB 12

Connections to Standards

Table 12.2 highlights how the investigation can be used to address learning objectives from AP Physics 1; learning objectives from AP Physics C: Mechanics, *Common Core State Standards*, in English language arts (*CCSS ELA*); and *Common Core State Standards*, Mathematics (*CCSS Mathematics*).

TABLE 12.2

Lab 12 alignment with standards

NGSS **performance expectations**	• None
AP Physics 1 learning objectives	• 3.A.3.1: The student is able to analyze a scenario and make claims (develop arguments, justify assertions) about the forces exerted on an object by other objects for different types of forces or components of forces. • 3.F.1.1: The student is able to use representations of the relationship between force and torque. • 3.F.1.4: The student is able to design an experiment and analyze data testing a question about torques in a balanced rigid system. • 3.F.1.5: The student is able to calculate torques on a two-dimensional system in static equilibrium by examining a representation or model (such as a diagram or physical construction). • 4.D.1.2: The student is able to plan data collection strategies designed to establish that torque, angular velocity, angular acceleration, and angular momentum can be predicted accurately when the variables are treated as being clockwise or counterclockwise with respect to a well-defined axis of rotation, and refine the research question based on the examination of data.
AP Physics C: Mechanics learning objectives	• I.E.2.a (1): Calculate the magnitude and direction of the torque associated with a given force. • I.E.2.d (3): Find the rotational inertia of a thin cylindrical shell about its axis, or an object that may be viewed as being made up of coaxial shells.
Literacy connections (*CCSS ELA*)	• *Reading:* Key ideas and details, craft and structure, integration of knowledge and ideas • *Writing:* Text types and purposes, production and distribution of writing, research to build and present knowledge, range of writing • *Speaking and listening:* Comprehension and collaboration, presentation of knowledge and ideas

Mathematics connections (*CCSS Mathematics*)	• *Mathematical practices:* Reason abstractly and quantitatively, construct viable arguments and critique the reasoning of others, use appropriate tools strategically, attend to precision • *Number and quantity:* Reason quantitatively and use units to solve problems, represent and model with vector quantities, perform operations on vectors • *Algebra:* Interpret the structure of expressions, understand solving equations as a process of reasoning and explain the reasoning, solve equations and inequalities in one variable • *Functions:* Understand the concept of a function and use function notation, interpret functions that arise in applications in terms of the context, interpret expressions for functions in terms of the situation they model • *Statistics and probability:* Summarize, represent, and interpret data on two categorical and quantitative variables; understand and evaluate random processes underlying statistical experiments; make inferences and justify conclusions from sample surveys, experiments, and observational studies

Reference

Wilson, E. O. 2013. *Letters to a young scientist.* New York: Liveright Publishing.

LAB 12

Lab Handout

Lab 12. Torque and Rotation: How Can Someone Predict the Amount of Force Needed to Open a Bottle Cap?

Introduction

With the invention of the modern twist cap, there have been major advances in the ways that food and beverages can be packaged; in fact, it is difficult to find a bottle that does not close by using a twist cap. Common twist caps are used to seal soda bottles, gallon milk jugs, jars of pasta sauce, and even small bottles of medicine. Twist caps lock onto bottles and jars using threads—like a screw—on the underside of the cap; the outside of the bottle opening also has a set of threads. When the cap is placed over the bottle opening, the threads match up, and twisting the cap causes the threads to lock together and form a tight seal on the bottle. The threads of the twist cap and the threads on the bottle fit together like two ramps sliding past each other. The friction between the thread surfaces keeps the cap from coming loose by accident. Figure L12.1 shows how the threads of the cap and bottle fit together.

FIGURE L12.1
Cross-section of a threaded twist cap being screwed onto a threaded bottle

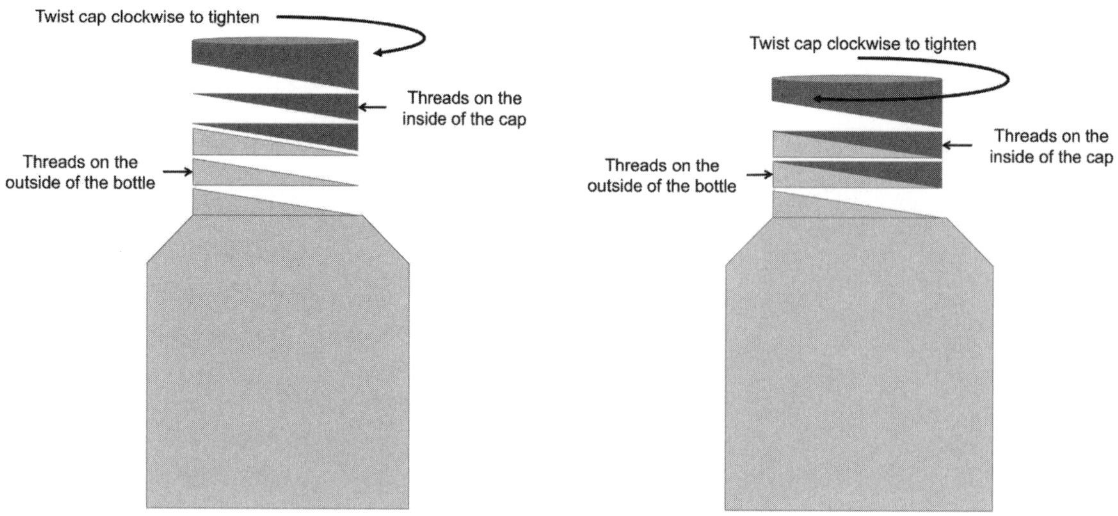

When companies are packaging different substances in bottles and jars that use this twist-top style of lid, they must ensure that the lids are installed with the proper amount of torque so that they form a good seal. A proper seal ensures that nothing leaks out and the food or drink stays fresh. When a net torque is applied to an object, the object will rotate.

The rotation of the object is related to two primary factors: (a) how strong of a force was applied to the object and (b) the location where the force was applied.

Think about closing the door to your science classroom. If you close the door by pushing on the handle, the door closes very easily. In this example, a small force is required when it is applied far away from the axis of rotation (the hinges of the door). When the same force is applied to the door, but halfway between the hinges and the handle, the door may not close. Because the location of the applied force is closer to the axis of rotation, the applied torque is decreased and the door may not close. If you want to close the door by applying a force at a location closer to the hinges, you likely need to apply a greater force.

Torque, and the relationship between the location of the applied force and the magnitude of the applied force, is important across many other relationships. Along with opening and closing bottle caps, torque plays an important role in tightening the lug nuts on your car's wheels, turning on a water faucet, and turning door handles. Part of the design process of tools and devices that twist is finding a balance between the location where a person will apply a force and the force needed to create a large enough torque. If you were to take apart a door handle, the part that actually turns to open your door is a small rod. The bulk of the handle is there to change the amount of force required to open the door.

Your Task

Use what you know about torque, proportional relationships, and systems and system models to design an investigation to determine the relationship between the force applied to open a bottle cap and the location from the axis of rotation the force is applied. Your goal is to create a mathematical model that will allow you to predict where a force must be applied on a bottle cap to open the cap for any specified force.

The guiding question of this investigation is, *How can someone predict the amount of force needed to open a bottle cap?*

Materials

You may use any of the following materials during your investigation:

Consumables
- String
- Tape

Equipment
- Safety glasses or goggles (required)
- Plastic bottle modified with ruler or meterstick attached to the cap
- Spring scales (or force sensor with interface)
- Meterstick
- Hanging mass set
- Table clamp

LAB 12

Safety Precautions

Follow all normal lab safety rules. In addition, take the following safety precautions:

1. Wear sanitized safety glasses or goggles during lab setup, hands-on activity, and takedown.

2. Wash hands with soap and water after completing the lab.

Investigation Proposal Required? ☐ Yes ☐ No

Getting Started

To answer the guiding question, you will need to design and carry out an experiment. To accomplish this task, you must determine what type of data you need to collect, how you will collect it, and how you will analyze it.

To determine *what type of data you need to collect,* think about the following questions:

- What are the boundaries and components of the system you are studying?
- How do the components of the system interact with each other?
- How could you keep track of changes in this system quantitatively?
- How will you know the torque required to unscrew the cap?
- What variables do you need to measure to calculate torque?

To determine *how you will collect the data,* think about the following questions:

- What scale or scales should you use when you take your measurements?
- What equipment will you need to collect the data?
- How will you make sure that your data are of high quality (i.e., how will you reduce error)?
- How will you keep track of the data you collect?
- How will you organize your data?

To determine *how you will analyze the data,* think about the following questions:

- What types of patterns might you look for as you analyze your data?
- Are there any proportional relationships that you can identify?
- What type of calculations will you need to make?
- What type of table or graph could you create to help make sense of your data?
- What types of mathematical relationships might you use to model the system under study?

Connections to the Nature of Scientific Knowledge and Scientific Inquiry

As you work through your investigation, you may want to consider

- the difference between data and evidence in science, and
- the role of imagination and creativity in science.

Initial Argument

Once your group has finished collecting and analyzing your data, your group will need to develop an initial argument. Your argument must include a claim, evidence to support your claim, and a justification of the evidence. The *claim* is your group's answer to the guiding question. The *evidence* is an analysis and interpretation of your data. Finally, the *justification* of the evidence is why your group thinks the evidence matters. The justification of the evidence is important because scientists can use different kinds of evidence to support their claims. Your group will create your initial argument on a whiteboard. Your whiteboard should include all the information shown in Figure L12.2.

FIGURE L12.2

Argument presentation on a whiteboard

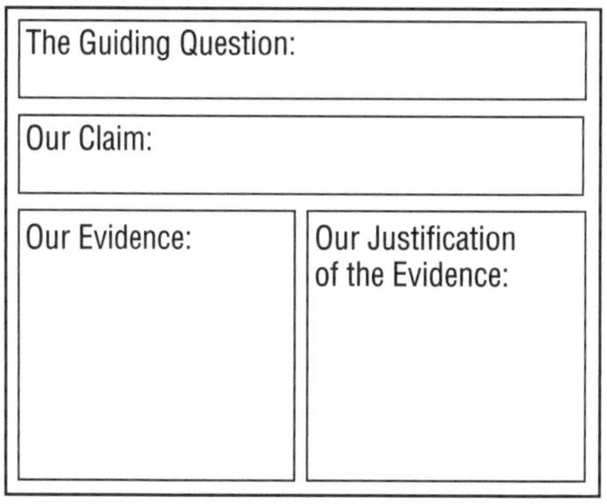

The Guiding Question:	
Our Claim:	
Our Evidence:	Our Justification of the Evidence:

Argumentation Session

The argumentation session allows all of the groups to share their arguments. One or two members of each group will stay at the lab station to share that group's argument, while the other members of the group go to the other lab stations to listen to and critique the other arguments. This is similar to what scientists do when they propose, support, evaluate, and refine new ideas during a poster session at a conference. If you are presenting your group's argument, your goal is to share your ideas and answer questions. You should also keep a record of the critiques and suggestions made by your classmates so you can use this feedback to make your initial argument stronger. You can keep track of specific critiques and suggestions for improvement that your classmates mention in the space below.

Critiques about our initial argument and suggestions for improvement:

If you are critiquing your classmates' arguments, your goal is to look for mistakes in their arguments and offer suggestions for improvement so these mistakes can be fixed. You should look for ways to make your initial argument stronger by looking for things that the other groups did well. You can keep track of interesting ideas that you see and hear during the argumentation in the space below. You can also use this space to keep track of any questions that you will need to discuss with your team.

Interesting ideas from other groups or questions to take back to my group:

Once the argumentation session is complete, you will have a chance to meet with your group and revise your initial argument. Your group might need to gather more data or design a way to test one or more alternative claims as part of this process. Remember, your goal at this stage of the investigation is to develop the best argument possible.

Report

Once you have completed your research, you will need to prepare an investigation report that consists of three sections. Each section should provide an answer to the following questions:

1. What question were you trying to answer and why?

2. What did you do to answer your question and why?

3. What is your argument?

Your report should answer these questions in two pages or less. This report must be typed, and any diagrams, figures, or tables should be embedded into the document. Be sure to write in a persuasive style; you are trying to convince others that your claim is acceptable or valid!

Checkout Questions

Lab 12. Torque and Rotation: How Can Someone Predict the Amount of Force Needed to Open a Bottle Cap?

1. A contestant on a game show was spinning a large wheel to try and win money. The contestant was spinning the wheel as hard as he could, but the wheel only spun around a couple of times. The contestant suggested moving the handles closer to the center so that he and other players could make the wheel spin more with each push. The suggested change is shown below.

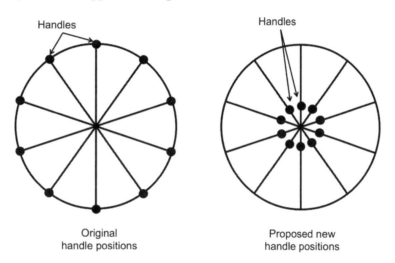

Original handle positions

Proposed new handle positions

Will moving the handles closer to the center of the wheel help the contestants get more spins per push?

a. Yes

b. No

Explain your answer, using what you know about torque and rotational motion.

2. The bones and muscles found in humans and chimpanzees are almost identical. Chimpanzees, however, are much stronger than humans even though they are smaller. One explanation for this difference in strength is that the muscles of

chimpanzees are attached to the bones in slightly different ways than those of humans; therefore, chimpanzees are able to generate a greater torque with smaller muscles. The diagram below shows where the bicep muscle is attached to the upper and lower limb in a human arm and in a chimpanzee arm. When thinking about this, remember that muscles can only contract, so a bicep muscle pulls the lower limb toward the upper limb.

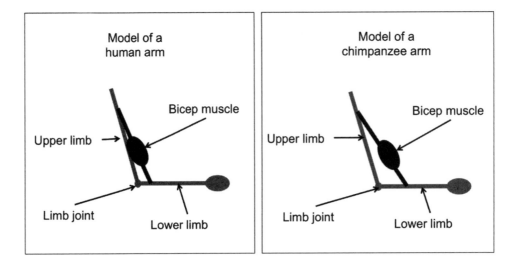

Use what you know about torque to explain why the difference could result in greater torque.

3. In science, it is possible for a variable to be proportionally related to two other variables.

   a. I agree with this statement.

   b. I disagree with this statement.

   Explain your answer, using an example from your investigation about torque and the rotation of bottle caps.

4. Science requires imagination and creativity.

   a. I agree with this statement.
   b. I disagree with this statement.

   Explain your answer, using an example from your investigation about torque and the rotation of bottle caps.

5. There is a difference between data and evidence in science. Explain what data and evidence are and how they are different from each other, using an example from your investigation about torque and the rotation of bottle caps.

6. In science, identifying the system under study is a prerequisite for being able to mathematically model the system. Explain why this statement is true, using an example from your investigation about torque and the rotation of bottle caps.

SECTION 5
Forces and Motion

Oscillations

Introduction Labs

LAB 13

Teacher Notes

Lab 13. Simple Harmonic Motion and Pendulums: What Variables Affect the Period of a Pendulum?

Purpose

The purpose of this lab is to *introduce* students to the core idea of forces and motion, part of the disciplinary core idea (DCI) of Motion and Stability: Forces and Interactions from the *NGSS*, by giving them an opportunity to explore simple harmonic motion and the behavior of pendulums. This lab also gives students an opportunity to learn about the crosscutting concepts (CCs) of (a) Patterns; (b) Cause and Effect: Mechanism and Explanation; and (c) Structure and Function from the *NGSS*. In addition, this lab can be used to help students understand two big ideas from AP Physics: (a) fields existing in space can be used to explain interactions and (b) the interactions of an object with other objects can be described by forces. As part of the explicit and reflective discussion, students will also learn about (a) the difference between data and evidence in science and (b) the nature and role of experiments in science.

Underlying Physics Concepts

Harmonic oscillation is a concept that appears throughout physics courses because it can be easily modeled using a sine function. Furthermore, harmonic oscillators have many uses in daily life, such as in clocks, metronomes, car shock absorbers, and even some electrical devices. There are certain features that all harmonic oscillators share. First, when the system is at rest, the oscillator is said to be at "equilibrium." When the oscillator is moved from the equilibrium position, a force acts on the oscillator to restore the oscillator to equilibrium. This force is called the *restoring force*. The further the oscillator is moved from the equilibrium position, the greater the restoring force becomes.

Newton's second law states that an unbalanced force will cause an object to accelerate. Thus, when the oscillator is disturbed from equilibrium, it accelerates in the general direction of the equilibrium position. When the oscillator reaches the equilibrium position, it has a non-zero velocity, while the sum of the forces acting on the oscillator is zero. Thus, the oscillator moves through the equilibrium position, at which point the restoring force increases until it reaches a maximum displacement from the equilibrium. At the maximum displacement, the restoring force is also at a maximum, while the velocity is equal to zero (0 m/s).

In this lab, students investigate pendulums and determine what variables affect the period of the pendulum. There are three major components of a pendulum, shown in Figure 13.1: the pivot point about which the pendulum swings, the mass (called a bob) at the end of the pendulum, and a rod (or string) that connects the pivot point and the bob. When the pendulum is not moving, the mass hangs vertically at the equilibrium position. When the pendulum swings back and forth, the pendulum is displaced from the equilibrium position by an angle θ. Finally, gravity is the restoring force that causes a pendulum disturbed from equilibrium to swing back and forth. The *period of the pendulum* is defined as the amount of time that it takes the pendulum to make one full swing (we define the initial position as the release point of the pendulum when it is moved from the equilibrium position). A full swing is the mass of the pendulum returning to its initial position. In other words, a full swing is the pendulum making one full back-and-forth swing.

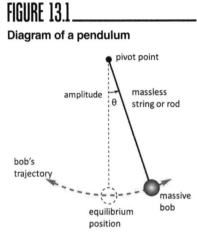

FIGURE 13.1

Diagram of a pendulum

The way a pendulum bob moves back and forth is described as simple harmonic motion. The motion of a pendulum bob is called *simple* because after the mass is pulled back from the equilibrium position to its initial position, the only other forces acting on the pendulum mass are the tension in the string and gravity. When friction also acts on a pendulum, the motion is called damped; if another force is exerted on the pendulum to help it keep swinging, the motion is called driven.

Students will discover that the only variable that affects the period of a pendulum is the length of string. Equation 13.1 describes the period of a pendulum, with *T* being the period of the pendulum, *L* being the length of the pendulum, and **g** being the acceleration due to gravity (9.8 m/s²; note that for mathematical purposes, we choose to write **g** as a positive number in this lab). In SI units, period is measured in seconds (s) and the length of the pendulum is measured in meters (m). This means that the other two variables that they are likely to test—the mass at the end of the pendulum and the angle of swing—do not influence the period of the pendulum.

$$\text{(Equation 13.1)} \quad T = 2\pi\sqrt{(L/\mathbf{g})},$$

Using the law of conservation of energy, it is easy to demonstrate why the mass at the end of the pendulum does not affect the period. When the mass is pulled from equilibrium, it will also raise height **h** above the equilibrium point. The law of conservation of energy says that the gravitational potential energy at the release point will equal the kinetic energy at the equilibrium point. We can then apply Equation 13.2 to the system, where *m* is the mass, **g** is the acceleration due to gravity (9.8 m/s²), **h** is the height above the equilibrium position, and **v** is the velocity of the pendulum bob. In SI units, mass is measured in kilograms (kg), height is measured in meters (m), and velocity is measured in meters per second (m/s).

$$\text{(Equation 13.2)} \quad m\mathbf{gh} = \tfrac{1}{2}m\mathbf{v}^2$$

LAB 13

When solving for velocity, the mass on each side divides to 1, thus the velocity at the bottom is shown in Equation 13.3. Notice how for any given height, the velocity at the bottom does not depend on the mass of the pendulum bob.

$$\textbf{(Equation 13.3)} \quad v = \sqrt{(2gh)}$$

The issue of the angle is a bit more complex. Equation 13.1 is valid when the angle of amplitude is small (see Figure 13.1). For larger angles, the amplitude will influence the period. However, this influence is small, even for relatively large angles; for example, an amplitude of 60° produces a deviation of only 7% between the true period and the predicted period. This difference is not large enough to affect the period in a meaningful way when using the suggested materials.

Timeline

The instructional time needed to complete this lab investigation is 170–230 minutes. Appendix 3 (p. 531) provides options for implementing this lab investigation over several class periods. Option C (230 minutes) should be used if students are unfamiliar with scientific writing, because this option provides extra instructional time for scaffolding the writing process. You can scaffold the writing process by modeling, providing examples, and providing hints as students write each section of the report. Option C can also be used if students are unfamiliar with any of the data collection and analysis tools. Option D (170 minutes) should be used if students are familiar with scientific writing and have developed the skills needed to write an investigation report on their own. In option D, students complete stage 6 (writing the investigation report) and stage 8 (revising the investigation report) as homework.

Materials and Preparation

The materials needed for this investigation are listed in Table 13.1. The consumables and equipment can be purchased from a science supply company such as Carolina, Flinn Scientific, PASCO, Vernier, or Ward's Science. You can also purchase many of these materials from a general retail store such as Wal-Mart or Target. Video analysis software can be purchased from Vernier (Logger *Pro*) or PASCO (SPARKvue or Capstone). These companies also have apps that can be used on Apple- or Android-based tablets and cell phones. We recommend consulting with your school's information technology coordinator to determine the best option for your students.

TABLE 13.1

Materials list for Lab 13

Item	Quantity
Consumables	
Tape	As needed
String	1 roll per group
Equipment and other materials	
Safety glasses or goggles	1 per student
Electronic or triple beam balance	1 per group
Washers	Several per group
Paper clips	5 per group
Protractor	1 per group
Ruler	1 per student
Meterstick	1 per group
Stopwatch	1 per student
Scissors	1 per group
Investigation Proposal C (optional)	3 per group
Whiteboard, 2' × 3'*	1 per group
Lab Handout	1 per student
Peer-review guide and teacher scoring rubric	1 per student
Checkout Questions	1 per student
Equipment for sensors (optional)	
Photogate and interface	1 per group
Computer, tablet, or graphing calculator with data collection and analysis software	1 per group
Equipment for video analysis (optional)	
Video camera	1 per group
Computer or tablet with video analysis software	1 per group

* As an alternative, students can use computer and presentation software such as Microsoft PowerPoint or Apple Keynote to create their arguments.

Be sure to use a set routine for distributing and collecting the materials during the lab investigation. One option is to set up the materials for each group at each group's lab station before class begins. This option works well when there is a dedicated section of the classroom for lab work and the materials are large and difficult to move (such as a dynamics track). A second option is to have all the materials on a table or cart at a central

LAB 13

location. You can then assign a member of each group to be the "materials manager." This individual is responsible for collecting all the materials his or her group needs from the table or cart during class and for returning all the materials at the end of the class. This option works well when the materials are small and easy to move (such as stopwatches, metersticks, or hanging masses). It also makes it easy to inventory the materials at the end of the class before students leave for the day.

Safety Precautions

Remind students to follow all normal lab safety rules. In addition, tell students to take the following safety precautions:

1. Wear sanitized safety glasses or goggles during lab setup, hands-on activity, and takedown.

2. Keep fingers and toes out of the way of moving objects.

3. Use caution when working with scissors. They are sharp and can cut or puncture skin.

4. Wash hands with soap and water after completing the lab.

Topics for the Explicit and Reflective Discussion

Reflecting on the Use of Core Ideas and Crosscutting Concepts During the Investigation

Teachers should begin the explicit and reflective discussion by asking students to discuss what they know about the core idea they used during the investigation. The following are some important concepts related to the core idea of forces and motion that students need to use to determine which variables do and which variables do not change the period of the pendulum:

- Force is considered a vector quantity because a force has both magnitude and direction.

- A restoring force can result in oscillatory motion.

- Simple harmonic motion occurs when a restoring force is directly proportional to the displacement of an object and acts in the direction opposite to that of displacement.

- Gravity can act as a restoring force.

To help students reflect on what they know about forces and motion, we recommend showing them two or three images using presentation software that help illustrate these important ideas. You can then ask the students the following questions to encourage them to share how they are thinking about these important concepts:

1. What do we see going on in this image?

2. Does anyone have anything else to add?

3. What might be going on that we can't see?

4. What are some things that we are not sure about here?

You can then encourage students to think about how CCs played a role in their investigation. There are at least three CCs that students need to use to determine which variables do and which variables do not change the period of the pendulum: (a) Patterns; (b) Cause and Effect: Mechanism and Explanation; and (c) Structure and Function (see Appendix 2 [p. 527] for a brief description of these CCs). To help students reflect on what they know about these CCs, we recommend asking them the following questions:

1. Why is it important to look for patterns during an investigation?

2. What patterns did you identify during your investigation? What did the identification of these patterns allow you to do?

3. Why is it important to identify causal relationships in science?

4. What did you have to do during your investigation to determine if a factor did or did not cause a change in the period of a pendulum? Why was that useful to do?

5. The way an object is shaped or structured determines many of its properties and how it functions. Why is it useful to think about the relationship between structure and function during an investigation?

6. Why was it useful to examine the structure of a pendulum when attempting to determine which variables do and which variables do not change the period of a pendulum?

You can then encourage the students to think about how they used all these different concepts to help answer the guiding question and why it is important to use these ideas to help justify their evidence for their final arguments. Be sure to remind your students to explain why they included the evidence in their arguments and make the assumptions underlying their analysis and interpretation of the data explicit in order to provide an adequate justification of their evidence.

Reflecting on Ways to Design Better Investigations

It is important for students to reflect on the strengths and weaknesses of the investigation they designed during the explicit and reflective discussion. Students should therefore be encouraged to discuss ways to eliminate potential flaws, measurement errors, or sources of uncertainty in their investigations. To help students be more reflective about the design of their investigation and what they can do to make their investigations more rigorous in the future, you can ask them the following questions:

1. What were some of the strengths of the way you planned and carried out your investigation? In other words, what made it scientific?

2. What were some of the weaknesses of the way you planned and carried out your investigation? In other words, what made it less scientific?

3. What rules can we make, as a class, to ensure that our next investigation is more scientific?

Reflecting on the Nature of Scientific Knowledge and Scientific Inquiry

This investigation can be used to illustrate two important concepts related to the nature of scientific knowledge and the nature of scientific inquiry: (a) the difference between data and evidence in science and (b) is the nature and role of experiments in science (see Appendix 2 for a brief description of these two concepts). Be sure to review these concepts during and at the end of the explicit and reflective discussion. To help students think about these concepts in relation to what they did during the lab, you can ask them the following questions:

1. You had to talk about data and evidence during your investigation. Can you give me some examples of data and evidence from your investigation?

2. Can you work with your group to come up with a rule that you can use to determine if a piece of information is data or evidence? Be ready to share in a few minutes.

3. I asked you to design and carry out an experiment as part of your investigation. Can you give me some examples of what experiments are used for in science?

4. Can you work with your group to come up with a rule that you can use to decide if an investigation is an experiment or not? Be ready to share in a few minutes.

You can also use presentation software or other techniques to encourage your students to think about these concepts. You can show examples of information from the investigation that are either data or evidence and ask students to classify each example and explain their thinking. You can also show images of different types of investigations (such as a physicist or an astronomer collecting data using a telescope as part of an observational or descriptive study, a person working in the library doing a literature review, a person working on a computer to analyze an existing data set, and an actual experiment) and ask students to indicate if they think each image represents an experiment and why or why not.

Be sure to remind your students that, to be proficient in science, it is important that they understand what counts as scientific knowledge and how that knowledge develops over time.

Hints for Implementing the Lab

- Allowing students to design their own procedures for collecting data gives students an opportunity to try, to fail, and to learn from their mistakes. However, you can scaffold students as they develop their procedure by having them fill out an investigation proposal. These proposals provide a way for you to offer students hints and suggestions without telling them how to do it. You can also check the proposals quickly during a class period. We suggest having them fill out an investigation proposal for each experiment they do. For this lab we suggest using Investigation Proposal C.

- Allow students to become familiar with the equipment and materials as part of the tool talk before they begin to design their investigation. Giving them 5–10 minutes to examine the equipment and materials allows them to see what they can and cannot do with them. This also gives students a chance to begin thinking about what variables to test and how to control for other variables.

- If you choose to have students mathematically model the relationship between the period and the length, it is important to realize that they will likely first run a linear regression and get a very high correlation coefficient for an equation of the form y = mx +b, where y corresponds to the period and x corresponds to the length of the pendulum. If this occurs, there are two ways to guide them toward a better equation. First, you can ask them to predict what the period of a pendulum with a length of 0.0 m would be. Their regression equation will give them a period greater than 0 s (because the value for b in their linear equation will be greater than zero), which is a nonsensical answer. Thus, they need an equation where the predicted period of a pendulum with length 0.0 m is 0 s. Second, you can ask them to collect data for a pendulum with a length greater than 1 m. This will provide a more pronounced data point, and the correlation coefficient will be reduced. They will then use other types of regressions and find that a power regression will give them the best model.

- If you choose to have students mathematically model the relationship between the period and the length, a power regression will give them an equation similar to $T = 2L^{\frac{1}{2}}$ or $T = 2\sqrt{(L)}$. This comes from recognizing that, first, it is possible to rewrite the full equation $T = 2\pi\sqrt{(L/\mathbf{g})}$ as $T = (2\pi/\sqrt{\mathbf{g}})(\sqrt{L})$; 2, π, and $\mathbf{g}$ are all constants. On Earth, the value of $\mathbf{g}$ is 9.8 m/s^2, so $2\pi/\sqrt{\mathbf{g}}$ resolves to be approximately 2. This is a nice opportunity for an extension, to ask students where the constant comes from in their regression equation.

- Be sure to allow students to go back and re-collect data at the end of the argumentation session. Students often realize that they made numerous mistakes when they were collecting data as a result of their discussions during the argumentation session. The students, as a result, will want a chance to re-collect data, and the re-collection of data should be encouraged when time allows.

This also offers an opportunity to discuss what scientists do when they realize a mistake is made inside the lab.

If students use a photogate

- Be sure that students record actual values (e.g., period in seconds) or save any graphs generated by a computer, rather than just attempting to hand draw what they see on the computer screen

If students use video analysis

- We suggest allowing students to familiarize themselves with the video analysis software before they finalize the procedure for the investigation, especially if they have not used such software previously. This gives students an opportunity to learn how to work with the software and to improve the quality of the video they take.

- Remind students to hold the video camera as still as possible. Any movement of the camera will introduce error into their analysis. If using actual camcorders, we recommend using a tripod to hold the camera steady. If students are using a camera on a cell phone or tablet, we recommend using a table to help steady the camera.

- Remind students to place a meterstick in the same field of view as the motion they are capturing with the video camera. Also, the meterstick should be approximately the same distance from the camera as the motion. Most video analysis software requires the user to define a scale in the video (this allows the software to establish distances and, subsequently, other variables dependent on distance and displacement).

Connections to Standards

Table 13.2 highlights how the investigation can be used to address learning objectives from AP Physics 1; learning objectives from AP Physics C: Mechanics, *Common Core State Standards*, in English language arts (*CCSS ELA*); and *Common Core State Standards*, Mathematics (*CCSS Mathematics*).

TABLE 13.2

Lab 13 alignment with standards

NGSS **performance expectations**	• None
AP Physics 1 learning objectives	• 3.A.1.1: The student is able to express the motion of an object using narrative, mathematical, and graphical representations. • 3.A.1.2: The student is able to design an experimental investigation of the motion of an object. • 3.A.1.3: The student is able to analyze experimental data describing the motion of an object and is able to express the results of the analysis using narrative, mathematical, and graphical representations. • 3.A.3.1: The student is able to analyze a scenario and make claims (develop arguments, justify assertions) about the forces exerted on an object by other objects for different types of forces or components of forces. • 3.B.3.1: The student is able to predict which properties determine the motion of a simple harmonic oscillator and what the dependence of the motion is on those properties. • 3.B.3.2: The student is able to design a plan and collect data in order to ascertain the characteristics of the motion of a system undergoing oscillatory motion caused by a restoring force.
AP Physics C: Mechanics learning objectives	• I.F.3: Students should be able to apply their knowledge of simple harmonic motion to the case of a pendulum, so they can: • I.F.3.a: Derive the expression for the period of a simple pendulum. • I.F.3.b: Apply the expression for the period of a simple pendulum. • I.F.3.d: Analyze the motion of a torsional pendulum or physical pendulum in order to determine the period of small oscillations.
Literacy connections (*CCSS ELA*)	• *Reading:* Key ideas and details, craft and structure, integration of knowledge and ideas • *Writing:* Text types and purposes, production and distribution of writing, research to build and present knowledge, range of writing • *Speaking and listening:* Comprehension and collaboration, presentation of knowledge and ideas
Mathematics connections (*CCSS Mathematics*)	• *Mathematical practices:* Reason abstractly and quantitatively, construct viable arguments and critique the reasoning of others, use appropriate tools strategically, attend to precision • *Number and quantity:* Reason quantitatively and use units to solve problems, represent and model with vector quantities, perform operations on vectors • *Algebra:* Interpret the structure of expressions, understand solving equations as a process of reasoning and explain the reasoning, solve equations and inequalities in one variable • *Functions:* Understand the concept of a function and use function notation, interpret functions that arise in applications in terms of the context, interpret expressions for functions in terms of the situation they model • *Statistics and probability:* Summarize, represent, and interpret data on two categorical and quantitative variables; understand and evaluate random processes underlying statistical experiments; make inferences and justify conclusions from sample surveys, experiments, and observational studies

LAB 13

Lab Handout

Lab 13. Simple Harmonic Motion and Pendulums: What Variables Affect the Period of a Pendulum?

Introduction

A pendulum, which is a mass swinging at the end of a rope, has a wide range of uses in our daily lives. One of the most frequent uses of a pendulum is in a clock. Christiaan Huygens built the first pendulum clock in 1656 (see Figure L13.1 and version available at *www.nsta.org/adi-physics1*), and his use of a pendulum to keep accurate time was considered a breakthrough in clock design. Pendulums can also be found as parts of amusement park rides, in religious ceremonies, and in tools that help musicians keep a beat. Most school-age children are also familiar with pendulums, because playground swings are just a pendulum with a person at one end.

Pendulums are part of a class of objects that undergo simple harmonic motion; such objects are called oscillators. Harmonic oscillators are objects that move about a point called the equilibrium position (see Figure L13.2). When a pendulum is not moving, the bob will rest (or hang motionless) at the equilibrium position. When an outside force moves the bob from its equilibrium position, a restoring force causes the object to move back toward its equilibrium position. This process is then repeated multiple times as the bob swings back and forth. This motion is referred to as *simple*, because after the initial force to move the bob from equilibrium, the only forces acting on the bob are the restoring force and the tension in the string. Other types of harmonic motion are called *damped*, when friction slows down the motion, or *driven*, when an outside force is repeatedly exerted on the oscillator. There

FIGURE L13.1

The original pendulum clock built by Christiaan Huygens in 1656

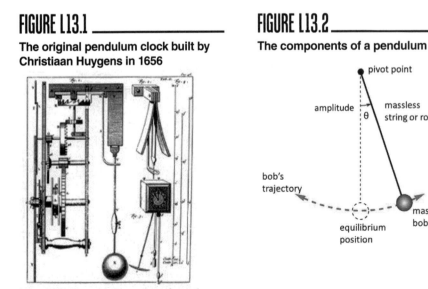

Note: This image is best viewed on the book's Extras page at *www.nsta.org/adi-physics1*.

FIGURE L13.2

The components of a pendulum

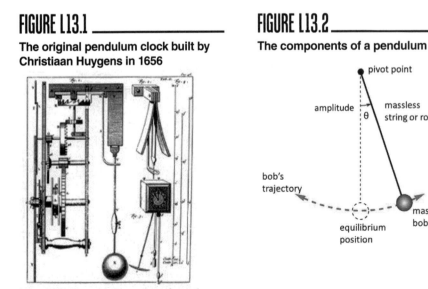

are many ways to describe the motion of a bob. The most frequent is the period (T), which is how long it takes a bob to make one full swing back and forth.

For most pendulums, the period does not change from one swing to the next. This makes the pendulum a particularly useful tool for timekeeping, such as in the pendulum clock shown in Figure L13.1. Early physicists recognized this and investigated the pendulum to understand what variables influence its period. This would allow them to more effectively use pendulums in clocks, as well as in other devices.

Your Task

Use what you know about simple harmonic motion, causal relationships, the relationship between structure and function in nature, and the importance of patterns to design and carry out a series of experiments to determine which variables do and which variables do not change the period of the pendulum.

The guiding question of this investigation is, **What variables affect the period of a pendulum?**

Materials

You may use any of the following materials during your investigation:

Consumables	Equipment	
• Tape	• Safety glasses or goggles (required)	• Protractor
• String	• Electronic or triple beam balance	• Ruler
	• Washers	• Meterstick
	• Paper clips	• Stopwatch
		• Scissors

To use a photogate system, you will need to have a sensor interface and a computer, tablet, or graphing calculator with data collection and analysis software. To use video analysis, you will need to have a video camera and a computer or tablet with video analysis software.

Safety Precautions

Follow all normal lab safety rules. In addition, take the following safety precautions:

1. Wear sanitized safety glasses or goggles during lab setup, hands-on activity, and takedown.

2. Keep fingers and toes out of the way of moving objects.

3. Use caution when working with scissors. They are sharp and can cut or puncture skin.

LAB 13

4. Wash hands with soap and water after completing the lab.

Investigation Proposal Required? ☐ Yes   ☐ No

Getting Started

To answer the guiding question, you will need to design and carry out several different experiments. Each experiment should look at one potential variable that may or may not affect the period of a pendulum. Some potential variables include the mass of the bob, the length of the pendulum, and the release angle. For each of your experiments, you must determine what type of data you need to collect, how you will collect it, and how you will analyze it before you begin.

To determine *what type of data you need to collect,* think about the following questions:

- What are the boundaries and components of the system you are studying?
- How can you describe the components of the system quantitatively?
- How could you keep track of changes in this system quantitatively?
- How might changes to the structure of pendulum change how it functions?
- What might be the underlying cause of a change in the period of a pendulum?
- What will be the independent variable and the dependent variable for each experiment?

To determine *how you will collect the data,* think about the following questions:

- How will you set up your pendulum?
- How will you measure the period of the pendulum?
- What will you need to hold constant during each experiment?
- What conditions need to be satisfied to establish a cause-and-effect relationship?
- What measurement scale or scales should you use to collect data?
- How will you make sure that your data are of high quality (i.e., how will you reduce error)?
- How will you keep track of and organize the data you collect?

To determine *how you will analyze the data,* think about the following questions:

- What type of calculations will you need to make?
- What types of patterns might you look for as you analyze your data?
- How could you use mathematics to show a cause-and-effect relationship?
- What type of table or graph could you create to help make sense of your data?

Connections to the Nature of Scientific Knowledge and Scientific Inquiry

As you work through your investigation, you may want to consider

- the difference between data and evidence in science, and
- the nature and role of experiments in science.

Initial Argument

Once your group has finished collecting and analyzing your data, your group will need to develop an initial argument. Your initial argument needs to include a claim, evidence to support your claim, and a justification of the evidence. The *claim* is your group's answer the guiding question. The *evidence* is an analysis and interpretation of your data. Finally, the *justification* of the evidence is why your group thinks the evidence matters. The justification of the evidence is important because scientists can use different kinds of evidence to support their claims. Your group will create your initial argument on a whiteboard. Your whiteboard should include all the information shown in Figure L13.3.

FIGURE L13.3 _____

Argument presentation on a whiteboard

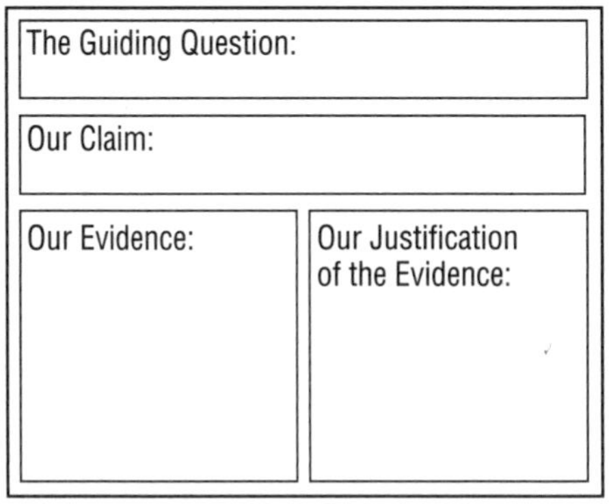

The Guiding Question:	
Our Claim:	
Our Evidence:	Our Justification of the Evidence:

Argumentation Session

The argumentation session allows all of the groups to share their arguments. One or two members of each group will stay at the lab station to share that group's argument, while the other members of the group go to the other lab stations to listen to and critique the other arguments. This is similar to what scientists do when they propose, support, evaluate, and refine new ideas during a poster session at a conference. If you are presenting your group's argument, your goal is to share your ideas and answer questions. You should also keep a record of the critiques and suggestions made by your classmates so you can use this feedback to make your initial argument stronger. You can keep track of specific critiques and suggestions for improvement that your classmates mention in the space below.

Critiques about our initial argument and suggestions for improvement:

If you are critiquing your classmates' arguments, your goal is to look for mistakes in their arguments and offer suggestions for improvement so these mistakes can be fixed. You should look for ways to make your initial argument stronger by looking for things that the other groups did well. You can keep track of interesting ideas that you see and hear during the argumentation in the space below. You can also use this space to keep track of any questions that you will need to discuss with your team.

Interesting ideas from other groups or questions to take back to my group:

Once the argumentation session is complete, you will have a chance to meet with your group and revise your initial argument. Your group might need to gather more data or design a way to test one or more alternative claims as part of this process. Remember, your goal at this stage of the investigation is to develop the best argument possible.

Report

Once you have completed your research, you will need to prepare an *investigation report* that consists of three sections. Each section should provide an answer to the following questions:

1. What question were you trying to answer and why?

2. What did you do to answer your question and why?

3. What is your argument?

Your report should answer these questions in two pages or less. This report must be typed, and any diagrams, figures, or tables should be embedded into the document. Be sure to write in a persuasive style; you are trying to convince others that your claim is acceptable or valid!

Reference

Jeremy Norman's HistoryofInformation.com. 2016. Huygens invents the pendulum clock, increasing accuracy sixty fold (1656). *www.historyofinformation.com/expanded.php?id=3506.*

Checkout Questions

Lab 13. Simple Harmonic Motion and Pendulums: What Variables Affect the Period of a Pendulum?

1. The equation for the period of a pendulum is $T = 2\pi\sqrt{(L/\mathbf{g})}$, where T is the period, L is the length of the pendulum, and $\mathbf{g}$ is the acceleration due to gravity. If a person were to take a pendulum to the Moon, which has a gravitation pull approximately one-sixth that of Earth, what would happen to the period of the pendulum?

   a. The period would increase.

   b. The period would decrease.

   c. The period would stay the same.

   How do you know?

2. Why does the mass of bob have no effect on the period of a pendulum?

3. It is equally important for scientists to identify variables that do have a cause-and-effect relationship and those variables that do not have a cause-and-effect relationship.

    a. I agree with this statement.

    b. I disagree with this statement.

Explain your answer, using an example from your investigation about pendulums.

4. Scientists use the term *data* when they are talking about observations and the term *evidence* when they are talking about measurements.

    a. I agree with this statement.

    b. I disagree with this statement.

Explain your answer, using an example from your investigation about pendulums.

5. Why is important to look for patterns in science? In your answer, be sure to include one example from your investigation on pendulums and at least one more example from another investigation you have conducted in either this class or another science class.

6. Scientists often examine the structure of an object or material during an investigation. Explain why it is useful to examine the structure of an object or material, using an example from your investigation about pendulums.

7. Experiments are one of the most powerful approaches to answering questions in science. Identify the components of an experiment and explain why they are so important in science, using an example from your investigation about pendulums.

LAB 14

Teacher Notes

Lab 14. Simple Harmonic Motion and Springs: What Is the Mathematical Model of the Simple Harmonic Motion of a Mass Hanging From a Spring?

Purpose

The purpose of this lab is to *introduce* students to the core idea of forces and motion, part of the disciplinary core idea (DCI) of Motion and Stability: Forces and Interactions from the *NGSS*, by having them explore the simple harmonic motion of a mass hanging on a spring. This lab also gives students an opportunity to learn about the crosscutting concepts (CCs) of (a) Patterns, (b) Systems and System Models, and (c) Stability and Change from the *NGSS*. In addition, this lab can be used to help students understand two big ideas from AP Physics: (a) the interactions of an object with other objects can be described by forces and (b) interactions between systems can result in changes in those systems. As part of the explicit and reflective discussion, students will also learn about (a) how scientists use different methods to answer different types of questions and (b) the role of imagination and creativity in science.

Underlying Physics Concepts

The position of a mass, $s(t)$, in terms of time, t, can be described in Equation 14.1, where A is the amplitude, B is the frequency, $\mathbf{C}$ is the initial displacement from equilibrium, and D is the phase shift. In SI units, position in terms of time and amplitude is measured in meters (m); frequency is measured either in hertz (Hz) or in radians per second (rad/s); initial displacement from equilibrium is measured in meters (m); and the phase shift is measured in seconds (s).

$$\textbf{(Equation 14.1)} \quad \mathbf{s}(t) = \mathbf{C} + A cos B\,(t - D)$$

Students will be able to develop a mathematical model of the simple harmonic motion of a mass hanging from a spring through two important steps: (1) collecting position-time data of a mass-spring system and (2) using graphical analysis software to graph the data they collected. The sinusoidal nature of the harmonic motion should suggest either a sine or cosine wave, and although it is commonplace to use cosine, a simple phase shift of the sine curve will yield the same result. In aiding students to find the parameters A, B, C, and D, consider asking them to compare the general cosine curve from Equation 14.1 with the data they collected. The following explanations and equations could be useful as students develop their mathematical model.

The amplitude, A, is equivalent to the distance from rest (equilibrium) to the maximum or minimum value, as shown in Equation 14.2.

$$(\text{Equation 14.2}) \quad A = \frac{s(t)_{max} - s(t)_{min}}{2}$$

The frequency, B, can be found by using Equation 14.3, which relates the observed period to the frequency, with observed period measured in seconds (s) and frequency measured in hertz (Hz) or radians per second (rad/s). To calculate the observed period, select two values of time, say t_1 and t_2, such that $s(t_1) = s(t_2)$, and calculate $t_2 - t_1$.

$$(\text{Equation 14.3}) \quad observed\ period = \frac{2\pi}{B}$$

The distance from rest or equilibrium, **C**, referred to mathematically as vertical shift, is calculated by averaging the maximum and minimum values and is shown in Equation 14.4, with **C**, $s(t)_{max}$ and $s(t)_{min}$ all measured in meters (m).

$$(\text{Equation 14.4}) \quad C = \frac{s(t)_{max} + s(t)_{min}}{2}$$

The phase shift, D, is the amount of time elapsed from $t = 0$ to the first maximum.

Timeline

The instructional time needed to complete this lab investigation is 200–280 minutes. Appendix 3 (p. 531) provides options for implementing this lab investigation over several class periods. Option E (280 minutes) should be used if students are unfamiliar with scientific writing, because this option provides extra instructional time for scaffolding the writing process. You can scaffold the writing process by modeling, providing examples, and providing hints as students write each section of the report. Option E can also be used if students are unfamiliar with any of the data collection and analysis tools. Option F (200 minutes) should be used if students are familiar with scientific writing and have developed the skills needed to write an investigation report on their own. In option F, students complete stage 6 (writing the investigation report) and stage 8 (revising the investigation report) as homework.

LAB 14

Materials and Preparation

The materials needed to implement this investigation are listed in Table 14.1. The equipment can be purchased from a science supply company such as Carolina, Flinn Scientific, PASCO Scientific, Vernier, or Ward's Science. Graphical analysis software can be purchased from Vernier (Logger *Pro*) or PASCO (SPARKvue or Capstone). These companies also have apps that can be used on Apple- or Android-based tablets. We recommend consulting with your school's information technology coordinator to determine the best option for your students.

TABLE 14.1

Materials list for Lab 14

Item	Quantity
Safety glasses or goggles	1 per student
Support stand	1 per group
Suspension hook clamp	1 per group
Hanging mass set	1 per group
Springs (variety)	Class set
Motion detector/sensor	1 per group
Interface for motion detector/sensor	1 per group
Computer, tablet, or graphing calculator with data collection and analysis software	1 per group
Investigation Proposal C (optional)	1 per group
Whiteboard, 2' × 3'*	1 per group
Lab Handout	1 per student
Peer-review guide and teacher scoring rubric	1 per student
Checkout Questions	1 per student

* As an alternative, students can use computer and presentation software such as Microsoft PowerPoint or Apple Keynote to create their arguments.

Be sure to use a set routine for distributing and collecting the materials during the lab investigation. One option is to set up the materials for each group at each group's lab station before class begins. This option works well when there is a dedicated section of the classroom for lab work and the materials are large and difficult to move (such as a dynamics track). A second option is to have all the materials on a table or cart at a central location. You can then assign a member of each group to be the "materials manager." This individual is responsible for collecting all the materials his or her group needs from the table or cart during class and for returning all the materials at the end of the class. This option works well when the materials are small and easy to move (such as stopwatches, metersticks, or hanging masses). It also makes it easy to inventory the materials at the end of the class before students leave for the day.

Safety Precautions

Remind students to follow all normal lab safety rules. In addition, tell students to take the following safety precautions:

1. Wear sanitized safety glasses or goggles during lab setup, hands-on activity, and takedown.

2. Keep fingers and toes out of the way of moving objects.

3. Wash hands with soap and water after completing the lab.

Topics for the Explicit and Reflective Discussion
Reflecting on the Use of Core Ideas and Crosscutting Concepts During the Investigation

Teachers should begin the explicit and reflective discussion by asking students to discuss what they know about the core idea they used during the investigation. The following are some important concepts related to the core idea of forces and motion that students will need to develop a mathematical model of the simple harmonic motion of a mass hanging from a spring:

• A reference frame is needed to determine the direction and the magnitude of the displacement, velocity, and acceleration of an object.

• Force is considered a vector quantity because a force has both magnitude and direction.

• A restoring force can result in oscillatory motion.

• Simple harmonic motion occurs when a restoring force is directly proportional to the displacement of an object and acts in the direction opposite to that of displacement.

To help students reflect on what they know about forces and motion, we recommend showing them two or three images using presentation software that help illustrate these important ideas. You can then ask the students the following questions to encourage them to share how they are thinking about these important concepts:

1. What do we see going on in this image?

2. Does anyone have anything else to add?

3. What might be going on that we can't see?

4. What are some things that we are not sure about here?

You can then encourage students to think about how CCs played a role in their investigation. There are at least three CCs that students need to use to develop a mathematical model of the simple harmonic motion of a mass hanging from a spring: (a) Patterns, (b) Systems and System Models, and (c) Stability and Change (see Appendix 2 [p. 527] for a brief description of these CCs). To help students reflect on what they know about these CCs, we recommend asking them the following questions:

1. Why is it important to look for patterns during an investigation?

2. What patterns did you identify during your investigation? What did the identification of these patterns allow you to do?

3. Why do scientists often need to define a system and then develop a model of it as part of an investigation?

4. How did you use a model during your investigation to understand the motion of a mass hanging from a spring?

5. Why is it important to think about what controls or affects the rate of change in system?

6. Which factors might have controlled the rate of change in the movement of the mass hanging from a spring? What did testing these factors systematically allow you to do?

You can then encourage the students to think about how they used all these different concepts to help answer the guiding question and why it is important to use these ideas to help justify their evidence for their final arguments. Be sure to remind your students to explain why they included the evidence in their arguments and make the assumptions underlying their analysis and interpretation of the data explicit in order to provide an adequate justification of their evidence.

Reflecting on Ways to Design Better Investigations

It is important for students to reflect on the strengths and weaknesses of the investigation they designed during the explicit and reflective discussion. Students should therefore be encouraged to discuss ways to eliminate potential flaws, measurement errors, or sources of uncertainty in their investigations. To help students be more reflective about the design of their investigation and what they can do to make their investigations more rigorous in the future, you can ask them the following questions:

1. What were some of the strengths of the way you planned and carried out your investigation? In other words, what made it scientific?

2. What were some of the weaknesses of the way you planned and carried out your investigation? In other words, what made it less scientific?

3. What rules can we make, as a class, to ensure that our next investigation is more scientific?

Reflecting on the Nature of Scientific Knowledge and Scientific Inquiry

This investigation can be used to illustrate two important concepts related to the nature of scientific knowledge and the nature of scientific inquiry: (a) how scientists use different methods to answer different types of questions and (b) the role of imagination and creativity in science (see Appendix 2 for a brief description of these two concepts). Be sure to review these concepts during and at the end of the explicit and reflective discussion. To help students think about these concepts in relation to what they did during the lab, you can ask them the following questions:

1. There is no universal step-by-step scientific method that all scientists follow. Why do you think there is no universal scientific method?

2. Think about what you did during this investigation. How would you describe the method you used to develop a mathematical model of the simple harmonic motion of a mass hanging from a spring? Why would you call it that?

3. Some people think that there is no room for imagination or creativity in science. What do you think?

4. Can you work with your group to come up with different ways that you needed to use your imagination or be creative during this investigation? Be ready to share in a few minutes.

You can also use presentation software or other techniques to encourage your students to think about these concepts. You can show one or more images of a "universal scientific method" that misrepresent the nature of scientific inquiry (see, e.g., *https://commons.wikimedia.org/wiki/File:The_Scientific_Method_as_an_Ongoing_Process.svg*) and ask students why each image is *not* a good representation of what scientists do to develop scientific knowledge. You can ask students to suggest revisions to the image that would make it more consistent with the way scientists develop scientific knowledge. You can also show students an image of the following quote by E. O. Wilson from *Letters to a Young Scientist* (2013) and ask them what they think he meant by it:

> The ideal scientist thinks like a poet and only later works like a bookkeeper. Keep in mind that innovators in both literature and science are basically dreamers and storytellers. In the early stages of the creation of both literature and science, everything in the mind is a story. There is an imagined ending, and usually an imagined beginning, and a selection of bits and pieces that might fit in between.

LAB 14

In works of literature and science alike, any part can be changed, causing a ripple among the other parts, some of which are discarded and new ones added. (p. 74)

Be sure to remind your students that, to be proficient in science, it is important that they understand what counts as scientific knowledge and how that knowledge develops over time.

Hints for Implementing the Lab

- Allowing students to design their own procedures for collecting data gives students an opportunity to try, to fail, and to learn from their mistakes. However, you can scaffold students as they develop their procedure by having them fill out an investigation proposal. These proposals provide a way for you to offer students hints and suggestions without telling them how to do it. You can also check the proposals quickly during a class period. We recommend having the students fill out a different investigation proposal for each experiment. For this lab we suggest using Investigation Proposal C.

- Learn how to use the motion detector/sensor and the data collection and analysis software that the students will use before the lab begins. It is important for you to know how to use the equipment so you can help students when technical issues arise.

- Allow the students to become familiar with the motion detector/sensor and other equipment as part of the tool talk before they begin to design their investigation. Giving them 5–10 minutes to examine the equipment allows them to see what they can and cannot do with it. This also gives students a chance to begin thinking about what variables to test and how to control for other variables.

- We recommend an experimental setup where the spring oscillates vertically, as opposed to horizontally. This will provide students with more robust data. Furthermore, the dampening effects will be minimal.

- When students are modeling the spring, they will want to allow the spring to come to rest at the equilibrium position before they begin collecting data. They can then pull the spring to a displacement from equilibrium position and measure the displacement before releasing the spring. The initial displacement will be equal to the amplitude.

- Be sure to allow students to go back and re-collect data at the end of the argumentation session. Students often realize that they made numerous mistakes when they were collecting data as a result of their discussions during the argumentation session. The students, as a result, will want a chance to re-collect data, and the re-collection of data should be encouraged when time allows. This also offers an opportunity to discuss what scientists do when they realize a mistake is made inside the lab.

Connections to Standards

Table 14.2 highlights how the investigation can be used to address learning objectives from AP Physics 1; learning objectives from AP Physics C: Mechanics, *Common Core State Standards,* in English language arts (*CCSS ELA*); and *Common Core State Standards,* Mathematics (*CCSS Mathematics*).

TABLE 14.2

Lab 14 alignment with standards

***NGSS* performance expectations**	• None
AP Physics 1 learning objectives	• 3.A.1.1: The student is able to express the motion of an object using narrative, mathematical, and graphical representations. • 3.A.1.2: The student is able to design an experimental investigation of the motion of an object. • 3.A.1.3: The student is able to analyze experimental data describing the motion of an object and is able to express the results of the analysis using narrative, mathematical, and graphical representations. • 3.B.3.1: The student is able to predict which properties determine the motion of a simple harmonic oscillator and what the dependence of the motion is on those properties. • 3.B.3.2: The student is able to design a plan and collect data in order to ascertain the characteristics of the motion of a system undergoing oscillatory motion caused by a restoring force.
AP Physics C: Mechanics learning objectives	• I.A.1.b(1): Write down expressions for velocity and position as functions of time, and identify or sketch graphs of these quantities. • IA.2.c(1): Write down expressions for the horizontal and vertical components of velocity and position as functions of time, and sketch or identify graphs of these components. • I.F.1.a: Sketch or identify a graph of displacement as a function of time, and determine from such a graph the amplitude, period, and frequency of the motion. • I.F.1.b: Write down an appropriate expression for displacement of the form $A \sin \omega t$ or $A \cos \omega t$ to describe the motion. • I.F.2.c: Analyze problems in which a mass hangs from a spring and oscillates vertically.

continued

Table 14.2 (*continued*)

Literacy connections (*CCSS ELA*)	• *Reading:* Key ideas and details, craft and structure, integration of knowledge and ideas • *Writing:* Text types and purposes, production and distribution of writing, research to build and present knowledge, range of writing • *Speaking and listening:* Comprehension and collaboration, presentation of knowledge and ideas
Mathematics connections (*CCSS Mathematics*)	• *Mathematical practices:* Make sense of problems and persevere in solving them, reason abstractly and quantitatively, construct viable arguments and critique the reasoning of others, model with mathematics, use appropriate tools strategically, attend to precision, look for and make use of structure, look for and express regularity in repeated reasoning • *Number and quantity:* Reason quantitatively and use units to solve problems, represent and model with vector quantities, perform operations on vectors • *Algebra:* Interpret the structure of expressions, create equations that describe numbers or relationships, understand solving equations as a process of reasoning and explain the reasoning, solve equations and inequalities in one variable, represent and solve equations and inequalities graphically. • *Functions:* Understand the concept of a function and use function notation, interpret functions that arise in applications in terms of the context, analyze functions using different representations, build a function that models a relationship between two quantities, construct and compare linear and exponential models and solve problems, interpret expressions for functions in terms of the situation they model, extend the domain of trigonometric functions using the unit circle, model periodic phenomena with trigonometric functions, prove and apply trigonometric identities. • *Statistics and probability:* Summarize, represent, and interpret data on two categorical and quantitative variables; understand and evaluate random processes underlying statistical experiments; make inferences and justify conclusions from sample surveys, experiments, and observational studies

Reference

Wilson, E. O. 2013. *Letters to a young scientist.* New York: Liveright Publishing.

Lab Handout

Lab 14. Simple Harmonic Motion and Springs: What Is the Mathematical Model of the Simple Harmonic Motion of a Mass Hanging From a Spring?

Introduction

A basic but important kind of motion is called simple harmonic motion. Simple harmonic motion is a type of periodic or oscillatory motion. An example of simple harmonic motion is the way a mass moves up and down when it is attached to a spring (see Figure L14.1). When the mass attached to a spring is not moving, the mass is said to be at equilibrium. When the mass is moved from the equilibrium position (such as when a person pulls down on the spring), a force from the spring (F_k) acts on the mass to restore it to the equilibrium position. This force is called the restoring force. The further the mass is moved from the equilibrium position, the greater the restoring force becomes. In fact, the magnitude of the restoring force is directly proportional to the distance the mass is moved away from the equilibrium position (the displacement) and acts in the direction opposite to the direction of displacement.

We can explain the underlying cause of simple harmonic motion using Newton's second law of motion. Newton's second law of motion indicates that an object will accelerate when acted on by an unbalanced force. Thus, when an oscillator (in this case a mass attached to a spring) is disturbed from equilibrium, it accelerates in the general direction of the equilibrium position. When the mass reaches the equilibrium position, it has a non-zero velocity and the sum of the forces acting on it is zero. The mass will therefore move through the equilibrium position, at which point the restoring force increases until the mass reaches a maximum displacement from the equilibrium position. At the maximum displacement position, the restoring force is also at a maximum, while the velocity is equal to zero (0 m/s). This process repeats over time, which results in periodic or oscillatory motion.

In this investigation you will have an opportunity to examine the simple harmonic motion of a mass hanging on a spring. Your goal is to create a mathematical model that you can use to describe the vertical position of the mass in terms of time. Therefore, you will need to investigate the effect of different masses, release points, and types of springs during your investigation. It is important to note that your model will ignore dampening, the effect of slowing the mass-spring system down to a stop by frictional forces.

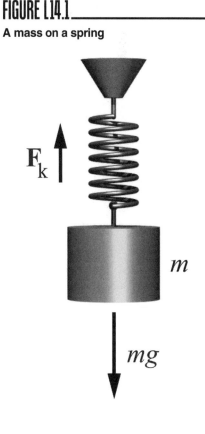

FIGURE L14.1

A mass on a spring

F_k

m

mg

LAB 14

Your Task

Use what you know about simple harmonic motion, patterns, and stability and change in systems to develop a function that will allow you to model the motion of a mass hanging from a spring. To develop a mathematical model, you will need to design and carry out several experiments to determine how (a) different masses, (b) different release points, and (c) different spring types affect the motion of a mass hanging from a spring. Once you have developed your model, you will need to test it to determine if allows you to make accurate predictions about the vertical position of the mass in terms of time.

The guiding question of this investigation is, *What is the mathematical model of the simple harmonic motion of a mass hanging from a spring?*

Materials

You may use any of the following materials during your investigation:

- Safety glasses or goggles (required)
- Support stand
- Suspension hook clamp
- Hanging mass set
- Springs (variety)

- Motion detector/sensor
- Interface for motion detector/sensor
- Computer, tablet, or graphing calculator with data collection and analysis software

Safety Precautions

Follow all normal lab safety rules. In addition, take the following safety precautions:

1. Wear sanitized safety glasses or goggles during lab setup, hands-on activity, and takedown.

2. Keep fingers and toes out of the way of moving objects.

3. Wash hands with soap and water after completing the lab.

Investigation Proposal Required? ☐ Yes ☐ No

Getting Started

The first step in developing your mathematical model is to design and carry out three experiments. In the first experiment, you will need to determine how changing the mass affects the motion of a mass-spring system. You will then need to determine how changing the release point of the mass affects the motion of the mass-spring system. Finally, you will need to determine how changing the type of spring affects the motion of the mass-spring system. Figure L14.2 illustrates how you can use the available equipment to study the motion of a mass-spring system in each experiment. Before you can design your experiments, however, you must determine what type of data you need to collect, how you will collect it, and how you will analyze it.

To determine *what type of data you need to collect,* think about the following questions:

- What are the boundaries and components of the mass-spring system you are studying?
- Which factor(s) might control the rate of change in the mass-spring system?
- How could you keep track of changes in this system quantitatively?
- Under what conditions is the system stable, and under what conditions does it change?
- How will you measure the vertical position of the mass over time?
- What will be the independent variables and the dependent variables for each experiment?

FIGURE L14.2 _____

One way to examine the motion of the mass-spring system using the available equipment

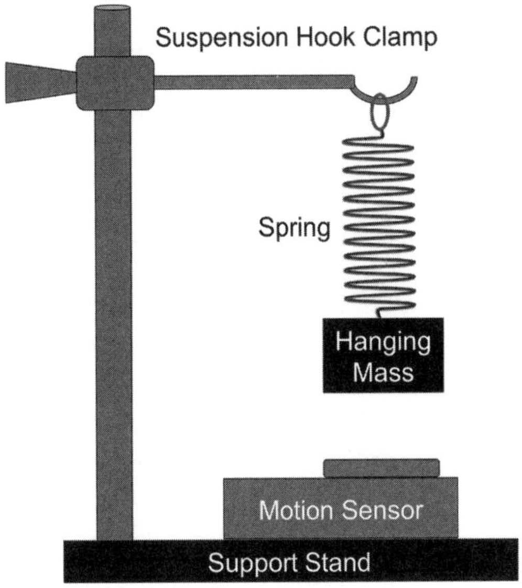

To determine *how you will collect the data,* think about the following questions:

- What other factors will you need to control during each experiment?
- What scale or scales should you use when you take your measurements?
- How will you make sure that your data are of high quality (i.e., how will you reduce error)?
- How will you keep track of and organize the data you collect?

To determine *how you will analyze the data,* think about the following questions:

- How much of your data is useful, given that you want to ignore dampening?
- What type of calculations will you need to make?
- What types of patterns might you look for as you analyze your data?
- What type of table or graph could you create to help make sense of your data?
- What types of equations can you use to describe motion that is periodic or harmonic?

Once you have determined how different masses, release points, and spring types affect the motion of a mass-spring system, your group will need to develop a mathematical model. The model must allow you to make accurate predictions about the vertical position of the mass in terms of time.

LAB 14

The last step in this investigation will be to test your model. To accomplish this goal, you can add different hanging masses (amounts that you did not test) to the end of one of the springs or try different release points (ones that you did not test) to determine whether your mathematical model helps you make accurate predictions. If you are able to use your model to make accurate predictions, then you will be able to generate the evidence you need to convince others that it is a valid and acceptable model of the simple harmonic motion of a mass hanging from a spring.

Connections to the Nature of Scientific Knowledge and Scientific Inquiry

As you work through your investigation, you may want to consider

- how scientists use different methods to answer different types of questions, and
- the role of imagination and creativity in science.

Initial Argument

Once your group has finished collecting and analyzing your data, your group will need to develop an initial argument. Your argument must include a claim, evidence to support your claim, and a justification of the evidence. The *claim* is your group's answer to the guiding question. The *evidence* is an analysis and interpretation of your data. Finally, the *justification* of the evidence is why your group thinks the evidence matters. The justification of the evidence is important because scientists can use different kinds of evidence to support their claims. Your group will create your initial argument on a whiteboard. Your whiteboard should include all the information shown in Figure L14.3.

FIGURE L14.3 _____

Argument presentation on a whiteboard

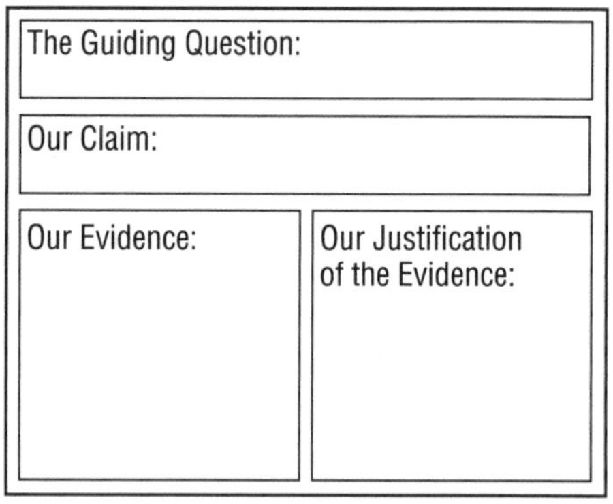

The Guiding Question:	
Our Claim:	
Our Evidence:	Our Justification of the Evidence:

Argumentation Session

The argumentation session allows all of the groups to share their arguments. One or two members of each group will stay at the lab station to share that group's argument, while the other members of the group go to the other lab stations to listen to and critique the other arguments. This is similar to what scientists do when they propose, support, evaluate, and refine new ideas during a poster session at a conference. If you are presenting your group's argument, your goal is to share your ideas and answer questions. You should also keep a record of the critiques and suggestions made by your classmates so you can use this feedback to make your initial argument stronger. You can keep track of specific critiques and suggestions for improvement that your classmates mention in the space below.

Critiques about our initial argument and suggestions for improvement:

If you are critiquing your classmates' arguments, your goal is to look for mistakes in their arguments and offer suggestions for improvement so these mistakes can be fixed. You should look for ways you to make your initial argument stronger by looking for things that the other groups did well. You can keep track of interesting ideas that you see and hear during the argumentation in the space below. You can also use this space to keep track of any questions that you will need to discuss with your team.

Interesting ideas from other groups or questions to take back to my group:

LAB 14

Once the argumentation session is complete, you will have a chance to meet with your group and revise your initial argument. Your group might need to gather more data or design a way to test one or more alternative claims as part of this process. Remember, your goal at this stage of the investigation is to develop the best argument possible.

Report

Once you have completed your research, you will need to prepare an *investigation report* that consists of three sections. Each section should provide an answer to the following questions:

1. What question were you trying to answer and why?

2. What did you do to answer your question and why?

3. What is your argument?

Your report should answer these questions in two pages or less. This report must be typed, and any diagrams, figures, or tables should be embedded into the document. Be sure to write in a persuasive style; you are trying to convince others that your claim is acceptable or valid!

Checkout Questions

Lab 14. Simple Harmonic Motion and Springs: What Is the Mathematical Model of the Simple Harmonic Motion of a Mass Hanging From a Spring?

In most physics textbooks, the position of an object in simple harmonic motion is described using the equation below:

$$x = A\cos(\omega t + \delta)$$

1. Given your model in terms of A, B, C, and D, define the parameters A, ω, and δ and discuss the meaning of each in terms of the position of the mass on a spring.

Use the graph below to answer questions 2 and 3.

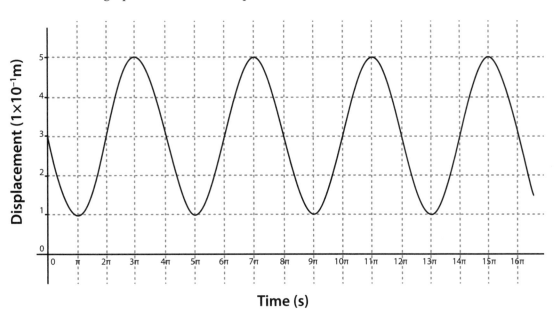

2. What are the parameters A, B, C, and D for the general model $\mathbf{s}(t) = \mathbf{C} + A \cos B \, (t - D)$?

3. What are the parameters for A, ω, and δ for the model $x = A\cos(\omega t + \delta)$?

4. Scientists do not need to be creative or have a good imagination.

   a. I agree with this statement.
   b. I disagree with this statement.

   Explain your answer, using an example from your investigation about simple harmonic motion and springs.

5. It is important to understand what makes a system stable or unstable and what contributes to the rates of change in a system.

   a. I agree with this statement.

   b. I disagree with this statement.

   Explain your answer, using an example from your investigation about simple harmonic motion and springs.

6. Scientists use different methods to answer different types of questions. Explain how the type of question a scientist asks affects the methods he or she uses to answer those questions, using an example from your investigation about simple harmonic motion and springs.

7. Scientists often look for and try to explain patterns in nature. Explain why it is useful to look for and explain patterns in nature, using an example from your investigation about simple harmonic motion and springs.

8. Models in science can be physical, conceptual, or mathematical. Explain the difference in these types of models and discuss the strengths and weaknesses of each type of model, using an example from your investigation about simple harmonic motion and springs.

Application Lab

LAB 15

Teacher Notes

Lab 15. Simple Harmonic Motion and Rubber Bands: Under What Conditions Do Rubber Bands Obey Hooke's Law?

Purpose

The purpose of this lab is for students to *apply* what they know about the core idea of forces and motion, part of the disciplinary core idea (DCI) of Motion and Stability: Forces and Interactions from the *NGSS,* to determine if a mass hanging from a rubber band obeys Hooke's law. This lab also gives students an opportunity to learn about the crosscutting concepts (CCs) of (a) Systems and System Models and (b) Stability and Change from the *NGSS.* In addition, this lab can be used to help students understand two big ideas from AP Physics: (a) fields existing in space can be used to explain interactions and (b) the interactions of an object with other objects can be described by forces. As part of the explicit and reflective discussion, students will also learn about (a) how the culture of science, societal needs, and current events influence the work of scientists; and (b) the role of imagination and creativity in science.

Underlying Physics Concepts

For harmonic oscillators that undergo compression and expansion (such as a spring), Hooke's law is a reliable description of the relationship between the restoring force and the displacement from equilibrium of the oscillator. Conceptually, Hooke's law states that the restoring force is directly proportional to the displacement from equilibrium. Mathematically, Hooke's law is described by Equation 15.1, where $\mathbf{F}$ is the restoring force, $\mathbf{x}$ is the displacement from equilibrium, and k is the spring constant. In SI units, force is measured in newtons (N), displacement is measured in meters (m), and the spring constant has units of newtons per meter (N/m).

$$\text{(Equation 15.1)} \quad \mathbf{F} = -k\mathbf{x}$$

Yet, Hooke's Law only applies to a small range of compressions or extensions. Above this range of compression or extension, the oscillator becomes deformed and will no longer return to its natural equilibrium position. That is, if too much force is used to compress or extend the oscillator, the oscillator will become deformed. In the case of a bungee cord or a rubber band, this means that it will be permanently "stretched out."

Figure 15.1 shows a graph of the applied force to stretch an oscillator (spring, bungee, or rubber band) versus the elongation (i.e., the distance an oscillator is stretched). There are several points that are important to identify on the graph in Figure 15.1. The first point is the *proportional limit,* which is the largest amount of force that can be applied to stretch the oscillator such that the oscillator will return to its normal shape upon release of the force. For any force less than or equal to the proportional limit, the oscillator will obey Hooke's law. For this lab, this means that when a mass is placed at the end of the rubber band, if $m\mathbf{g}$ is less than or equal to the proportional limit, then the rubber band will obey Hooke's law and oscillate around an equilibrium point.

The second point on the graph is the *elastic limit*. The elastic limit and the proportional limit define a range of forces that have unique properties with respect to the oscillator. In this range of forces, the linear relationship between the force applied to stretch the oscillator (e.g., the spring or rubber band) and the elongation no longer

FIGURE 15.1
Stress curve for a mass on a rubber band

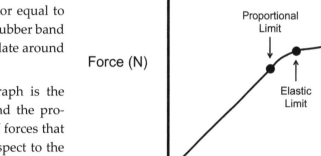

holds. Instead, a very complex relationship between the force applied and the elongation exists, and the mathematics required to model this relationship is beyond the scope of a high school physics course. Although no linear relationship exists in this range of forces, as long as the force used to stretch the oscillator is less than or equal to the elastic limit, the oscillator will return to its normal shape upon release of the applied force. For this lab, it means the rubber band will return to its original shape and size.

When a force greater than the elastic limit is applied to the oscillator, this force results in the permanent deformation of the oscillator. With respect to this lab, when a large enough mass is placed on the end of a rubber band, the rubber band will become "permanently stretched out." There is, however, an upper limit on the amount of force that can be applied to stretch the oscillator. At this limit, the force is strong enough to break the oscillator, represented as the *breaking point* on the stress curve in Figure 15.1. With respect to a rubber band, if you put too much mass on the end, it will tear the rubber band.

There is one caveat to Hooke's law for a rubber band or a bungee cord oscillating around an equilibrium point. When a mass is placed on the rubber band, a new equilibrium position is created where the restoring force upward due to the stretch of the rubber band is equal to the gravitational force on the mass. Mathematically, this can be expressed as the elongation ($\mathbf{x}$) where $\Sigma\mathbf{F} = 0$, such that $m\mathbf{g} + k\mathbf{x} = 0$. Any additional force applied to stretch the rubber band will cause the mass to begin to oscillate.

Unlike springs, bungee cords and rubber bands do not have a restoring force when $\mathbf{x}$ is positive. If the mass is pulled slightly down, the restoring force will act to bring the mass back to the equilibrium point. The mass will return to the equilibrium point where the sum of the forces is zero. However, the velocity of the mass at the equilibrium point is in the positive direction, which results in the mass passing through the equilibrium point. Unlike springs, rubber bands are not easily compressed. When the mass moves through the equilibrium point to a positive displacement, there is no restoring force exerted on the mass by the rubber band. During this portion of the motion, the only force acting on the mass is the force due to gravity. Because this force is *not* proportional to the displacement from equilibrium, the rubber band does not obey Hooke's law during this portion of its motion.

Thus, when students are conducting their investigation, there are two potential answers to the question. Some groups may create investigations to determine the elastic limit, with the answer to the question being that the rubber band obeys Hooke's law up to a certain mass limit. Other groups may design their investigation such that they realize that a rubber band only obeys Hooke's law when the mass has negative displacement relative to the equilibrium point.

Timeline

The instructional time needed to complete this lab investigation is 200–280 minutes. Appendix 3 (p. 531) provides options for implementing this lab investigation over several class periods. Option E (280 minutes) should be used if students are unfamiliar with scientific writing, because this option provides extra instructional time for scaffolding the writing process. You can scaffold the writing process by modeling, providing examples, and providing hints as students write each section of the report. Option E can also be used if you are introducing students to the video analysis programs. Option F (200 minutes) should be used if students are familiar with scientific writing and have developed the skills needed to write an investigation report on their own. In option F, students complete stage 6 (writing the investigation report) and stage 8 (revising the investigation report) as homework.

Materials and Preparation

The materials needed for this investigation are listed in Table 15.1. The consumables and equipment can be purchased from a science supply company such as Carolina, Flinn Scientific, PASCO, Vernier, or Ward's Science. You can also purchase many of these materials from a general retail store such as Wal-Mart or Target. Video analysis software can be purchased from Vernier (Logger *Pro*) or PASCO (SPARKvue or Capstone). These companies also have apps that can be used on Apple- or Android-based tablets and cell phones. We recommend consulting with your school's information technology coordinator to determine the best option for your students.

TABLE 15.1

Materials list for lab 15

Item	Quantity
Consumables	
Rubber (nonlatex) bands	10 per group
Tape	As needed
Equipment and other materials	
Safety glasses or goggles	1 per student
Hanging mass set	1 per group
Paper clips	As needed
Ruler	1 per student
Support stand	1 per group
Electronic or triple beam balance	1 per group
Stopwatch	1 per student
Investigation Proposal A (optional)	1 per group
Whiteboard, 2' × 3'*	1 per group
Lab Handout	1 per student
Peer-review guide and teacher scoring rubric	1 per student
Checkout Questions	1 per student
Equipment for video analysis (optional)	
Video camera	1 per group
Computer or tablet with video analysis software	1 per group

* As an alternative, students can use computer and presentation software such as Microsoft PowerPoint or Apple Keynote to create their arguments.

Be sure to use a set routine for distributing and collecting the materials during the lab investigation. One option is to set up the materials for each group at each group's lab station before class begins. This option works well when there is a dedicated section of the classroom for lab work and the materials are large and difficult to move (such as a dynamics track). A second option is to have all the materials on a table or cart at a central location. You can then assign a member of each group to be the "materials manager." This individual is responsible for collecting all the materials his or her group needs from the table or cart during class and for returning all the materials at the end of the class. This option works well when the materials are small and easy to move (such as stopwatches,

metersticks, or hanging masses). It also makes it easy to inventory the materials at the end of the class before students leave for the day.

Safety Precautions

Remind students to follow all normal lab safety rules. In addition, tell students to take the following safety precautions:

1. Wear sanitized safety glasses or goggles during lab setup, hands-on activity, and takedown.

2. Keep fingers and toes out of the way of moving objects.

3. Wash hands with soap and water after completing the lab.

Topics for the Explicit and Reflective Discussion

Reflecting on the Use of Core Ideas and Crosscutting Concepts During the Investigation

Teachers should begin the explicit and reflective discussion by asking students to discuss what they know about the core idea they used during the investigation. The following are some important concepts related to the core idea of forces and motion that students will need to determine the conditions in which rubber bands obey Hooke's law:

- A reference frame is needed to determine the direction and the magnitude of the displacement, velocity, and acceleration of an object.

- Force is considered a vector quantity because a force has both magnitude and direction.

- A restoring force can result in oscillatory motion.

- Simple harmonic motion occurs when a restoring force is directly proportional to the displacement of an object and acts in the direction opposite to that of displacement.

- Hooke's law indicates that the strain (deformation) of an elastic object or material is proportional to the stress applied to it.

To help students reflect on what they know about forces and motion, we recommend showing them two or three images using presentation software that help illustrate these important ideas. You can then ask the students the following questions to encourage them to share how they are thinking about these important concepts:

1. What do we see going on in this image?

2. Does anyone have anything else to add?

3. What might be going on that we can't see?

4. What are some things that we are not sure about here?

You can then encourage students to think about how CCs played a role in their investigation. There are at least two CCs that students need to determine the conditions in which rubber bands obey Hooke's law: (a) Systems and System Models and (b) Stability and Change (see Appendix 2 [p. 527] for a brief description of these CCs). To help students reflect on what they know about these CCs, we recommend asking them the following questions:

1. Why do scientists often need to define a system and then develop a model of it as part of an investigation?

2. How did you use a model during your investigation to understand the motion of a mass hanging from a rubber band?

3. Why is it important to think about the conditions under which the behavior of a system is stable?

4. Under what conditions was the rubber band–mass system stable (i.e., under what conditions does the system obey Hooke's law)? Why is it important to understand the conditions under which a system is stable and when the system is not stable?

You can then encourage the students to think about how they used all these different concepts to help answer the guiding question and why it is important to use these ideas to help justify their evidence for their final arguments. Be sure to remind your students to explain why they included the evidence in their arguments and make the assumptions underlying their analysis and interpretation of the data explicit in order to provide an adequate justification of their evidence.

Reflecting on Ways to Design Better Investigations
It is important for students to reflect on the strengths and weaknesses of the investigation they designed during the explicit and reflective discussion. Students should therefore be encouraged to discuss ways to eliminate potential flaws, measurement errors, or sources of uncertainty in their investigations. To help students be more reflective about the design of their investigation and what they can do to make their investigations more rigorous in the future, you can ask them the following questions:

1. What were some of the strengths of the way you planned and carried out your investigation? In other words, what made it scientific?

2. What were some of the weaknesses of the way you planned and carried out your investigation? In other words, what made it less scientific?

3. What rules can we make, as a class, to ensure that our next investigation is more scientific?

LAB 15

Reflecting on the Nature of Scientific Knowledge and Scientific Inquiry

This investigation can be used to illustrate two important concepts related to the nature of scientific knowledge and the nature of scientific inquiry: (a) how the culture of science, societal needs, and current events influence the work of scientists; and (b) the role of imagination and creativity in science (see Appendix 2 for a brief description of these two concepts). Be sure to review these concepts during and at the end of the explicit and reflective discussion. To help students think about these concepts in relation to what they did during the lab, you can ask them the following questions:

1. People view some types of research as being more important than other types of research because of cultural values and current events. Can you come up with some examples of how cultural values and current events have influenced the work of scientists?

2. Scientists share a set of values, norms, and commitments that shape what counts as knowing, how to represent or communicate information, and how to interact with other scientists. Can you work with your group to come up with a rule that you can use to decide if something is science or not science? Be ready to share in a few minutes.

3. Some people think that there is no room for imagination or creativity in science. What do you think?

4. Can you work with your group to come up different ways that you needed to use your imagination or be creative during this investigation? Be ready to share in a few minutes.

You can also use presentation software or other techniques to encourage your students to think about these concepts. You can show examples of research projects that were influenced by cultural values and current events and ask students to think about was going on at the time and why that research was viewed as being important for the greater good. You can also show students an image of the following quote by E. O. Wilson from *Letters to a Young Scientist* (2013) and ask them what they think he meant by it:

> The ideal scientist thinks like a poet and only later works like a bookkeeper. Keep in mind that innovators in both literature and science are basically dreamers and storytellers. In the early stages of the creation of both literature and science, everything in the mind is a story. There is an imagined ending, and usually an imagined beginning, and a selection of bits and pieces that might fit in between. In works of literature and science alike, any part can be changed, causing a ripple among the other parts, some of which are discarded and new ones added. (p. 74)

Be sure to remind your students that, to be proficient in science, it is important that they understand what counts as scientific knowledge and how that knowledge develops over time.

Hints for Implementing the Lab

- Allowing students to design their own procedures for collecting data gives students an opportunity to try, to fail, and to learn from their mistakes. However, you can scaffold students as they develop their procedure by having them fill out an investigation proposal. These proposals provide a way for you to offer students hints and suggestions without telling them how to do it. You can also check the proposals quickly during a class period. For this lab we suggest using Investigation Proposal A.

- We suggest that you try this lab on your own before introducing it to the students. Different size and widths of rubber bands will have slightly different stress curves. Trying the lab yourself with different rubber bands will allow you to give students a set of masses that are appropriate for the rubber bands you are using.

- Allow the students to become familiar with the equipment and materials as part of the tool talk before they begin to design their investigation. Giving them 5–10 minutes to examine the equipment and materials will let them see what they can and cannot do with them. This also gives students a chance to begin thinking about what variables to test and how to control for other variables.

- For students who may need additional guidance, we suggest having them create force diagrams with respect to the oscillating mass. This will help students identify when the restoring force is obeying Hooke's law and when it is not.

- Be sure to allow students to go back and re-collect data at the end of the argumentation session. Students often realize that they made numerous mistakes when they were collecting data as a result of their discussions during the argumentation session. The students, as a result, will want a chance to re-collect data, and the re-collection of data should be encouraged when time allows. This also offers an opportunity to discuss what scientists do when they realize a mistake is made inside the lab.

If students use video analysis

- We suggest allowing students to familiarize themselves with the video analysis software before they finalize the procedure for the investigation, especially if they have not used such software previously. This gives students an opportunity to learn how to work with the software and to improve the quality of the video they take.

- Remind students to hold the video camera as still as possible. Any movement of the camera will introduce error into their analysis. If using actual camcorders, we recommend using a tripod to hold the camera steady. If students are using a camera on a cell phone or tablet, we recommend using a table to help steady the camera.

- Remind students to place a meterstick in the same field of view as the motion they are capturing with the video camera. Also, the meterstick should be approximately the same distance from the camera as the motion. Most video analysis software requires the user to define a scale in the video (this allows the software to establish distances and, subsequently, other variables dependent on distance and displacement).

Connections to Standards

Table 15.2 highlights how the investigation can be used to address learning objectives from AP Physics 1; learning objectives from AP Physics C: Mechanics, *Common Core State Standards*, in English language arts (*CCSS ELA*); and *Common Core State Standards*, Mathematics (*CCSS Mathematics*).

TABLE 15.2

Lab 15 alignment with standards

NGSS performance expectations	• None
AP Physics 1 learning objectives	• 3.A.1.2: The student is able to design an experimental investigation of the motion of an object. • 3.A.1.3: The student is able to analyze experimental data describing the motion of an object and is able to express the results of the analysis using narrative, mathematical, and graphical representations. • 3.B.1.2: The student is able to design a plan to collect and analyze data for motion (static, constant, or accelerating) from force measurements and carry out an analysis to determine the relationship between the net force and the vector sum of the individual forces. • 3.B.3.c: Students should be able to calculate force and acceleration for any given displacement for an object oscillating on a spring. • 3.B.3.2: The student is able to design a plan and collect data in order to ascertain the characteristics of the motion of a system undergoing oscillatory motion caused by a restoring force. • 3.B.3.4: The student is able to construct a qualitative and/or a quantitative explanation of oscillatory behavior given evidence of a restoring force.

AP Physics C: Mechanics learning objectives	• I.C.2.b(4): Write an expression for the force exerted by an ideal spring and for the potential energy of a stretched or compressed spring. • I. F.2.b: Apply the expression for the period of oscillation of a mass on a spring. • I.F.2.c: Analyze problems in which a mass hangs from a spring and oscillates vertically.
Literacy connections (CCSS ELA)	• *Reading:* Key ideas and details, craft and structure, integration of knowledge and ideas • *Writing:* Text types and purposes, production and distribution of writing, research to build and present knowledge, range of writing • *Speaking and listening:* Comprehension and collaboration, presentation of knowledge and ideas
Mathematics connections (CCSS Mathematics)	• *Mathematical practices:* Make sense of problems and persevere in solving them, reason abstractly and quantitatively, construct viable arguments and critique the reasoning of others, model with mathematics, use appropriate tools strategically, attend to precision, look for and make use of structure, look for and express regularity in repeated reasoning • *Number and quantity:* Reason quantitatively and use units to solve problems, represent and model with vector quantities, perform operations on vectors • *Algebra:* Interpret the structure of expressions, write expressions in equivalent forms to solve problems, understand solving equations as a process of reasoning and explain the reasoning, solve equations and inequalities in one variable, represent and solve equations and inequalities graphically • *Functions:* Understand the concept of a function and use function notation, interpret functions that arise in applications in terms of the context, analyze functions using different representations, interpret expressions for functions in terms of the situation they model • *Statistics and probability:* Summarize, represent, and interpret data on two categorical and quantitative variables; interpret linear models; understand and evaluate random processes underlying statistical experiments; make inferences and justify conclusions from sample surveys, experiments, and observational studies

Reference

Wilson, E. O. 2013. *Letters to a young scientist.* New York: Liveright Publishing.

LAB 15

Lab Handout

Lab 15. Simple Harmonic Motion and Rubber Bands: Under What Conditions Do Rubber Bands Obey Hooke's Law?

Introduction

Harmonic oscillators are objects that move about an equilibrium point due to a restoring force. Two of the most common harmonic oscillators are pendulums and springs. Many objects that we encounter on a daily basis either contain a pendulum, such as a grandfather clock, or are a pendulum, such as the swings found on playgrounds. There are also many other objects that we use that incorporate springs. The shock absorbers found on cars, for example, are just large springs. Pendulums and springs, however, are not the only harmonic oscillators that are found in the world around us. For example, tire swings are a type of harmonic oscillator called a torsional oscillator because the restoring force from the rope causes the tire swing to twist back and forth (torsional motion is the motion of twisting).

Another object that oscillates is a bungee jump ride (see Figure L15.1). When a person makes a bungee jump, he or she is attached to bungee cord and then dropped from a considerable height. That person then bounces up and down for the duration of the ride.

Many oscillators obey the equations that describe simple harmonic motion. One of those equations is called Hooke's law, which states that the farther the oscillator is moved from its equilibrium point, the greater the restoring force on the oscillator. More specifically, the restoring force is directly proportional to the displacement the oscillator is from equilibrium. The equation for Hooke's law is $\mathbf{F} = -k\mathbf{x}$, where $\mathbf{F}$ is the restoring force, $\mathbf{x}$ is the displacement from equilibrium, and k is a constant of proportionality. This equation includes a negative sign because the restoring force always acts in the direction opposite the direction of displacement. For example, if the bungee cord is pulled in the down direction, then the restoring force acts in the up direction.

It is important to note, however, that Hooke's law is not valid for all ranges of possible displacements. For example, if a spring is pulled too far from its equilibrium point by a disturbing force, the spring will actually deform and remain stretched out, instead of returning to the equilibrium position. This caveat to Hooke's law is particularly important for those who incorporate an oscillator into the design of an amusement park ride. If the displacement exceeds the allowable range, the restoring force can no longer return the oscillator to equilibrium. For a bungee ride, this issue can pose safety risks, because the riders would continue to fall toward the ground. It is therefore very important to know how much mass can be added to a bungee cord before it deforms or breaks. One way to determine the range of mass that can safely be added to a bungee is to use a smaller-scale model of the system to explore the relationship between mass and displacement. In addition, we can use a rubber band to approximate the behavior of a bungee cord because rubber bands are much smaller and much less expensive. This type of modeling allows engineers to determine parameters they will need to consider when doing tests using actual bungee cords.

Your Task

Use what you know about forces, oscillation, systems and system models, and stability and change to design and carry out an investigation to determine the range of mass that can be added to a bungee cord so that it still obeys Hooke's law.

The guiding question of this investigation is, **Under what conditions do rubber bands obey Hooke's law?**

Materials

You may use any of the following materials during your investigation:

Consumables	Equipment	
• Rubber bands	• Safety glasses or goggles (required)	• Ruler
• Tape	• Hanging mass set	• Support stand
	• Paper clips	• Electronic or triple beam balance
		• Stopwatch

If you have access to the following equipment, you may also consider using a video camera and a computer or tablet with video analysis software.

Safety Precautions

Follow all normal lab safety rules. In addition, take the following safety precautions:

1. Wear sanitized safety glasses or goggles during lab setup, hands-on activity, and takedown.

2. Keep fingers and toes out of the way of moving objects.

3. Wash hands with soap and water after completing the lab.

LAB 15

Getting Started

To answer the guiding question, you will need to design and carry out an investigation to determine the range of mass that can be added to a rubber band so that it still obeys Hooke's law. Figure L15.2 illustrates how you can use the available equipment to study the motion of a mass–rubber band system. Before you can design your investigation, however, you must determine what type of data you need to collect, how you will collect it, and how you will analyze it.

To determine *what type of data you need to collect,* think about the following questions:

- What are the boundaries and components of the mass–rubber band system?

- How do the components of the system interact with each other?

- When is this system stable, and under which conditions does it change?

- How could you keep track of changes in this system quantitatively?

- How will you measure the motion of the hanging mass?

- What are the independent variables and the dependent variables for each experiment?

To determine *how you will collect the data,* think about the following questions:

- What other factors will you need to control during each experiment?

- What scale or scales should you use when you take your measurements?

- How will you make sure that your data are of high quality (i.e., how will you reduce error)?

- How will you keep track of and organize the data you collect?

To determine *how you will analyze the data,* think about the following questions:

- What types of patterns might you look for as you analyze your data?

- What type of table or graph could you create to help make sense of your data?

- How will you model the system to indicate under what parameters the harmonic motion is stable?

FIGURE L15.2 _____

One way to examine the motion of the mass–rubber band system using the available equipment

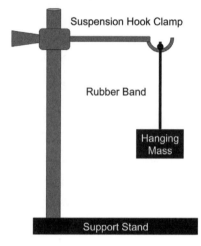

Connections to the Nature of Scientific Knowledge and Scientific Inquiry

As you work through your investigation, you may want to consider

- how the culture of science, societal needs, and current events influence the work of scientists; and
- the role that imagination and creativity play in scientific research.

Initial Argument

Once your group has finished collecting and analyzing your data, your group will need to develop an initial argument. Your initial argument needs to include a claim, evidence to support your claim, and a justification of the evidence. The *claim* is your group's answer to the guiding question. The *evidence* is an analysis and interpretation of your data. Finally, the *justification* of the evidence is why your group thinks the evidence matters. The justification of the evidence is important because scientists can use different kinds of evidence to support their claims. Your group will create your initial argument on a whiteboard. Your whiteboard should include all the information shown in Figure L15.3.

FIGURE L15.3

Argument presentation on a whiteboard

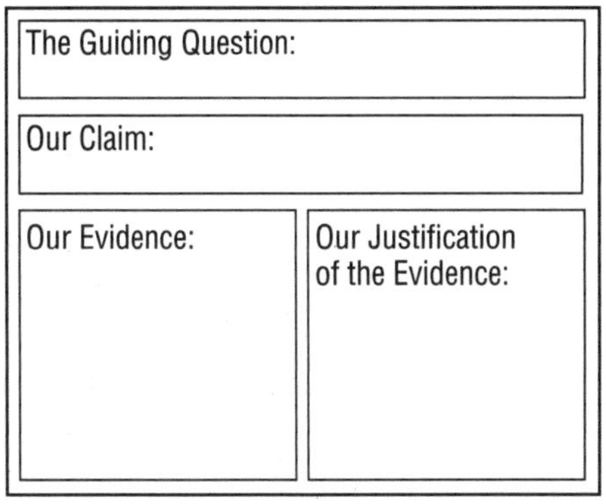

The Guiding Question:	
Our Claim:	
Our Evidence:	Our Justification of the Evidence:

Argumentation Session

The argumentation session allows all of the groups to share their arguments. One or two members of each group will stay at the lab station to share that group's argument, while the other members of the group go to the other lab stations to listen to and critique the other arguments. This is similar to what scientists do when they propose, support, evaluate, and refine new ideas during a poster session at a conference. If you are presenting your group's argument, your goal is to share your ideas and answer questions. You should also keep a record of the critiques and suggestions made by your classmates so you can use this feedback to make your initial argument stronger. You can keep track of specific critiques and suggestions for improvement that your classmates mention in the space below.

Critiques about our initial argument and suggestions for improvement:

If you are critiquing your classmates' arguments, your goal is to look for mistakes in their arguments and offer suggestions for improvement so these mistakes can be fixed. You should look for ways to make your initial argument stronger by looking for things that the other groups did well. You can keep track of interesting ideas that you see and hear during the argumentation in the space below. You can also use this space to keep track of any questions that you will need to discuss with your team.

Interesting ideas from other groups or questions to take back to my group:

Once the argumentation session is complete, you will have a chance to meet with your group and revise your initial argument. Your group might need to gather more data or design a way to test one or more alternative claims as part of this process. Remember, your goal at this stage of the investigation is to develop the best argument possible.

Report

Once you have completed your research, you will need to prepare an *investigation report* that consists of three sections. Each section should provide an answer to the following questions:

1. What question were you trying to answer and why?

2. What did you do to answer your question and why?

3. What is your argument?

Your report should answer these questions in two pages or less. This report must be typed, and any diagrams, figures, or tables should be embedded into the document. Be sure to write in a persuasive style; you are trying to convince others that your claim is acceptable or valid!

Checkout Questions

Lab 15. Simple Harmonic Motion and Rubber Bands: Under What Conditions Do Rubber Bands Obey Hooke's Law?

1. What is the mathematical relationship between the force acting on the rubber band and the elongation of the rubber band?

    Is the function linear as the force increases?

2. In springs, the spring constant k is a function of both the material the spring is made from and the shape of the spring. What factors do you think might affect the constant of proportionality relating the force on a rubber band to the elongation of the rubber band?

Explain your answer, based on what you observed during your investigation about Hooke's law and rubber bands.

3. The imagination and creativity of a scientist play an important role in planning and carrying out investigations.

   a. I agree with this statement.
   b. I disagree with this statement.

   Explain your answer, using an example from your investigation about Hooke's law and rubber bands.

4. The research done by a scientist is often influenced by what is important in society.

   a. I agree with this statement.
   b. I disagree with this statement.

Explain your answer, using an example from your investigation about Hooke's law and rubber bands.

5. Models are used to understand complex phenomena across the different scientific disciplines. Explain why models are so important, using an example from your investigation about Hooke's law and rubber bands.

6. Scientists often seek to identify the parameters under which a system is stable and what happens to the system when those parameters are exceeded. Explain why this is such an important research aim, using an example from your investigation about Hooke's law and rubber bands.

SECTION 6
Forces and Motion

Systems of Particles and Linear Momentum

Introduction Labs

LAB 16

Teacher Notes

Lab 16. Linear Momentum and Collisions: When Two Objects Collide and Stick Together, How Do the Initial Velocity and Mass of One of the Moving Objects Affect the Velocity of the Two Objects After the Collision?

Purpose

The purpose of this lab is to *introduce* students to the core idea of forces and motion, part of the disciplinary core idea (DCI) of Motion and Stability: Forces and Interactions from the *NGSS*, by giving them the opportunity to explore the conservation of momentum during a collision. This lab also gives students an opportunity to learn about the crosscutting concepts (CCs) of (a) Scale, Proportion, and Quantity; and (b) Energy and Matter: Flows, Cycles, and Conservation from the *NGSS*. In addition, this lab can be used to help students understand three big ideas from AP Physics: (a) the interactions of an object with other objects can be described by forces, (b) interactions between systems can result in changes in those systems, and (c) changes that occur as result of interactions are constrained by conservation laws. As part of the explicit and reflective discussion, students will also learn about (a) the difference between observations and inferences in science and (b) how the culture of science, societal needs, and current events influence the work of scientists.

Underlying Physics Concepts

Linear momentum is defined as the product of an object's mass (m) and velocity ($\mathbf{v}$). The mathematical relationship is shown in Equation 16.1. Because we already use m to represent mass in equations, physicists have agreed to use $\mathbf{p}$ to symbolize linear momentum. In SI units, momentum is measured in kilogram-meters per second (kg·m/s), mass is measured in kilograms (kg), and velocity is measured in meters per second (m/s).

$$\text{(Equation 16.1)} \quad \mathbf{p} = m\mathbf{v}$$

Among other quantities, linear momentum obeys a conservation law. The law of conservation of momentum is that the total momentum of a system of isolated objects remains constant. An *isolated system of objects* is a system of objects in which no outside forces act on the system as a whole or on the individual objects within the system. When analyzing the motion of objects to understand their momentum, we often define the system in such a way as to ignore the force of gravity on the objects, thereby treating the system as an isolated system. For example, when two pool balls collide on a pool table, we define the

Linear Momentum and Collisions

When Two Objects Collide and Stick Together, How Do the Initial Velocity and Mass of One of the Moving Objects Affect the Velocity of the Two Objects After the Collision?

system such that we ignore the gravitational and normal forces acting on the balls and only include the force of the two balls on each other when they collide.

Although the law of conservation of momentum states that the total momentum of the system must be conserved, the momentum of each object can change within the system, such as when two objects collide. Mathematically, this is represented in Equation 16.2 (This equation is for linear motion; there are rotational analogues, but they can be ignored in this investigation). The subscripts 1 and 2 are used to denote each object, so $\mathbf{p}_1$ is the momentum of the first object and $\mathbf{p}_2$ is the momentum of the second object. The left side of the equation is the momentum of each object before the collision, and the right side of the equation is the momentum of each object after the collision. The symbol (') on the right side of the equation is used to denote that those values represent the momentum after the collision, and it is pronounced "prime" (e.g., $\mathbf{p}_1{}'$ is read as "p one prime").

$$\text{(Equation 16.2)} \quad \mathbf{p}_1 + \mathbf{p}_2 = \mathbf{p}_1{}' + \mathbf{p}_2{}'$$

Using Equation 16.1, we can replace the value of momentum in Equation 16.2 with the product of each object's mass and velocity both before and after the collision, such that we get Equation 16.3:

$$\text{(Equation 16.3)} \quad m_1\mathbf{v}_1 + m_2\mathbf{v}_2 = m_1\mathbf{v}_1{}' + m_2\mathbf{v}_2{}'$$

Notice that on the right side of Equation 16.3 the prime symbol is only on the velocity for each object, respectively. That is, after the collision, we may say "v one prime." We do not, however, place a prime symbol on the mass, because we generally assume that the mass of each object does not change during the collision. Finally, Equation 16.3 is not restricted to two objects. If, for example, there was a collision between three objects, we could add the third object's momentum to both the right and left side of the equation.

In this lab, students explore a *perfectly inelastic collision,* or a collision where the objects stick together after they collide. Mathematically, this type of collision is represented in Equation 16.4.

$$\text{(Equation 16.4)} \quad m_1\mathbf{v}_1 + m_2\mathbf{v}_2 = (m_1 + m_2)\mathbf{v}'$$

Because the objects stick together, we can treat them as one object with a mass equal to the sum of the masses of the two individual objects. Finally, they will both move together after the collision with the same velocity. If we assume that the second object is stationary before the collision, then Equation 16.4 becomes Equation 16.5.

$$\text{(Equation 16.5)} \quad m_1\mathbf{v}_1 = (m_1 + m_2)\mathbf{v}'$$

This means that before the collision only the first object has momentum. After the collision, when both objects stick together, they both have momentum and move together with the same velocity. We can rearrange this relationship to find the answer to the guiding question for the two relevant variables at the heart of this investigation.

The first task requires students to determine the effect of increased velocity for object 1 on the velocity of both objects after the collision. Using Equation 16.5, we find a linear relationship between the two velocities, given by the equation $\mathbf{v}' = (\frac{m_1}{m_1 + m_2})\mathbf{v_1}$. As the velocity of the first object increases, the velocity of both objects after the collision will increase linearly, with the slope of the line being equal to $(\frac{m_1}{m_1 + m_2})$.

The second task requires students to determine the effect of increasing the mass of the incoming object on the resultant velocity of the two objects after the collision. The equation governing this relationship remains $\mathbf{v}' = (\frac{m_1}{m_1 + m_2})\mathbf{v_1}$. However, in this case, $\mathbf{v}_1$ is constant. When increasing the mass of the incoming object, the resultant velocity of the two objects after the collision will also increase, but this relationship is not linear. If students were to graph this relationship, they would find that $\mathbf{v}'$ approaches $\mathbf{v}_1$ but that there is a horizontal asymptote at this point.

Timeline

The instructional time needed to complete this lab investigation is 220–280 minutes. Appendix 3 (p. 531) provides options for implementing this lab investigation over several class periods. Option A (280 minutes) should be used if students are unfamiliar with scientific writing, because this option provides extra instructional time for scaffolding the writing process. You can scaffold the writing process by modeling, providing examples, and providing hints as students write each section of the report. Option A can also be used if students are unfamiliar with any of the data collection and analysis tools. Option B (220 minutes) should be used if students are familiar with scientific writing and have developed the skills needed to write an investigation report on their own. In option F, students complete stage 6 (writing the investigation report) and stage 8 (revising the investigation report) as homework.

Materials and Preparation

The materials needed to implement this investigation are listed in Table 16.1. The items can be purchased from a science supply company such as Carolina, Flinn Scientific, PASCO, Vernier, or Ward's Science. The data collection and analysis software and/or the video analysis software) can be purchased from Vernier (Logger *Pro*) or PASCO Scientific (SPARKvue or Capstone). These companies also have apps that can be used on Apple- or Android-based tablets and cell phones. We recommend consulting with your school's information technology coordinator to determine the best option for your students.

TABLE 16.1

Materials list for Lab 16

Item	Quantity
Safety glasses or goggles	1 per student
Dynamics cart (with Velcro or magnetic bumpers)	2 per group
Dynamics track	1 per group
Motion detector/sensor	2 per group
Interface for motion detector/sensor (if used)	1 per group
Video camera	1 per group
Computer or tablet with data collection and analysis software (for use with motion detector/sensor) or video analysis software (for use with the video camera)	1 per group
Electronic or triple beam balance	1 per group
Cart picket fence	1 per group
Mass set	1 per group
Stopwatch	2–3 per group
Meterstick or ruler	1 per group
Whiteboard, 2' × 3'*	1 per group
Lab Handout	1 per student
Investigation Proposal C (optional)	2 per group
Peer-review guide and teacher scoring rubric	1 per student
Checkout Questions	1 per student

* As an alternative, students can use computer and presentation software such as Microsoft PowerPoint or Apple Keynote to create their arguments.

Be sure to use a set routine for distributing and collecting the materials during the lab investigation. One option is to set up the materials for each group at each group's lab station before class begins. This option works well when there is a dedicated section of the classroom for lab work and the materials are large and difficult to move (such as a dynamics track). A second option is to have all the materials on a table or cart at a central location. You can then assign a member of each group to be the "materials manager." This individual is responsible for collecting all the materials his or her group needs from the table or cart during class and for returning all the materials at the end of the class. This option works well when the materials are small and easy to move (such as stopwatches,

metersticks, or hanging masses). It also makes it easy to inventory the materials at the end of the class before students leave for the day.

Safety Precautions

Remind students to follow all normal lab safety rules. In addition, tell students to take the following safety precautions:

1. Wear sanitized safety glasses or goggles during lab setup, hands-on activity, and takedown.

2. Keep fingers and toes out of the way of moving objects.

3. Wash hands with soap and water after completing the lab.

Topics for the Explicit and Reflective Discussion

Reflecting on the Use of Core Ideas and Crosscutting Concepts During the Investigation

Teachers should begin the explicit and reflective discussion by asking students to discuss what they know about the core idea they used during the investigation. The following are some important concepts related to the core idea of forces and motion that students will need to explain what happens when two objects collide and stick together:

- A reference frame is needed to determine the direction and the magnitude of the displacement, velocity, and acceleration of an object.

- Force is considered a vector quantity because a force has both magnitude and direction.

- A *system* is an object or a collection of objects. An object is treated as if it has no internal structure.

- The boundary between a system and its environment is a decision made by the person considering the situation in order to simplify or otherwise assist in analysis.

- Conserved quantities are constant in an isolated or a closed system. An open system is one that exchanges any conserved quantity with its surroundings.

- Linear momentum is conserved in all systems.

To help students reflect on what they know about forces and motion, we recommend showing them two or three images using presentation software that help illustrate these important ideas. You can then ask the students the following questions to encourage them to share how they are thinking about these important concepts:

1. What do we see going on in this image?

2. Does anyone have anything else to add?

3. What might be going on that we can't see?

4. What are some things that we are not sure about here?

You can then encourage students to think about how CCs played a role in their investigation. There are at least two CCs that students need to explain what happens when two objects collide and stick together: (a) Scale, Proportion, and Quantity; and (b) Energy and Matter: Flows, Cycles, and Conservation (see Appendix 2 [p. 527] for a brief description of these CCs). To help students reflect on what they know about these CCs, we recommend asking them the following questions:

1. Why is it important keep track of changes in a system quantitatively during an investigation?

2. What did you keep track of quantitatively during your investigation? What did that allow you to do?

3. Scientists often attempt to track how energy and matter move into, out of, and within systems as part of an investigation. Why is this useful?

4. How did you attempt to track how matter moves within the system you were studying? What did tracking the movement of matter allow you to do during your investigation?

You can then encourage the students to think about how they used all these different concepts to help answer the guiding question and why it is important to use these ideas to help justify their evidence for their final arguments. Be sure to remind your students to explain why they included the evidence in their arguments and make the assumptions underlying their analysis and interpretation of the data explicit in order to provide an adequate justification of their evidence.

Reflecting on Ways to Design Better Investigations

It is important for students to reflect on the strengths and weaknesses of the investigation they designed during the explicit and reflective discussion. Students should therefore be encouraged to discuss ways to eliminate potential flaws, measurement errors, or sources of uncertainty in their investigations. To help students be more reflective about the design of their investigation and what they can do to make their investigations more rigorous in the future, you can ask them the following questions:

1. What were some of the strengths of the way you planned and carried out your investigation? In other words, what made it scientific?

2. What were some of the weaknesses of the way you planned and carried out your investigation? In other words, what made it less scientific?

3. What rules can we make, as a class, to ensure that our next investigation is more scientific?

Reflecting on the Nature of Scientific Knowledge and Scientific Inquiry

This investigation can be used to illustrate two important concepts related to the nature of scientific knowledge and the nature of scientific inquiry: (a) the difference between observations and inferences in science and (b) how the culture of science, societal needs, and current events influence the work of scientists (see Appendix 2 for a brief description of these two concepts). Be sure to review these concepts during and at the end of the explicit and reflective discussion. To help students think about these concepts in relation to what they did during the lab, you can ask them the following questions:

1. You had to make observations and inferences during your investigation. Can you give me some examples of these observations and inferences?

2. Can you work with your group to come up with a rule that you can use to decide if a piece of information is an observation or an inference? Be ready to share in a few minutes.

3. People view some types of research as being more important than other types of research because of cultural values and current events. Can you come up with some examples of how cultural values and current events have influenced the work of scientists?

4. Scientists share a set of values, norms, and commitments that shape what counts as knowing, how to represent or communicate information, and how to interact with other scientists. Can you work with your group to come up with a rule that you can use to decide if something is science or not science? Be ready to share in a few minutes.

You can also use presentation software or other techniques to encourage your students to think about these concepts. You can show examples of information from the investigation that are either observations or inferences and ask students to classify each example and explain their thinking. You can also show examples of research projects that were influenced by cultural values and current events and ask students to think about what was going on at the time and why that research was viewed as being important for the greater good.

Be sure to remind your students that, to be proficient in science, it is important that they understand what counts as scientific knowledge and how that knowledge develops over time.

Linear Momentum and Collisions

When Two Objects Collide and Stick Together, How Do the Initial Velocity and Mass of One of the Moving Objects Affect the Velocity of the Two Objects After the Collision?

Hints for Implementing the Lab

- In this lab, students will have to conduct two different experiments to determine (1) how changing the velocity of a moving object affects the velocity of that object and a second object after a collision, when holding mass constant; and (2) how changing the mass of the moving object affects the velocity of both objects after the collision, when holding the incoming velocity of the moving object constant.

- Allowing students to design their own procedures for collecting data gives students an opportunity to try, to fail, and to learn from their mistakes. However, you can scaffold students as they develop their procedure by having them fill out an investigation proposal. These proposals provide a way for you to offer students hints and suggestions without telling them how to do it. You can also check the proposals quickly during a class period. We recommend having students fill out a proposal for each experiment. For this lab we suggest using Investigation Proposal C.

- To determine the velocity of the carts, we recommend using a motion detector/ sensor attached to the track as shown in Figure L16.3 in the Lab Handout (p. 357); as an alternative, students can conduct a video analysis of the motion of the carts.

- If students use motion detectors/sensors, they can use either one or two motion detectors (see Figure L16.3 in the Lab Handout). If they use only one motion detector, they will want to make sure that it is on the side of the track with the cart that has a non-zero initial velocity (see the next hint). If students use two motion detectors (one for each cart), after the collision, they will record equal but opposite values for the velocity of the system. This is because the velocity after the collision will be away from the first motion detector (i.e., a positive direction) and toward the second motion detector (i.e., a negative direction). This provides an opportunity to reinforce the importance of defining positive and negative reference frames.

- Although it is not necessary, we recommend having the second object remain stationary during the collision. This way, the value of $m_2\mathbf{v}_2$ in equation 16.4 is zero. In other words, the easiest way to control for the momentum of the second object is for the momentum of the second object to be equal to zero. Because the second object has a mass greater than zero, the velocity of the second object must be zero if the momentum of the object is zero.

- When adding masses to the cart to investigate the relationship between the increased mass and the resulting velocity, it is important to add sufficient mass relative to the mass of the cart. For example, if the mass of each cart is 100 g, adding an additional 1 g mass will not have much of an effect on the outcome.

- Learn how to use the dynamics track before the lab begins. It is important for you to know how to use the equipment so you can help students when technical issues arise. Some collision tracks require pressurized air; others use magnetic fields, which require access to power outlets, and others just have carts on wheels that have a very small coefficient of friction.

- Allow the students to become familiar with the dynamics track as part of the tool talk before they begin to design their investigation. Giving them 5–10 minutes to examine the equipment will let them see what they can and cannot do with it. This also gives students a chance to begin thinking about what variables to test and how to control for other variables.

- Be sure to allow students to go back and re-collect data at the end of the argumentation session. Students often realize that they made numerous mistakes when they were collecting data as a result of their discussions during the argumentation session. The students, as a result, will want a chance to re-collect data, and the re-collection of data should be encouraged when time allows. This also offers an opportunity to discuss what scientists do when they realize a mistake is made inside the lab.

If students use a motion detector/sensor

- Learn how to use the motion detector/sensor and the data collection and analysis software before the lab begins. It is important for you to know how to use the equipment and software so you can help students when technical issues arise.

- Allow the students to become familiar with the motion detector/sensor and data collection and analysis software as part of the tool talk before they begin to design their investigation. Giving them 5–10 minutes to examine the equipment and software will let them see what they can and cannot do with it.

If students use video analysis

- We suggest allowing students to familiarize themselves with the video analysis software before they finalize the procedure for the investigation, especially if they have not used such software previously. This gives students an opportunity to learn how to work with the software and to improve the quality of the video they take.

- Remind students to hold the video camera as still as possible. Any movement of the camera will introduce error into their analysis. If using actual camcorders, we recommend using a tripod to hold the camera steady. If students are using a camera on a cell phone or tablet, we recommend using a table to help steady the camera.

- Remind students to place a meterstick in the same field of view as the motion they are capturing with the video camera. Also, the meterstick should be approximately the same distance from the camera as the motion. Most video analysis software

Linear Momentum and Collisions

When Two Objects Collide and Stick Together, How Do the Initial Velocity and Mass of One of the Moving Objects Affect the Velocity of the Two Objects After the Collision?

requires the user to define a scale in the video (this allows the software to establish distances and, subsequently, other variables dependent on distance and displacement).

Connections to Standards

Table 16.2 highlights how the investigation can be used to address specific performance expectations from the *NGSS;* learning objectives from AP Physics 1; learning objectives from AP Physics C: Mechanics, *Common Core State Standards,* in English language arts (*CCSS ELA*); and *Common Core State Standards,* Mathematics (*CCSS Mathematics*).

TABLE 16.2

Lab 16 alignment with standards

NGSS performance expectation	• HS-PS2-2: Use mathematical representations to support the claim that the total momentum of a system of objects is conserved when there is no net force on the system.
AP Physics 1 learning objectives	• 3.A.1.1: The student is able to express the motion of an object using narrative, mathematical, and graphical representations. • 3.A.1.2: The student is able to design an experimental investigation of the motion of an object. • 3.A.1.3: The student is able to analyze experimental data describing the motion of an object and is able to express the results of the analysis using narrative, mathematical, and graphical representations. • 4.B.1.1: The student is able to calculate the change in linear momentum of a two-object system with constant mass in linear motion from a representation of the system (data, graphs, etc.). • 4.B.1.2: The student is able to analyze data to find the change in linear momentum for a constant-mass system using the product of the mass and the change in velocity of the center of mass. • 5.A.2.1: The student is able to define open and closed/isolated systems for everyday situations and apply conservation concepts for energy, charge, and linear momentum to those situations. • 5.D.1.1: The student is able to make qualitative predictions about natural phenomena based on conservation of linear momentum and restoration of kinetic energy in elastic collisions. • 5.D.1.4: The student is able to design an experimental test of an application of the principle of the conservation of linear momentum, predict an outcome of the experiment using the principle, analyze data generated by that experiment whose uncertainties are expressed numerically, and evaluate the match between the prediction and the outcome. • 5.D.2.2: The student is able to plan data collection strategies to test the law of conservation of momentum in a two-object collision that is elastic or inelastic and analyze the resulting data graphically. • 5.D.2.4: The student is able to analyze data that verify conservation of momentum in collisions with and without an external friction force.

Table 16.2 (*continued*)

AP Physics C: Mechanics learning objectives	• I.D.2.a: Relate mass, velocity, and linear momentum for a moving object, and calculate the total linear momentum of a system of objects. • I.D.3.a(2): Identify situations in which linear momentum, or a component of the linear momentum vector, is conserved. • I.D.3.a(3): Apply linear momentum conservation to one-dimensional elastic and inelastic collisions and two-dimensional completely inelastic collisions.
Literacy connections (*CCSS ELA*)	• *Reading:* Key ideas and details, craft and structure, integration of knowledge and ideas • *Writing:* Text types and purposes, production and distribution of writing, research to build and present knowledge, range of writing • *Speaking and listening:* Comprehension and collaboration, presentation of knowledge and ideas
Mathematics connections (*CCSS Mathematics*)	• *Mathematical practices:* Reason abstractly and quantitatively, construct viable arguments and critique the reasoning of others, use appropriate tools strategically, attend to precision • *Number and quantity:* Reason quantitatively and use units to solve problems, represent and model with vector quantities, perform operations on vectors • *Algebra:* Interpret the structure of expressions, understand solving equations as a process of reasoning and explain the reasoning, solve equations and inequalities in one variable, represent and solve equations and inequalities graphically • *Functions:* Understand the concept of a function and use function notation, interpret functions that arise in applications in terms of the context, analyze functions using different representations, interpret expressions for functions in terms of the situation they model • *Statistics and probability:* Summarize, represent, and interpret data on two categorical and quantitative variables; understand and evaluate random processes underlying statistical experiments; make inferences and justify conclusions from sample surveys, experiments, and observational studies

Linear Momentum and Collisions

When Two Objects Collide and Stick Together, How Do the Initial Velocity and Mass of One of the Moving Objects Affect the Velocity of the Two Objects After the Collision?

Lab Handout

Lab 16. Linear Momentum and Collisions: When Two Objects Collide and Stick Together, How Do the Initial Velocity and Mass of One of the Moving Objects Affect the Velocity of the Two Objects After the Collision?

Introduction

The incidence of traumatic brain injury (TBI) is on the rise in the United States (CDC 2016). At least 1.7 million TBIs occur every year in this country, and these injuries are a contributing factor in about a third (30.5%) of all injury-related deaths (Faul et al. 2010). Adolescents (ages 15–19 years), older adults (ages 65 years and older), and males across all age groups are most likely to sustain a TBI (Faul et al. 2010). A TBI is caused by a bump, blow, or jolt to the head, although not all such events result in a TBI. TBIs can range from mild to severe, but most TBIs are mild and are commonly called concussions (CDC 2016). A single concussion, however, can cause temporary memory loss, confusion, and impaired vision or hearing. It can also cause depression, anxiety, and mood swings. Multiple concussions can make these symptoms permanent.

A concussion is caused when a person's brain hits the inside of the skull. This can occur when a moving person hits a stationary object (like running into a glass door), when a stationary person is hit by a moving object (like a person getting hit in the head by a baseball), or when two people collide with each other (like two people running into each other while playing a sport). As can be seen in Figure L16.1, when the skull is jolted too fast or is impacted by something, the brain shifts and hits against the skull. The "harder" the brain collides with the inside of the skull, the more severe the concussion.

Scientists have shown that the momentum of a collision is related to the severity of a concussion. Consider for example, what happens when a person is hit in the head with a ball. This often happens to softball players, baseball players, and soccer players. When a person is hit in the head by a moving ball, the severity of the concussion will be related to the momentum of that moving ball. Momentum is a function of both the mass of an object and its velocity. As an object accelerates, its

FIGURE L16.1

Movement of the brain during a concussive event

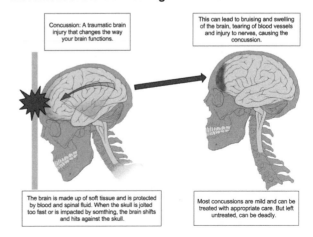

Concussion: A traumatic brain injury that changes the way your brain functions.

This can lead to bruising and swelling of the brain, tearing of blood vessels and injury to nerves, causing the concussion.

The brain is made up of soft tissue and is protected by blood and spinal fluid. When the skull is jolted too fast or is impacted by somthing, the brain shifts and hits against the skull.

Most concussions are mild and can be treated with appropriate care. But left untreated, can be deadly.

momentum increases. And, for two objects moving at the same velocity, the object with the greater mass will have a greater momentum.

Scientists have been studying collisions between two objects, such as cars, for some time. More and more scientists, however, are now studying the types of collisions that happen during different kinds of sports to better protect athletes from concussions. One type of collision that appears to be related to a high incidence of concussion is tackling in football. Tackling often results in two bodies staying together after the collision; Figure L16.2 shows one example of this. In this example, a defensive player (who is moving) collides with the quarterback (who is stationary). The defensive player holds on to the quarterback after the initial collision so they stay together as they fall to the ground. One goal of this line of research is to determine how the velocity of the moving object (or an athlete) before the collision affects the velocity of the two objects stuck together after the collision. In this investigation, you will have an opportunity to explore this relationship.

A collision between two football players

Your Task

Use what you know about momentum, collisions, systems, tracking the movement of matter within systems, and the importance of considering issues related to scale, proportion, and quantity to design and carry out an investigation that will allow you to understand what happens when two objects collide and stick together.

The guiding question of this investigation is, *When two objects collide and stick together, how does the initial velocity and mass of one of the moving objects affect the velocity of the two objects after the collision?*

Linear Momentum and Collisions

When Two Objects Collide and Stick Together, How Do the Initial Velocity and Mass of One of the Moving Objects Affect the Velocity of the Two Objects After the Collision?

Materials

You may use any of the following materials during your investigation (some items may not be available):

- Safety glasses or goggles (required)
- 2 Dynamics carts (with Velcro or magnetic bumpers)
- Dynamics track
- Motion detector/sensor and interface
- Video camera
- Computer or tablet with data collection and analysis software and/or video analysis software

- Electronic or triple beam balance
- Cart picket fence
- Mass set
- Stopwatch
- Meterstick or ruler

Safety Precautions

Follow all normal lab safety rules. In addition, take the following safety precautions:

1. Wear sanitized safety glasses or goggles during lab setup, hands-on activity, and takedown.

2. Keep fingers and toes out of the way of moving objects.

3. Wash your hands with soap and water after completing the lab.

Investigation Proposal Required?   ☐ Yes   ☐ No

Getting Started

To answer the guiding question, you will need to design and carry out at least two different experiments. First, you will need to determine how changing the initial velocity of a moving object affects the velocity of the two objects after the collision. Next, you will need to determine how changing the mass of the moving object affects the velocity of the two objects after the collision. Figure L16.3 shows how you can use motion detectors/sensors to measure the velocity of a moving object (in this case, a dynamics cart) before and after a collision. The velocity of the moving object can also be measured by using a video camera and video analysis software. Before you can design your two experiments, however, you must determine what type of data you need to collect, how you will collect it, and how you will analyze it.

FIGURE L16.3 _____

One way to measure the velocity of a moving object before and after a collision

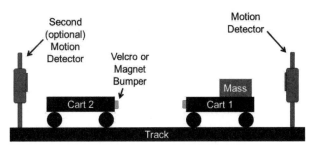

To determine *what type of data you need to collect*, think about the following questions:

- What are the boundaries and components of the system you are studying?
- How do components of the system under study interact?
- How will you track the movement of matter within this system?
- How could you keep track of changes in this system quantitatively?
- What factors affects the momentum of an object?
- How will you determine the velocity of each object?
- What will be the independent variable and the dependent variable for each experiment?

To determine *how you will collect the data*, think about the following questions:

- What other factors will you need to control or measure during each experiment?
- Which quantities are vectors, and which quantities are scalars?
- For any vector quantities, which directions are positive and which directions are negative?
- What scale or scales should you use to take your measurements?
- What equipment will you need to collect the measurements you need?
- How will you make sure that your data are of high quality (i.e., how will you reduce error)?
- How will you keep track of and organize the data you collect?

To determine *how you will analyze the data*, think about the following questions:

- What type of calculations will you need to do?
- What types of patterns might you look for as you analyze your data?
- Are there any proportional relationships you can identify?
- What types of comparisons will be useful to make?
- What type of table or graph could you create to help make sense of your data?

Connections to the Nature of Scientific Knowledge and Scientific Inquiry

As you work through your investigation, you may want to consider

- the difference between observations and inferences in science, and
- how the culture of science, societal needs, and current events influence the work of scientists.

Linear Momentum and Collisions

When Two Objects Collide and Stick Together, How Do the Initial Velocity and Mass of One of the Moving Objects Affect the Velocity of the Two Objects After the Collision?

Initial Argument

Once your group has finished collecting and analyzing your data, your group will need to develop an initial argument. Your initial argument needs to include a claim, evidence to support your claim, and a justification of the evidence. The *claim* is your group's answer to the guiding question. The *evidence* is an analysis and interpretation of your data. Finally, the *justification* of the evidence is why your group thinks the evidence matters. The justification of the evidence is important because scientists can use different kinds of evidence to support their claims. Your group will create your initial argument on a whiteboard. Your whiteboard should include all the information shown in Figure L16.4.

FIGURE L16.4

Argument presentation on a whiteboard

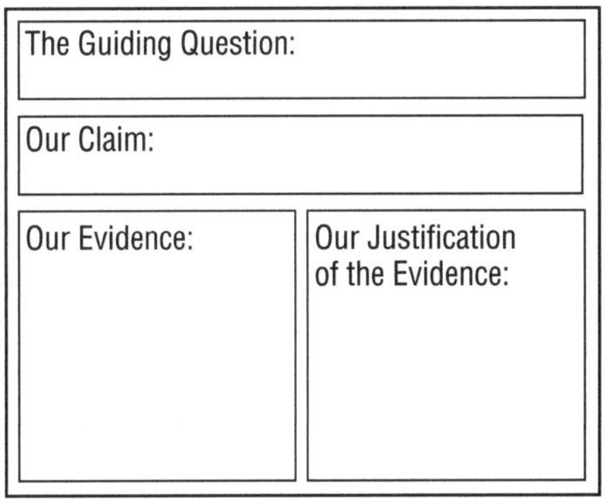

The Guiding Question:	
Our Claim:	
Our Evidence:	Our Justification of the Evidence:

Argumentation Session

The argumentation session allows all of the groups to share their arguments. One or two members of each group will stay at the lab station to share that group's argument, while the other members of the group go to the other lab stations to listen to and critique the other arguments. This is similar to what scientists do when they propose, support, evaluate, and refine new ideas during a poster session at a conference. If you are presenting your group's argument, your goal is to share your ideas and answer questions. You should also keep a record of the critiques and suggestions made by your classmates so you can use this feedback to make your initial argument stronger. You can keep track of specific critiques and suggestions for improvement that your classmates mention in the space below.

Critiques about our initial argument and suggestions for improvement:

LAB 16

If you are critiquing your classmates' arguments, your goal is to look for mistakes in their arguments and offer suggestions for improvement so these mistakes can be fixed. You should look for ways to make your initial argument stronger by looking for things that the other groups did well. You can keep track of interesting ideas that you see and hear during the argumentation in the space below. You can also use this space to keep track of any questions that you will need to discuss with your team.

Interesting ideas from other groups or questions to take back to my group:

Once the argumentation session is complete, you will have a chance to meet with your group and revise your initial argument. Your group might need to gather more data or design a way to test one or more alternative claims as part of this process. Remember, your goal at this stage of the investigation is to develop the best argument possible.

Report

Once you have completed your research, you will need to prepare an *investigation report* that consists of three sections. Each section should provide an answer to the following questions:

1. What question were you trying to answer and why?

2. What did you do to answer your question and why?

3. What is your argument?

Linear Momentum and Collisions

When Two Objects Collide and Stick Together, How Do the Initial Velocity and Mass of One of the Moving Objects Affect the Velocity of the Two Objects After the Collision?

Your report should answer these questions in two pages or less. This report must be typed, and any diagrams, figures, or tables should be embedded into the document. Be sure to write in a persuasive style; you are trying to convince others that your claim is acceptable or valid!

References

Centers for Disease Control and Prevention. 2016. TBI: Get the facts. *www.cdc.gov/traumaticbraininjury/get_the_facts.html*.

Faul, M., L. Xu, M. M. Wald, and V. G. Coronado. 2010. Traumatic brain injury in the United States: Emergency department visits, hospitalizations and deaths. Atlanta, GA: Centers for Disease Control and Prevention, National Center for Injury Prevention and Control.

LAB 16

Lab 16. Linear Momentum and Collisions: When Two Objects Collide and Stick Together, How Do the Initial Velocity and Mass of One of the Moving Objects Affect the Velocity of the Two Objects After the Collision?

The images below show the motion of two carts on a track before they collide with each other. Assume that both carts stick together after the collision. Use this information to answer questions 1 and 2.

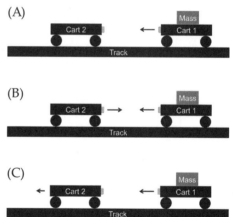

1. How would the magnitude of the velocity of the carts after the collision in situation A compare with the magnitude of the velocity of the carts after the collision in situation B? For situation B, assume the magnitude of the velocity for cart 1 equals the magnitude of the velocity for cart 2.

   a. The velocity will be greater in A than in B.
   b. The velocity will be less in A than in B.
   c. The velocity will be equal in A and B.

   How do you know?

2. How would the magnitude of the velocity of the carts after the collision in situation A compare with the magnitude of the velocity of the carts after the collision in situation C? For situation C, assume the magnitude of the velocity for cart 1 is greater than the magnitude of the velocity for cart 2.

   a. The velocity will be greater in A than in C.
   b. The velocity will be less in A than in C.
   c. The velocity will be equal in A and C.

How do you know?

3. The mass of the carts did not change while they were moving during your investigation. Are there instances where the mass of a moving object changes as it moves?

   a. Yes

   b. No

Explain your answer using an example.

4. How does decreasing the mass of a moving object as it moves affect the momentum of that object?

   a. It decreases the momentum of the object.

   b. It increases the momentum of the object.

   c. It has no effect on the momentum of the object.

How do you know?

5. In science, there is a difference between inferences and observations.

   a. I agree with this statement.
   b. I disagree with this statement.

   Explain your answer, using an example from your investigation about linear momentum and collisions.

6. Scientists share a set of values, norms, and commitments that shape what counts as knowing, how to represent or communicate information, and how to interact with other scientists.

   a. I agree with this statement.
   b. I disagree with this statement.

   Explain your answer, using an example from your investigation about linear momentum and collisions.

7. Scientists often need to need to define the system under study as part of the investigation. Explain why this is useful to do, using an example from your investigation about linear momentum and collisions.

8. Scientists often need to track how matter moves within a system. Explain why this is useful to do, using an example from your investigation about linear momentum and collisions.

9. Scientists often focus on proportional relationships. Explain what a proportional relationship is and why these relationships are useful, using an example from your investigation about linear momentum and collisions.

LAB 17

Teacher Notes

Lab 17. Impulse and Momentum: How Does Changing the Magnitude and Duration of a Force Acting on an Object Affect the Momentum of That Object?

Purpose

The purpose of this lab is to *introduce* students to the core idea of forces and motion, part of the disciplinary core idea (DCI) of Motion and Stability: Forces and Interactions from the *NGSS*, by giving students an opportunity to explore impulses and momentum. This lab also gives students an opportunity to learn about the crosscutting concepts (CCs) of (a) Scale, Proportion, and Quantity; and (b) Energy and Matter: Flows, Cycles, and Conservation from the *NGSS*. In addition, this lab can be used to help students understand three big ideas from AP Physics: (a) the interactions of an object with other objects can be described by forces, (b) interactions between systems can result in changes in those systems, and (c) changes that occur as result of interactions are constrained by conservation laws. As part of the explicit and reflective discussion, students will also learn about (a) the difference between laws and theories in science and (b) the difference between data and evidence in science.

Underlying Physics Concepts

When a force, $\mathbf{F}$, is applied to an object, it acts for some time, Δt. The product, $\mathbf{F}\Delta t$, is defined as an *impulse*, $\mathbf{J}$, in Equation 17.1. In SI units, force is measured in newtons (N), time in seconds (s), and impulse in newton seconds (N·s).

$$\text{(Equation 17.1)} \quad \mathbf{J} = \mathbf{F}\Delta t$$

As we know from Newton's second law, an unbalanced force causes an object to accelerate. An impulse can cause a change in an object's momentum, as shown in Equation 17.2, where $\mathbf{J}$ is impulse and $\mathbf{p}$ is momentum (we use $\mathbf{p}$ for momentum because m is already used for mass). In SI units, momentum is measured in kilogram-meters per second (kg·m/s).

$$\text{(Equation 17.2)} \quad \mathbf{J} = \Delta\mathbf{p}$$

An object's momentum is equal to the product of its mass, m, and velocity, $\mathbf{v}$, as shown in Equation 17.3. In SI units, mass is measured in kilograms (kg) and velocity is measured in meters per second (m/s).

$$\text{(Equation 17.3)} \quad \mathbf{p} = m\mathbf{v}$$

Taken together, Equations 17.2 and 17.3 imply that, for an object of constant mass, an impulse has the effect of changing the velocity. We can combine Equations 17.1, 17.2, and 17.3 to write an equation in terms of observable quantities, shown in Equation 17.4.

$$\textbf{(Equation 17.4)} \quad \textbf{F}\Delta t = m\Delta \textbf{v}$$

Dividing both sides by Δt will yield the familiar version of Newton's second law (i.e., $\textbf{F} = m\textbf{a}$), though something similar to Equation 17.4 was how Newton actually presented the second law in his famous work *Principia Mathematica* (1687). That is, instead of saying that a force will cause an object to accelerate, Newton's description of his second law is more accurately translated from Latin as stating that the change in momentum of a body is proportional to the impulse acting on the body.

Note that impulse and momentum are vector quantities. As such, the direction of an impulse causes a momentum change in the same direction (see Equation 17.2). Equation 17.4 illustrates this more clearly by showing that the direction of the velocity's *change* comes from the direction of the force applied. Therefore, a force directed to the left can cause an increase in leftward velocity (i.e., increase in momentum) *or* a decrease in rightward velocity (i.e., decreased momentum).

In this investigation, students must determine the effect of increased force on the change in the momentum of the object and the effect of a longer duration of an applied force on the momentum of an object. Assuming that the initial momentum and the force are in the same direction, then an increased force will cause a larger change in momentum over a constant amount of time. And, if the force is held constant, a longer duration will lead to a greater change in momentum.

Timeline

The instructional time needed to complete this lab investigation is 200–280 minutes. Appendix 3 (p. 531) provides options for implementing this lab investigation over several class periods. Option E (280 minutes) should be used if students are unfamiliar with scientific writing, because this option provides extra instructional time for scaffolding the writing process. You can scaffold the writing process by modeling, providing examples, and providing hints as students write each section of the report. Option E can also be used if students are unfamiliar with any of the data collection and analysis tools. Option F (200 minutes) should be used if students are familiar with scientific writing and have developed the skills needed to write an investigation report on their own. In option F, students complete stage 6 (writing the investigation report) and stage 8 (revising the investigation report) as homework.

LAB 17

Materials and Preparation

The materials needed to implement this investigation are listed in Table 17.1. The items can be purchased from a science supply company such as Carolina, Flinn Scientific, PASCO, Vernier, or Ward's Science. The data collection and analysis software and/or the video analysis software) can be purchased from Vernier (Logger *Pro*) or PASCO (SPARKvue or Capstone). These companies also have apps that can be used on Apple- or Android-based tablets and cell phones. We recommend consulting with your school's information technology coordinator to determine the best option for your students.

TABLE 17.1

Materials list for Lab 17

Item	Quantity
• Safety glasses or goggles	1 per student
• Dynamics cart	1 per group
• Fan attachment for cart, with variable speed and duration settings	1 per group
• Dynamics track	1 per group
• Motion detector/sensor	1 per group
• Interface for motion detector/sensor (if used)	1 per group
• Video camera	1 per group
• Computer or tablet with data collection and analysis software (for use with motion detector/sensor) or video analysis software (for use with video camera)	1 per group
• Electronic or triple beam balance	1 per group
• Stopwatch	1 per group
• Meterstick or ruler	1 per group
• Investigation Proposal C (optional)	2 per group
• Whiteboard, 2' × 3'*	1 per group
• Lab Handout	1 per student
• Peer-review guide and teacher scoring rubric	1 per student
• Checkout Questions	1 per student

* As an alternative, students can use computer and presentation software such as Microsoft PowerPoint or Apple Keynote to create their arguments.

Be sure to use a set routine for distributing and collecting the materials during the lab investigation. One option is to set up the materials for each group at each group's lab

station before class begins. This option works well when there is a dedicated section of the classroom for lab work and the materials are large and difficult to move (such as a dynamics track). A second option is to have all the materials on a table or cart at a central location. You can then assign a member of each group to be the "materials manager." This individual is responsible for collecting all the materials his or her group needs from the table or cart during class and for returning all the materials at the end of the class. This option works well when the materials are small and easy to move (such as stopwatches, metersticks, or hanging masses). It also makes it easy to inventory the materials at the end of the class before students leave for the day.

Safety Precautions

Remind students to follow all normal lab safety rules. In addition, tell students to take the following safety precautions:

1. Wear sanitized safety glasses or goggles during lab setup, hands-on activity, and takedown.

2. Keep fingers and toes out of the way of moving objects.

3. Wash hands with soap and water after completing the lab.

Topics for the Explicit and Reflective Discussion

Reflecting on the Use of Core Ideas and Crosscutting Concepts During the Investigation

Teachers should begin the explicit and reflective discussion by asking students to discuss what they know about the core idea they used during the investigation. The following are some important concepts related to the core idea of forces and motion that students will need to explain how the momentum of an object will change in response to an impulse:

- A reference frame is needed to determine the direction and the magnitude of the displacement, velocity, and acceleration of an object.

- Force is considered a vector quantity because a force has both magnitude and direction.

- A *system* is an object or a collection of objects. An object is treated as if it has no internal structure.

- The boundary between a system and its environment is a decision made by the person considering the situation in order to simplify or otherwise assist in analysis.

- An *impulse* is the product of the average force acting on an object and the time interval during which the interaction occurred.

LAB 17

To help students reflect on what they know about forces and motion, we recommend showing them two or three images using presentation software that help illustrate these important ideas. You can then ask the students the following questions to encourage them to share how they are thinking about these important concepts:

1. What do we see going on in this image?

2. Does anyone have anything else to add?

3. What might be going on that we can't see?

4. What are some things that we are not sure about here?

You can then encourage students to think about how CCs played a role in their investigation. There are at least two CCs that students need to explain how the momentum of an object will change in response to an impulse: (a) Scale, Proportion, and Quantity; and (b) Energy and Matter: Flows, Cycles, and Conservation (see Appendix 2 [p. 527] for a brief description of these CCs). To help students reflect on what they know about these CCs, we recommend asking them the following questions:

1. Why is it important to keep track of changes in a system quantitatively during an investigation?

2. What did you keep track of quantitatively during your investigation? What did that allow you to do?

3. Scientists often attempt to track how energy and matter move into, out of, and within systems as part of an investigation. Why is this useful?

4. How did you attempt to track how matter moves within the system you were studying? What did tracking the movement of matter allow you to do during your investigation?

You can then encourage the students to think about how they used all these different concepts to help answer the guiding question and why it is important to use these ideas to help justify their evidence for their final arguments. Be sure to remind your students to explain why they included the evidence in their arguments and make the assumptions underlying their analysis and interpretation of the data explicit in order to provide an adequate justification of their evidence.

Reflecting on Ways to Design Better Investigations

It is important for students to reflect on the strengths and weaknesses of the investigation they designed during the explicit and reflective discussion. Students should therefore be encouraged to discuss ways to eliminate potential flaws, measurement errors, or sources of uncertainty in their investigations. To help students be more reflective about the design

of their investigation and what they can do to make their investigations more rigorous in the future, you can ask them the following questions:

1. What were some of the strengths of the way you planned and carried out your investigation? In other words, what made it scientific?

2. What were some of the weaknesses of the way you planned and carried out your investigation? In other words, what made it less scientific?

3. What rules can we make, as a class, to ensure that our next investigation is more scientific?

Reflecting on the Nature of Scientific Knowledge and Scientific Inquiry

This investigation can be used to illustrate two important concepts related to the nature of scientific knowledge and the nature of scientific inquiry: (a) the difference between laws and theories in science and (b) the difference between data and evidence in science (see Appendix 2 for a brief description of these two concepts. Be sure to review these concepts during and at the end of the explicit and reflective discussion. To help students think about these concepts in relation to what they did during the lab, you can ask them the following questions:

1. Laws and theories are different in science. Is $\mathbf{F}\Delta t = m\Delta\mathbf{v}$ an example of a theory or a law? Why?

2. Can you work with your group to come up with a rule that you can use to decide if something is a theory or a law? Be ready to share in a few minutes.

3. You had to talk about data and evidence during your investigation. Can you give me some examples of data and evidence from your investigation?

4. Can you work with your group to come up with a rule that you can use to decide if a piece of information is data or evidence? Be ready to share in a few minutes.

You can also use presentation software or other techniques to encourage your students to think about these concepts. You can show examples of either a law (such as $\mathbf{g} = GM/\mathbf{r}^2$) or a theory (such as *gravity is the curvature of four-dimensional space-time due to the presence of mass*) and ask students to indicate if they think it is a law or a theory and why. You can also show examples of information from the investigation that are either data or evidence and ask students to classify each example and explain their thinking.

Be sure to remind your students that, to be proficient in science, it is important that they understand what counts as scientific knowledge and how that knowledge develops over time.

LAB 17

Hints for Implementing the Lab

- Allowing students to design their own procedures for collecting data gives students an opportunity to try, to fail, and to learn from their mistakes. However, you can scaffold students as they develop their procedure by having them fill out an investigation proposal. These proposals provide a way for you to offer students hints and suggestions without telling them how to do it. You can also check the proposals quickly during a class period. We recommend having them fill out a proposal for each experiment. For this lab we suggest using Investigation Proposal C.

- We recommend using motion sensors to collect data about the velocity of the carts. However, students can also determine the velocity of the carts by conducting a video analysis of the motion of the carts.

- We recommend using a fan attachment with variable speed and duration settings. This type of fan attachment makes it easier for students to design more informative experiments. The variable duration setting is especially useful.

- Learn how to use the fan attachment and dynamics track before the lab begins. It is important for you to know how to use the equipment so you can help students when technical issues arise.

- Allow students to become familiar with the fan attachment and dynamics track as part of the tool talk before they begin to design their investigation. Giving them 5–10 minutes to examine the available equipment will let them see what they can and cannot do with it. This also gives students a chance to begin thinking about what variables to test and how to control for other variables.

- Be sure to allow students to go back and re-collect data at the end of the argumentation session. Students often realize that they made numerous mistakes when they were collecting data as a result of their discussions during the argumentation session. The students, as a result, will want a chance to re-collect data, and the re-collection of data should be encouraged when time allows. This also offers an opportunity to discuss what scientists do when they realize a mistake is made inside the lab.

If students use a motion detector/sensor

- Learn how to use the motion detector/sensor and the data collection and analysis software before the lab begins. It is important for you to know how to use the equipment and software so you can help students when technical issues arise.

- Allow the students to become familiar with the motion detector/sensor and data collection and analysis software as part of the tool talk before they begin to design their investigation. Giving them 5–10 minutes to examine the equipment and software will let them see what they can and cannot do with it.

If students use video analysis

- We suggest allowing students to familiarize themselves with the video analysis software before they finalize the procedure for the investigation, especially if they have not used such software previously. This gives students an opportunity to learn how to work with the software and to improve the quality of the video they take.

- Remind students to hold the video camera as still as possible. Any movement of the camera will introduce error into their analysis. If using actual camcorders, we recommend using a tripod to hold the camera steady. If students are using a camera on a cell phone or tablet, we recommend using a table to help steady the camera.

- Remind students to place a meterstick in the same field of view as the motion they are capturing with the video camera. Also, the meterstick should be approximately the same distance from the camera as the motion. Most video analysis software requires the user to define a scale in the video (this allows the software to establish distances and, subsequently, other variables dependent on distance and displacement).

Connections to Standards

Table 17.2 (p. 374) highlights how the investigation can be used to address specific performance expectations from the *NGSS*; learning objectives from AP Physics 1; learning objectives from AP Physics C: Mechanics, *Common Core State Standards*, in English language arts (*CCSS ELA*); and *Common Core State Standards*, Mathematics (*CCSS Mathematics*).

TABLE 17.2

Lab 17 alignment with standards

NGSS performance expectation	• HS-PS2-2: Use mathematical representation to support the claim that the total momentum of a system of objects is conserved when there is no net force on the system.
AP Physics 1 learning objectives	• 3.D.1.1: The student is able to justify the selection of data needed to determine the relationship between the direction of the force acting on an object and the change in momentum caused by that force. • 3.D.2.1: The student is able to justify the selection of routines for the calculation of the relationships between changes in momentum of an object, average force, impulse, and time of interaction. • 3.D.2.2: The student is able to predict the change in momentum of an object from the average force exerted on the object and the interval of time during which the force is exerted. • 3.D.2.3: The student is able to analyze data to characterize the change in momentum of an object from the average force exerted on the object and the interval of time during which the force is exerted. • 3.D.2.4: The student is able to design a plan for collecting data to investigate the relationship between changes in momentum and the average force exerted on an object over time.
AP Physics C: Mechanics learning objectives	• I.D.2.a: Relate mass, velocity, and linear momentum for a moving object, and calculate the total linear momentum of a system of objects. • I.D.2.b: Relate impulse to the change in linear momentum and the average force acting on an object. • I.D.2.e: Calculate the change in momentum of an object given a function $\mathbf{F}(t)$ for the net force acting on the object.
Literacy connections (_CCSS ELA_)	• _Reading:_ Key ideas and details, craft and structure, integration of knowledge and ideas • _Writing:_ Text types and purposes, production and distribution of writing, research to build and present knowledge, range of writing • _Speaking and listening:_ Comprehension and collaboration, presentation of knowledge and ideas

Mathematics connections (*CCSS Mathematics*)	• *Mathematical practices:* Reason abstractly and quantitatively, construct viable arguments and critique the reasoning of others, use appropriate tools strategically, attend to precision • *Number and quantity:* Reason quantitatively and use units to solve problems, represent and model with vector quantities, perform operations on vectors • *Algebra:* Interpret the structure of expressions, understand solving equations as a process of reasoning and explain the reasoning, solve equations and inequalities in one variable, represent and solve equations and inequalities graphically • *Functions:* Understand the concept of a function and use function notation, interpret functions that arise in applications in terms of the context, analyze functions using different representations, interpret expressions for functions in terms of the situation they model • *Statistics and probability:* Summarize, represent, and interpret data on two categorical and quantitative variables; understand and evaluate random processes underlying statistical experiments; make inferences and justify conclusions from sample surveys, experiments, and observational studies

Reference

Newton, I. 1687. *Philosophiae naturalis principia mathematica* [Mathematical principles of natural philosophy]. London: S. Pepys.

LAB 17

Lab Handout

Lab 17. Impulse and Momentum: How Does Changing the Magnitude and Duration of a Force Acting on an Object Affect the Momentum of That Object?

Introduction

Forces are responsible for all changes in motion and momentum. Regardless of how quickly a force is applied, it can still change the motion of an object. Consider what happens when someone hits a ball with a bat. The time that the bat and ball are in contact is very short, but the force from this collision is strong enough to significantly change the motion of a ball. The force that results from a bat hitting a ball can even be strong enough to change the shape of the ball (see Figure L17.1) or the bat (see Figure L17.2).

FIGURE L17.1 _____

A ball deforming from the force that results from the collision of the bat and the ball

FIGURE L17.2 _____

The bat breaking from the force that results from the collision of the bat and the ball

Momentum is defined as the mass of an object multiplied by its velocity. Momentum is a vector quantity because it has both a magnitude and a direction. As a result, the momentum of an object can be positive or negative, depending on the direction an object is moving. Force is also a vector quantity because forces have both a magnitude and a direction. When an unbalanced force acts on an object, the momentum of that object will change. In other words, an unbalanced force will change the momentum of an object.

The amount of time that an unbalanced force acts on an object is also important to consider when examining the change in momentum of an object. Sometimes the amount of time a force is applied to an object is very short, such as when a bat hits a ball, and other times it is applied over long periods, such as when the thrusters attached to a satellite are fired for several minutes to launch that satellite into orbit. The term *impulse* is used to describe the product of the magnitude and the duration of a force that acts on an object. In this investigation you will have an opportunity to examine how the nature of an impulse can change the momentum of a cart moving in one dimension. Your goal is to create a conceptual model that you can use to explain how the magnitude and duration of a force affects the change in momentum of a cart.

Your Task

Use what you know about momentum, impulse, the movement of matter within a system, and scale, proportional relationships, and quantity to develop a conceptual model that will enable you to explain how the momentum of an object will change in response to an impulse. To develop this conceptual model, you will need to design and carry out two different experiments to determine how (a) the magnitude of a force affects the momentum of an object and (b) the duration of a force affects the momentum of an object. Once you have developed your model, you will need to test it to determine if allows you to make accurate predictions about the change of momentum of an object over time in response to different types of impulse.

The guiding question of this investigation is, *How does changing the magnitude and the duration of a force acting on an object affect the momentum of that object?*

Materials

You may use any of the following materials during your investigation (some items may not be available):

- Safety glasses or goggles (required)
- Dynamics cart with fan attachment
- Dynamics track
- Motion detector/sensor and interface
- Video camera
- Computer or tablet with data collection and analysis software and/or video analysis software
- Electronic or triple beam balance
- Stopwatch
- Meterstick or ruler

Safety Precautions

Follow all normal lab safety rules. In addition, take the following safety precautions:

1. Wear sanitized safety glasses or goggles during lab setup, hands-on activity, and takedown.

LAB 17

2. Keep fingers and toes out of the way of the moving objects.

3. Wash hands with soap and water after completing the lab.

Investigation Proposal Required? ☐ Yes ☐ No

Getting Started

The first step in developing your conceptual model is to design and carry out two experiments. In the first experiment, you will need to determine how changing the magnitude of a force will affect the momentum of a cart. In the second experiment, you will need to determine how changing the duration of the force affects the momentum of a cart. Figure L17.3 illustrates how you can use the available equipment to study the momentum of cart moving in one dimension. Before you can design your experiments, however, you must determine what type of data you need to collect, how you will collect it, and how you will analyze it.

FIGURE L17.3

One way to study the change in momentum of an object

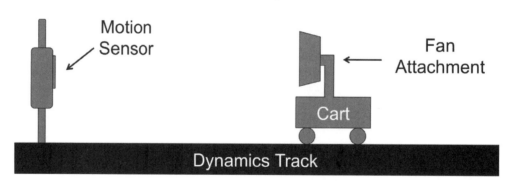

To determine *what type of data you need to collect,* think about the following questions:

- What are the boundaries and components of the system you are studying?
- How do components of the system under study interact?
- How will you track the movement of matter within this system?
- How could you keep track of changes in this system quantitatively?
- What factors affects the momentum of an object?
- How will you determine the velocity of each object?
- What will be the independent variable and the dependent variable for each experiment?

To determine *how you will collect the data,* think about the following questions:

- What other factors will you need to control or measure during each experiment?
- Which quantities are vectors, and which quantities are scalars?
- For any vector quantities, which directions are positive and which directions are negative?
- What scale or scales should you use when you take your measurements?
- What equipment will you need to collect the measurements you need?
- How will you make sure that your data are of high quality (i.e., how will you reduce error)?
- How will you keep track of and organize the data you collect?

To determine *how you will analyze the data,* think about the following questions:

- What type of calculations will you need to do?
- What types of patterns might you look for as you analyze your data?
- Are there any proportional relationships you can identify?
- What types of comparisons will be useful to make?
- What type of table or graph could you create to help make sense of your data?

Once you have determined how the magnitude of a force and the duration of a force affect the momentum of a cart in one dimension, your group will need to develop a conceptual model. Your model must include the various forces acting on the cart and allow you to make accurate predictions about how the momentum of cart changes over time in response to different forces.

The last step in this investigation will be to test your model. To accomplish this goal, you can apply different impulses (ones that you did not test) to the cart to determine if your model enables you to make accurate predictions about how the momentum of the cart changes over time. If you are able to use your model to make accurate predictions, then you will be able to generate the evidence you need to convince others that your model is a valid and acceptable. The fan attached to the cart you will use in this investigation may have a limited number of different speeds, so it will be important to reserve at least one speed setting for this step of your investigation.

Connections to the Nature of Scientific Knowledge and Scientific Inquiry

As you work through your investigation, you may want to consider

- the difference between laws and theories in science, and
- the difference between data and evidence in science.

LAB 17

Initial Argument

Once your group has finished collecting and analyzing your data, your group will need to develop an initial argument. Your initial argument needs to include a claim, evidence to support your claim, and a justification of the evidence. The *claim* is your group's answer to the guiding question. The *evidence* is an analysis and interpretation of your data. Finally, the *justification* of the evidence is why your group thinks the evidence matters. The justification of the evidence is important because scientists can use different kinds of evidence to support their claims. Your group will create your initial argument on a whiteboard. Your whiteboard should include all the information shown in Figure L17.4.

FIGURE L17.4 _____

Argument presentation on a whiteboard

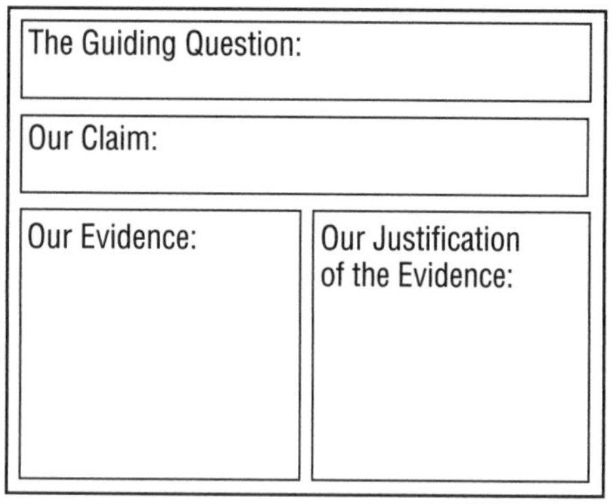

The Guiding Question:	
Our Claim:	
Our Evidence:	Our Justification of the Evidence:

Argumentation Session

The argumentation session allows all of the groups to share their arguments. One or two members of each group will stay at the lab station to share that group's argument, while the other members of the group go to the other lab stations to listen to and critique the other arguments. This is similar to what scientists do when they propose, support, evaluate, and refine new ideas during a poster session at a conference. If you are presenting your group's argument, your goal is to share your ideas and answer questions. You should also keep a record of the critiques and suggestions made by your classmates so you can use this feedback to make your initial argument stronger. You can keep track of specific critiques and suggestions for improvement that your classmates mention in the space below.

Critiques about our initial argument and suggestions for improvement:

If you are critiquing your classmates' arguments, your goal is to look for mistakes in their arguments and offer suggestions for improvement so these mistakes can be fixed. You should look for ways to make your initial argument stronger by looking for things that the other groups did well. You can keep track of interesting ideas that you see and hear during the argumentation in the space below. You can also use this space to keep track of any questions that you will need to discuss with your team.

Interesting ideas from other groups or questions to take back to my group:

Once the argumentation session is complete, you will have a chance to meet with your group and revise your initial argument. Your group might need to gather more data or design a way to test one or more alternative claims as part of this process. Remember, your goal at this stage of the investigation is to develop the best argument possible.

Report

Once you have completed your research, you will need to prepare an *investigation report* that consists of three sections. Each section should provide an answer to the following questions:

1. What question were you trying to answer and why?

2. What did you do to answer your question and why?

3. What is your argument?

Your report should answer these questions in two pages or less. This report must be typed, and any diagrams, figures, or tables should be embedded into the document. Be sure to write in a persuasive style; you are trying to convince others that your claim is acceptable or valid!

LAB 17

Lab 17. Impulse and Momentum: How Does Changing the Magnitude and Duration of a Force Acting on an Object Affect the Momentum of That Object?

1. How can the shape of a force versus time graph be used to determine an object's momentum?

   Use the following information to answer questions 2–4. Consider a cart starting from rest with a fan attachment that applies a constant force. Assume that there is no friction acting on the cart as it moves.

2. What would the momentum versus time graph look like if the fan force doubled halfway through the trial?

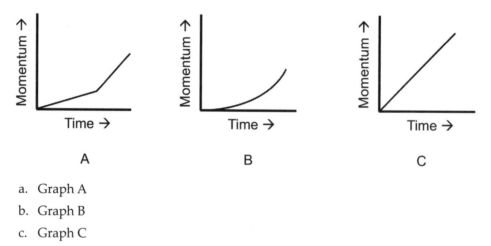

   a. Graph A
   b. Graph B
   c. Graph C

   How do you know?

3. What could cause the slope to become zero?

    a. Doubling the fan force a second time

    b. Turning the fan off

    c. Leaving the fan as is

    How do you know?

4. Draw a momentum versus time graph showing the change in momentum if the fan stayed on for twice as long.

    Why did you draw your momentum versus time graph like that?

5. There is a difference between a scientific law and a scientific theory.

    a. I agree with this statement.

    b. I disagree with this statement.

    Explain your answer, using an example from your investigation about impulse and momentum.

6. There is a difference between data and evidence in science.

    a. I agree with this statement.
    b. I disagree with this statement.

    Explain your answer, using an example from your investigation about impulse and momentum.

7. In physics, it is important to classify something as either a vector quantity or a scalar quantity. Explain what a vector is and why it is important to identify vector quantities in physics, using an example from your investigation about impulse and momentum.

8. Scientists often need to track how energy or matter moves into, out of, or within a system during an investigation. Explain why tracking energy and matter is such an important part of science, using an example from your investigation about impulse and momentum.

Application Labs

LAB 18

Teacher Notes

Lab 18. Elastic and Inelastic Collisions: Which Properties of a System Are Conserved During a Collision?

Purpose

The purpose of this lab is for students to *apply* what they know about the core idea of forces and motion, part of the disciplinary core idea (DCI) of Motion and Stability: Forces and Interactions from the *NGSS,* to determine which properties of a system are conserved in a collision. This lab also gives students an opportunity to learn about the crosscutting concepts (CCs) of (a) Systems and System Models and (b) Energy and Matter: Flows, Cycles, and Conservation from the *NGSS.* In addition, this lab can be used to help students understand three big ideas from AP Physics: (a) the interactions of an object with other objects can be described by forces, (b) interactions between systems can result in changes in those systems, and (c) changes that occur as result of interactions are constrained by conservation laws. As part of the explicit and reflective discussion, students will also learn about (a) the difference between laws and theories in science and (b) the difference between data and evidence in science.

Underlying Physics Concepts

Scientists have identified a number of conservation laws. A conservation law states that some property of an isolated system is the same before and after some specific interaction takes place within that system. Some of the conservation laws that have been identified include conservation of linear momentum, rotational momentum, and total energy within a system. Mathematically, we can represent conservation laws using Equation 18.1, where $\sum$ is "the sum of" and x is a generic symbol for some property of constituent parts of the system; x is the value of the property before the collision and x' is the value of the property after the collision.

$$\text{(Equation 18.1)} \quad \sum x = \sum x'$$

If a property is conserved, then it will obey Equation 18.1. In this investigation, the two properties of the system that will be conserved are mass and momentum. Thus, for each property students are testing in this investigation, they are asking if Equation 18.1 holds true for that property when two carts collide.

The law of conservation of mass is expressed in Equations 18.2 and 18.3 relative to the collision between two carts. In Equation 18.2, the law of conservation of mass is shown as the sum of the masses of the two carts before the collision, respectively, will be equal to the sum of the masses of the two carts after the collision. In Equation 18.3, the conservation of mass is shown where each mass is treated separately. In this equation, m_1 is the mass of cart 1 before the collision, m'_1 is the mass of cart 1 after the collision, m_2 is the mass of cart 2 before the collision, and m'_2 is the mass of cart 2 after the collision. In SI units, mass is measured in kilograms (kg).

$$\text{(Equation 18.2)} \quad \sum m = \sum m'$$

$$\text{(Equation 18.3)} \quad m_1 + m_2 = m'_1 + m'_2$$

Equations 18.4 and 18.5 (which is derived from Equation 18.4) show the law of conservation of momentum expressed in slightly different mathematical terms. In Equation 18.4, the law of conservation of momentum is shown as the sum of the momentum of each cart, respectively, before and after the collision, where **p** is used to represent momentum. In Equation 18.5, the law of conservation of momentum is shown with the momentum of each cart treated separately both before and after the collision. In SI units, momentum is measured in kilogram-meters per second (kg·m/s).

$$\text{(Equation 18.4)} \quad \sum \mathbf{p} = \sum \mathbf{p}'$$

$$\text{(Equation 18.5)} \quad \mathbf{p}_1 + \mathbf{p}_2 = \mathbf{p}'_1 + \mathbf{p}'_2$$

Because momentum of an object is the product of the mass of the object times the velocity of the object, we can rewrite Equation 18.5 by substituting in the mass and velocity values for each cart, as shown in Equation 18.6. In this equation, m_1 represents the mass of cart 1 before the collision, $\mathbf{v}_1$ represents the mass of cart 1 before the collision, m'_1 represents the mass of cart 1 after the collision, and $\mathbf{v}'_1$ represents the velocity of cart 1 after the collision. In SI units, velocity is measured in meters per second (m/s).

$$\text{(Equation 18.6)} \quad m_1\mathbf{v}_1 + m_2\mathbf{v}_2 = m_1\mathbf{v}'_1 + m_2\mathbf{v}'_2$$

With regard to the total momentum of the system, it is important to keep in mind the sign conventions for velocity. Traditionally, things in the upward, right, north, or east direction get a positive sign, and things in the downward, left, south, or west direction get a negative sign. For example, in Figure 18.1 (p. 388), the truck moving to the right would have a positive value for velocity. The car moving to the left would have a negative velocity. This means that the total momentum of the system can be smaller in magnitude than the momentum of two constituent parts. For example, if the truck has a mass of 5,000 kg and is moving to the right at 10 m/s and the car has a mass of 2,000 kg and is moving to

LAB 18

the left at 20 m/s, then the total momentum of the two-vehicle system before the collision is 10,000 kg·m/s, as shown in Equation 18.7.

(Equation 18.7) *Total momentum* = (5,000 kg · 10 m/s) + (2,000 kg · −20 m/s) =
10,000 kg·m/s

Note that the magnitude of the momentum of the truck is 50,000 kg·m/s and the magnitude of the momentum of the small car is 40,000 kg·m/s. Yet the total momentum of the system is only 10,000 kg·m/s to the right.

FIGURE 18.1

A truck and a car before a collision

Finally, physicists distinguish between elastic and inelastic collisions. In an elastic collision, both momentum and kinetic energy are conserved. In an inelastic collision, only momentum is conserved. Mathematically, an inelastic collision only obeys Equation 18.4, whereas an elastic collision obeys both Equation 18.4 and Equation 18.8, where KE is used to represent the kinetic energy of the system. Equation 18.9 shows the mathematical expression for kinetic energy, where m is the mass of an object and $\mathbf{v}$ is the velocity of the object. In SI units, kinetic energy is measured in joules (J).

(Equation 18.8) $\sum KE = \sum KE'$

(Equation 18.9) $KE = \frac{1}{2}m\mathbf{v}^2$

If we substitute Equation 18.9 into Equation 18.8, we can derive Equation 18.10, which shows the conservation of kinetic energy for an elastic collision.

(Equation 18.10) $\frac{1}{2}m_1\mathbf{v}_1^2 + \frac{1}{2}m_2\mathbf{v}_2^2 = \frac{1}{2}m_1\mathbf{v}_1'^2 + \frac{1}{2}m_2\mathbf{v}_2'^2$

Most collisions that occur on a daily basis are inelastic, where kinetic energy is not conserved. In fact, the only elastic collisions that have been observed are collisions between atoms. This does not mean that inelastic collisions violate the law of conservation of energy, because the law of conservation of energy states that the total energy is conserved, not just the kinetic energy. During a collision, some energy is transformed into sound or heat energy. Finally, some collisions that we can observe (such as two billiard balls colliding) are very close to being elastic, so we often treat them as if they are elastic, while recognizing this is not the case.

Timeline

The instructional time needed to complete this lab investigation is 170–230 minutes. Appendix 3 (p. 531) provides options for implementing this lab investigation over several class periods. Option C (230 minutes) should be used if students are unfamiliar with scientific writing, because this option provides extra instructional time for scaffolding the writing process. You can scaffold the writing process by modeling, providing examples, and providing hints as students write each section of the report. Option C can also be used if students are unfamiliar with any of the data collection and analysis tools. Option D (170 minutes) should be used if students are familiar with scientific writing and have developed the skills needed to write an investigation report on their own. In option D, students complete stage 6 (writing the investigation report) and stage 8 (revising the investigation report) as homework.

Materials and Preparation

The materials needed to implement this investigation are listed in Table 18.1 (p. 390). The items can be purchased from a science supply company such as Carolina, Flinn Scientific, PASCO, Vernier, or Ward's Science. The video analysis software can be purchased from Vernier (Logger *Pro*) or PASCO (SPARKvue or Capstone). These companies also have apps that can be used on Apple- or Android-based tablets and cell phones. We recommend consulting with your school's information technology coordinator to determine the best option for your students.

LAB 18

TABLE 18.1 _____

Materials list for Lab 18

Item	Quantity
Safety glasses or goggles	1 per student
Dynamics carts	2 per group
Dynamics track	1 per group
Bumper kit for the carts (includes hoops, magnets or Velcro, rubber, and clay)	1 per group
Video camera	1 per group
Computer or tablet with video analysis software	1 per group
Electronic or triple beam balance	1 per group
Mass set	1 per group
Stopwatches	2–3 per group
Ruler	1 per student
Whiteboard, 2' × 3'*	1 per group
Lab Handout	1 per student
Investigation Proposal B (optional)	2 per group
Peer-review guide and teacher scoring rubric	1 per student
Checkout Questions	1 per student

* As an alternative, students can use computer and presentation software such as Microsoft PowerPoint or Apple Keynote to create their arguments.

Be sure to use a set routine for distributing and collecting the materials during the lab investigation. One option is to set up the materials for each group at each group's lab station before class begins. This option works well when there is a dedicated section of the classroom for lab work and the materials are large and difficult to move (such as a dynamics track). A second option is to have all the materials on a table or cart at a central location. You can then assign a member of each group to be the "materials manager." This individual is responsible for collecting all the materials his or her group needs from the table or cart during class and for returning all the materials at the end of the class. This option works well when the materials are small and easy to move (such as stopwatches, metersticks, or hanging masses). It also makes it easy to inventory the materials at the end of the class before students leave for the day.

Safety Precautions

Remind students to follow all normal lab safety rules. In addition, tell students to take the following safety precautions:

1. Wear sanitized safety glasses or goggles during lab setup, hands-on activity, and takedown.

2. Keep fingers and toes out of the way of moving objects.

3. Wash hands with soap and water after completing the lab.

Topics for the Explicit and Reflective Discussion

Reflecting on the Use of Core Ideas and Crosscutting Concepts During the Investigation

Teachers should begin the explicit and reflective discussion by asking students to discuss what they know about the core idea they used during the investigation. The following are some important concepts related to the core idea of forces and motion that students will need to determine which properties of a system are conserved during a collision:

- A reference frame is needed to determine the direction and the magnitude of the displacement, velocity, and acceleration of an object.

- Force is considered a vector quantity because a force has both magnitude and direction.

- A *system* is an object or a collection of objects. An object is treated as if it has no internal structure.

- The boundary between a system and its environment is a decision made by the person considering the situation in order to simplify or otherwise assist in analysis.

- Conserved quantities are constant in an isolated or a closed system. An open system is one that exchanges any conserved quantity with its surroundings.

- Linear momentum is conserved in all systems.

- The change in momentum of an object occurs over a time interval.

- The force that one object exerts on a second object changes the momentum of the second object (in the absence of other forces on the second object).

To help students reflect on what they know about forces and motion, we recommend showing them two or three images using presentation software that help illustrate these important ideas. You can then ask the students the following questions to encourage them to share how they are thinking about these important concepts:

1. What do we see going on in this image?

2. Does anyone have anything else to add?

3. What might be going on that we can't see?

4. What are some things that we are not sure about here?

You can then encourage students to think about how CCs played a role in their investigation. There are at least two CCs that students need to determine which properties of a system are conserved during a collision: (a) Systems and System Models and (b) Energy and Matter: Flows, Cycles, and Conservation (see Appendix 2 [p.527] for a brief description of these CCs). To help students reflect on what they know about these CCs, we recommend asking them the following questions:

1. Scientists often need to define the system under study and then make a model during an investigation. Why is developing a model of system so useful in science?

2. What were the boundaries and components of the system you studied during this investigation? What were the strengths and limitations of the model you developed?

3. Scientists often attempt to track how energy and matter move into, out of, and within systems as part of an investigation. Why is this useful?

4. How did you attempt to track how matter moves within the system you were studying? What did tracking the movement of matter allow you to do during your investigation?

You can then encourage the students to think about how they used all these different concepts to help answer the guiding question and why it is important to use these ideas to help justify their evidence for their final arguments. Be sure to remind your students to explain why they included the evidence in their arguments and make the assumptions underlying their analysis and interpretation of the data explicit in order to provide an adequate justification of their evidence.

Reflecting on Ways to Design Better Investigations

It is important for students to reflect on the strengths and weaknesses of the investigation they designed during the explicit and reflective discussion. Students should therefore be encouraged to discuss ways to eliminate potential flaws, measurement errors, or sources of uncertainty in their investigations. To help students be more reflective about the design of their investigation and what they can do to make their investigations more rigorous in the future, you can ask them the following questions:

1. What were some of the strengths of the way you planned and carried out your investigation? In other words, what made it scientific?

2. What were some of the weaknesses of the way you planned and carried out your investigation? In other words, what made it less scientific?

3. What rules can we make, as a class, to ensure that our next investigation is more scientific?

Reflecting on the Nature of Scientific Knowledge and Scientific Inquiry

This investigation can be used to illustrate two important concepts related to the nature of scientific knowledge and the nature of scientific inquiry: (a) the difference between laws and theories in science and (b) the difference between data and evidence in science (see Appendix 2 for a brief description of these two concepts). Be sure to review these concepts during and at the end of the explicit and reflective discussion. To help students think about these concepts in relation to what they did during the lab, you can ask them the following questions:

1. Laws and theories are different in science. Is $\mathbf{F} \Delta t = m\Delta \mathbf{v}$ an example of a theory or a law? Why?

2. Can you work with your group to come up with a rule that you can use to decide if something is a law or a theory? Be ready to share in a few minutes.

3. You had to talk about data and evidence during your investigation. Can you give me some examples of data and evidence from your investigation?

4. Can you work with your group to come up with a rule that you can use to decide if a piece of information is data or evidence? Be ready to share in a few minutes.

You can also use presentation software or other techniques to encourage your students to think about these concepts. You can show examples of either a law (such as $\mathbf{g} = GM/\mathbf{r}^2$) or a theory (such as *gravity is the curvature of four-dimensional space-time due to the presence of mass*) and ask students to indicate if they think it is a law or a theory and why. You can also show examples of information from the investigation that are either data or evidence and ask students to classify each example and explain their thinking.

Be sure to remind your students that, to be proficient in science, it is important that they understand what counts as scientific knowledge and how that knowledge develops over time.

Hints for Implementing the Lab

- This investigation can be used at several points during the year. You can use this investigation after introducing momentum but before introducing the law of conservation of momentum. You can also use it after teaching about conservation of momentum but before introducing the concept of elastic and inelastic collisions. If you choose to use this investigation after introducing momentum but before introducing the law of conservation of momentum, we suggest not discussing energy during the explicit and reflective discussion. Finally, you can use this lab after teaching about momentum and energy but before teaching about elastic and inelastic collisions.

- Allowing students to design their own procedures for collecting data gives students an opportunity to try, to fail, and to learn from their mistakes. However, you can scaffold students as they develop their procedure by having them fill out an investigation proposal. These proposals provide a way for you to offer students hints and suggestions without telling them how to do it. You can also check the proposals quickly during a class period. We recommend that students fill out a proposal for each experiment they design. For this lab we suggest using Investigation Proposal B.

- Be sure to remind students to use different types of bumpers on the cart to model different types of collisions. Bumper kits for carts usually include hoops, Velcro or magnets, rubber, and clay.

- Learn how to use the dynamics cart and track, the bumper kit, and the video analysis software before the lab begins. It is important for you to know how to use the equipment so you can help students when technical issues arise.

- Allow the students to become familiar with the dynamics cart and track, the bumper kit, and the video analysis software as part of the tool talk before they begin to design their investigation. Giving them 5–10 minutes to examine the equipment and materials will let them see what they can and cannot do with them. This also gives students a chance to begin thinking about what variables to test and how to control for other variables.

- Remind students to hold the video camera as still as possible. Any movement of the camera will introduce error into their analysis. If using actual camcorders, we recommend using a tripod to hold the camera steady. If students are using a camera on a cell phone or tablet, we recommend using a table to help steady the camera.

- Remind students to place a meterstick in the same field of view as the motion they are capturing with the video camera. Also, the meterstick should be approximately the same distance from the camera as the motion. Most video analysis software requires the user to define a scale in the video (this allows the software to establish distances and, subsequently, other variables dependent on distance and displacement).

- During this lab, students may falsify a conservation law for velocity, acceleration, and force. That is, the data will show that these quantities are not conserved. This is an important finding, and we recommend that you suggest to students that they make these findings part of their argument.

- Be sure to allow students to go back and re-collect data at the end of the argumentation session. Students often realize that they made numerous mistakes when they were collecting data as a result of their discussions during the argumentation session. The students, as a result, will want a chance to re-collect data, and the re-collection of data should be encouraged when time allows. This also offers an opportunity to discuss what scientists do when they realize a mistake is made inside the lab.

Topic Connections

Table 18.2 (p. 396) highlights how the investigation can be used to address specific performance expectations from the *NGSS*; learning objectives from AP Physics 1; learning objectives from AP Physics C: Mechanics, *Common Core State Standards,* in English language arts (*CCSS ELA*); and *Common Core State Standards,* Mathematics (*CCSS Mathematics*).

LAB 18

TABLE 18.2 _____

Lab 18 alignment with standards

NGSS performance expectation	• HS-PS2-2: Use mathematical representations to support the claim that the total momentum of a system of objects is conserved when there is no net force on the system.
AP Physics 1 learning objectives	• 1.A.5.1: The student is able to model verbally or visually the properties of a system based on its substructure and to relate this to changes in the system properties over time as external variables are changed. • 3.A.1.2: The student is able to design an experimental investigation of the motion of an object. • 3.A.1.3: The student is able to analyze experimental data describing the motion of an object and is able to express the results of the analysis using narrative, mathematical, and graphical representations. • 4.B.1.1: The student is able to calculate the change in linear momentum of a two-object system with constant mass in linear motion from a representation of the system (data, graphs, etc.). • 4.B.1.2: The student is able to analyze data to find the change in linear momentum for a constant-mass system using the product of the mass and the change in velocity of the center of mass. • 5.A.2.1: The student is able to define open and closed/isolated systems for everyday situations and apply conservation concepts for energy, charge, and linear momentum to those situations. • 5.D.1.4: The student is able to design an experimental test of an application of the principle of the conservation of linear momentum, predict an outcome of the experiment using the principle, analyze data generated by that experiment whose uncertainties are expressed numerically, and evaluate the match between the prediction and the outcome. • 5.D.2.2: The student is able to plan data collection strategies to test the law of conservation of momentum in a two-object collision that is elastic or inelastic and analyze the resulting data graphically. • 5.D.2.4: The student is able to analyze data that verify conservation of momentum in collisions with and without an external friction force.
AP Physics C: Mechanics learning objectives	• I.D.2.a: Relate mass, velocity, and linear momentum for a moving object, and calculate the total linear momentum of a system of objects. • I.D.3.a(2): Identify situations in which linear momentum, or a component of the linear momentum vector, is conserved. • I.D.3.a(3): Apply linear momentum conservation to one-dimensional elastic and inelastic collisions and two-dimensional completely inelastic collisions.

Literacy connections (*CCSS ELA*)	• *Reading:* Key ideas and details, craft and structure, integration of knowledge and ideas • *Writing:* Text types and purposes, production and distribution of writing, research to build and present knowledge, range of writing • *Speaking and listening:* Comprehension and collaboration, presentation of knowledge and ideas
Mathematics connections (*CCSS Mathematics*)	• *Mathematical practices:* Reason abstractly and quantitatively, construct viable arguments and critique the reasoning of others, use appropriate tools strategically, attend to precision • *Number and quantity:* Reason quantitatively and use units to solve problems, represent and model with vector quantities, perform operations on vectors • *Algebra:* Interpret the structure of expressions, understand solving equations as a process of reasoning and explain the reasoning, solve equations and inequalities in one variable, represent and solve equations and inequalities graphically • *Functions:* Understand the concept of a function and use function notation, interpret functions that arise in applications in terms of the context, analyze functions using different representations, interpret expressions for functions in terms of the situation they model • *Statistics and probability:* Summarize, represent, and interpret data on two categorical and quantitative variables; understand and evaluate random processes underlying statistical experiments; make inferences and justify conclusions from sample surveys, experiments, and observational studies

LAB 18

Lab Handout

Lab 18. Elastic and Inelastic Collisions: Which Properties of a System Are Conserved During a Collision?

Introduction

Physics is the scientific study of time, space, and matter. Some branches of physics, such as cosmology, investigate questions regarding the entire universe (e.g., how old is it, how did it begin). Most branches of physics, however, investigate questions related to smaller scales and systems. When studying a system, a physicist will identify the system and then ask questions about (1) the matter contained in that system, (2) the interactions between matter contained in the system, and (3) how the matter moves in the system. When doing this, the physicist ignores any influences from outside the system during the investigation, while recognizing that those influences are still there. For example, when studying acceleration of an object in free fall due to gravity, a physicist might ignore the influence of air resistance when its effects are less than the uncertainty in the measurements. Systems come in all sizes. Astronomers study systems as large as galaxies. Chemists study systems that can be as small as a few atoms.

Scientists have identified several laws of conservation that are the same across all systems. A conservation law states that some property of an isolated system is the same before and after some specific interaction takes place within that system. The properties of each object in the system, however, do not need to be the same before and after the interaction in order for some property of the system to stay the same. For example, the law of conservation of energy indicates that the total amount of energy in a system stays the same before and after any interaction that takes place between one or more components of that system. To illustrate what the conservation of energy means, consider what happens when a person places a hot metal spoon into a cold cup of water. When the hot metal spoon is placed into the cup, some heat energy will transfer from the spoon into the water, but the total amount of energy of the system remains constant because the energy that transferred from the spoon into the water is still a part of the system. The transfer of energy caused the temperature of the water to increase and the temperature of the metal to decrease but the total amount of energy in the system did not change, so energy is conserved within the system.

When studying the interactions between two or more objects in a system, physicists often try to identify which properties are conserved during the interaction. In the example of putting a hot metal spoon into cold water, energy is conserved but temperature is not. Another type of interaction that physicists often study is a collision. Collisions are a common experience, from billiard balls colliding on a pool table to an asteroid hitting a planet, or, as shown in Figure L18.1, a collision between two cars.

There are a number of properties that could be conserved during a collision. Some examples include acceleration, velocity, force, energy, and momentum. There are likely other properties that might also be conserved during a collision. In this investigation, you will have an opportunity to determine which properties of a system are conserved during a two-car collision.

Your Task

Use what you know about momentum, velocity, acceleration, the conservation of energy and matter, and systems and system models to design and carry out an investigation that will allow you to understand what happens to the different properties of a two-car system before and after a collision.

The guiding question of this investigation is, *Which properties of a system are conserved during a collision?*

FIGURE L18.1

A collision between two vehicles

Materials

You may use any of the following materials during your investigation:

- Safety glasses or goggles (required)
- Dynamics carts
- Dynamics track
- Bumper kit for the carts
- Video camera
- Computer or tablet with video analysis software
- Electronic or triple beam balance
- Mass set
- Stopwatches
- Ruler

Safety Precautions

Follow all normal lab safety rules. In addition, take the following safety precautions:

1. Wear sanitized safety glasses or goggles during lab setup, hands-on activity, and takedown.

2. Keep fingers and toes out of the way of moving objects.

3. Wash hands with soap and water after completing the lab.

Investigation Proposal Required?   ☐ Yes   ☐ No

LAB 18

Getting Started

To answer the guiding question you will need to design and carry out two experiments. Figure L18.2 shows how you can set up two carts on a track to examine changes in the velocity, acceleration, and position of each cart before and after a collision. You can also change the nature of the collision by changing the bumpers on the carts. You can use a hoop bumper to make the carts bounce apart after they collide or a Velcro or magnet bumper to make them stick together. To measure changes in velocity, acceleration, and the position of the two carts at the same time, you will need to use a video camera and video analysis software. Before you can design your two experiments, however, you must first determine what type of data you need to collect, how you will collect it, and how you will analyze it.

FIGURE L18.2

One way to measure the velocity, acceleration, or position of a moving object before and after a collision

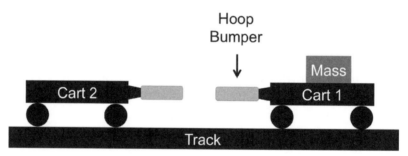

To determine *what type of data you need to collect,* think about the following questions:

- What are the boundaries and components of the system under study?
- How do the components of the system interact?
- What properties of the system might be conserved?
- What properties of the system are directly measurable?
- What properties of the system will you need to calculate from other measurements?
- What types of collisions will you need to model?
- How can you track how matter and energy flows into, out of, or within this system?
- What will be the independent variable and the dependent variable for each experiment?

To determine *how you will collect the data,* think about the following questions:

- What other factors will you need to control during each experiment?

- Which quantities are vectors, and which quantities are scalars?

- What scale or scales should you use when you take your measurements?

- What equipment will you need to collect the data you need?

- How will you make sure that your data are of high quality (i.e., how will you reduce error)?

- How will you keep track of and organize the data you collect?

To determine *how you will analyze the data*, think about the following questions:

- How will you determine if a property has been conserved during a collision?

- What type of calculations will you need to make?

- What types of comparison will be useful for you to make?

- How could you use mathematics to describe a relationship between variables?

- What type of table or graph could you create to help make sense of your data?

Connections to the Nature of Scientific Knowledge and Scientific Inquiry

As you work through your investigation, you may want to consider

- the difference between laws and theories in science, and

- the difference between data and evidence in science.

Initial Argument

Once your group has finished collecting and analyzing your data, your group will need to develop an initial argument. Your initial argument needs to include a claim, evidence to support your claim, and a justification of the evidence. The *claim* is your group's answer to the guiding question. The *evidence* is an analysis and interpretation of your data. Finally, the *justification* of the evidence is why your group thinks the evidence matters. The justification of the evidence is important because scientists can use different kinds of evidence to support their claims. Your group will create your initial argument on a whiteboard. Your whiteboard should include all the information shown in Figure L18.3.

FIGURE L18.3

Argument presentation on a whiteboard

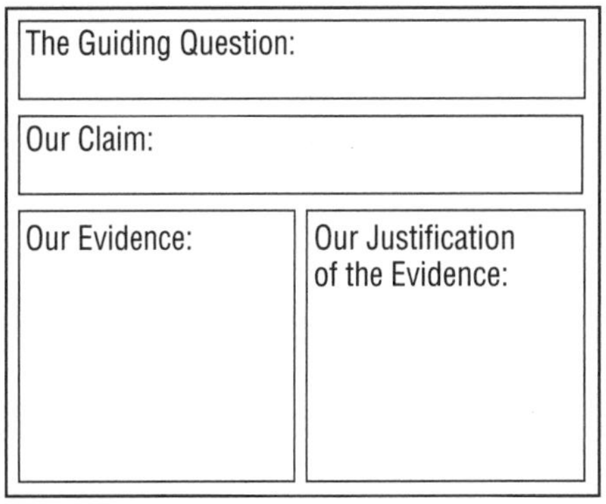

The Guiding Question:	
Our Claim:	
Our Evidence:	Our Justification of the Evidence:

LAB 18

Argumentation Session

The argumentation session allows all of the groups to share their arguments. One or two members of each group will stay at the lab station to share that group's argument, while the other members of the group go to the other lab stations to listen to and critique the other arguments. This is similar to what scientists do when they propose, support, evaluate, and refine new ideas during a poster session at a conference. If you are presenting your group's argument, your goal is to share your ideas and answer questions. You should also keep a record of the critiques and suggestions made by your classmates so you can use this feedback to make your initial argument stronger. You can keep track of specific critiques and suggestions for improvement that your classmates mention in the space below.

Critiques about our initial argument and suggestions for improvement:

If you are critiquing your classmates' arguments, your goal is to look for mistakes in their arguments and offer suggestions for improvement so these mistakes can be fixed. You should look for ways you to make your initial argument stronger by looking for things that the other groups did well. You can keep track of interesting ideas that you see and hear during the argumentation in the space below. You can also use this space to keep track of any questions that you will need to discuss with your team.

Interesting ideas from other groups or questions to take back to my group:

National Science Teachers Association

Once the argumentation session is complete, you will have a chance to meet with your group and revise your initial argument. Your group might need to gather more data or design a way to test one or more alternative claims as part of this process. Remember, your goal at this stage of the investigation is to develop the best argument possible.

Report

Once you have completed your research, you will need to prepare an *investigation report* that consists of three sections. Each section should provide an answer to the following questions:

1. What question were you trying to answer and why?

2. What did you do to answer your question and why?

3. What is your argument?

Your report should answer these questions in two pages or less. This report must be typed, and any diagrams, figures, or tables should be embedded into the document. Be sure to write in a persuasive style; you are trying to convince others that your claim is acceptable or valid!

Checkout Questions

Lab 18. Elastic and Inelastic Collisions: Which Properties Are Conserved During a Collision?

1. In your investigation, the kinetic energy of the system before the collision was greater than the kinetic energy of the system after the collision. Did your investigation violate the law of conservation of energy?

    a. Yes, it violated the law of conservation of energy.
    b. No, it did not violate the law of conservation of energy.

    Explain why or why not.

    Use the following information to answer questions 2 and 3. A truck has a mass of 5,000 kg and is moving to the right at 10 m/s. A small car has a mass of 2,000 kg and is moving to the left at 20 m/s. The two vehicles collide head on.

2. Is this collision elastic or inelastic?

    a. Elastic
    b. Inelastic

    How do you know?

3. What is the total momentum of the two-vehicle system before the collision?

    a. 10,000 kg·m/s

    b. 40,000 kg·m/s

    c. 50,000 kg·m/s

    d. 90,000 kg·m/s

    How do you know?

4. The terms *data* and *evidence* mean the same thing in science.

    a. I agree with this statement.

    b. I disagree with this statement.

    Explain your answer, using an example from your investigation about collisions.

LAB 18

5. It is important to track how energy and matter move into, out of, and within a system and to determine if any of properties are conserved within the system.

    a. I agree with this statement.

    b. I disagree with this statement.

Explain your answer, using an example from your investigation about collisions.

6. In science, there is a difference between a law and a theory. What is the difference between a law and a theory? Explain why this distinction is important, using an example from your investigation about collisions.

7. Scientists often need to identify the system under study before they start collecting data. Explain why defining the system under study is so important in science, using an example from your investigation about collisions.

LAB 19

Teacher Notes

Lab 19. Impulse and Materials: Which Material Is Most Likely to Provide the Best Protection for a Phone That Has Been Dropped?

Purpose

The purpose of this lab is for students to *apply* what they know about the core idea of forces and motion, part of the disciplinary core idea (DCI) of Motion and Stability: Forces and Interactions from the *NGSS,* to identify a suitable material for a new protective cell phone case. This lab also gives students an opportunity to learn about the crosscutting concepts (CCs) of (a) Scale, Proportion, and Quantity; and (b) Structure and Function from the *NGSS.* In addition, this lab can be used to help students understand three big ideas from AP Physics: (a) the interactions of an object with other objects can be described by forces, (b) interactions between systems can result in changes in those systems, and (c) changes that occur as result of interactions are constrained by conservation laws. As part of the explicit and reflective discussion, students will also learn about (a) how the culture of science, societal needs, and current events influence the work of scientists; and (b) the nature and role of experiments in science.

Underlying Physics Concepts

Newton's second law is typically written to describe the relationship between the forces acting on an object and the acceleration of that object. This formulation of Newton's second law is shown in Equation 19.1, where $\sum F$ is the net force acting on an object, m is the mass of that object, and **a** is the acceleration of the object. In SI units, force is measured in newtons (N), mass is measured in kilograms (kg), and acceleration is measured in meters per second squared (m/s^2).

$$\text{(Equation 19.1)} \quad \sum F = m\mathbf{a}$$

It turns out, however, that this formulation of the second law does not appear in Newton's major work, *Philosophiae Naturalis Principia Mathematica* (published in 1687). Instead, Newton's second law states that a net force on an object will result in a change in momentum of that object. This is shown mathematically in Equation 19.2, where $\sum F$ is the net force, Δ (pronounced "delta") means a change in the subsequent quantity, **p** is the momentum of the object, and t is the time in which the force acts on the object. In SI units, momentum is measured in kilogram-meters per second (kg·m/s) and time is measured in seconds (s).

(Equation 19.2) $\sum \mathbf{F} = \Delta \mathbf{p}/\Delta t$

Momentum is equal to the mass of an object times the velocity of the object. If we assume that the mass of an object does not change when a net force acts on it (a valid assumption at non-relativistic velocities), then Equation 19.2 can be rewritten as Equation 19.3, where **v** is velocity. In SI units, velocity is measured in meters per second (m/s).

(Equation 19.3) $\sum \mathbf{F} = m\Delta \mathbf{v}/\Delta t$

The quantity $\Delta \mathbf{v}/\Delta t$ is equal to the acceleration of an object. This is why Newton's second law is often written as Equation 19.1, as opposed to Equation 19.2 or 19.3. Because the mass remains constant, an applied force on an object must cause an acceleration. If we multiply both sides of Equation 19.2 by Δt, then we get Equation 19.4:

(Equation 19.4) $\sum \mathbf{F}\Delta t = m\Delta \mathbf{v}$

Newton defined the left side of the equation, $\sum \mathbf{F}\Delta t$, as the "impulse," often given the symbol **J**, measured in units of newton seconds (N·s); this mathematical relationship is shown in Equation 19.5.

(Equation 19.5) $\sum \mathbf{F}\Delta t = \mathbf{J}$

Finally, assuming that the mass of an object remains constant, then Newton's second law implies that an impulse acting on an object will cause a change in the object's velocity. This is shown in Equation 19.6.

(Equation 19.6) $\mathbf{J} = m\Delta \mathbf{v}$

Students will roll a cart with a force probe down a ramp and crash the cart into various materials during this investigation. If the cart's velocity is the same each time the cart collides with a specific material, then the total impulse required to bring the car to rest will remain the same. In other words, the mass of the cart and the velocity of the cart will be the same for each collision. Impulse is a function of both the force acting on an object and the time the force acts on that object. These two variables are inversely related, so that the larger the force, the shorter the amount of time the force acts on the object for a given impulse.

We can represent this graphically. Figure 19.1 (p. 410) shows a force versus time graph for a collision. In this collision, a constant force of 4 N acts on the object and brings it to rest in 2 seconds. Because impulse is equal to force times time, the total impulse acting on the object is 8 N·s. This also happens to be the area underneath the curve for a force versus time graph.

FIGURE 19.1

A force versus time graph, with a constant force of 4 N acting on the object

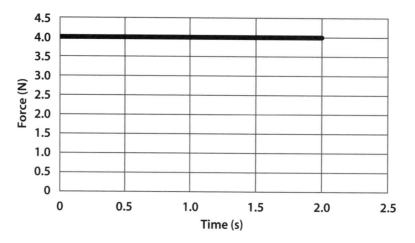

Figure 19.2 shows another force versus time graph. In this case, the force acting on the object is 8 N. However, this force only needs to act on the object for 1 second to bring the object to rest. In this case, the impulse is also 8 N·s.

FIGURE 19.2

A force versus time graph, with a constant force of 8 N acting on the object

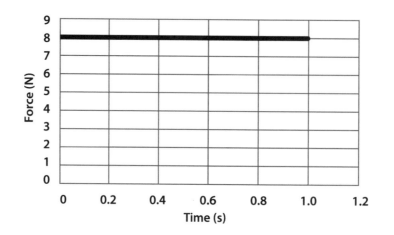

Students will be able to create force versus time graphs for each collision. If the velocity and mass of the cart before impact is constant for all materials tested, then the total impulse required to bring the cart to rest will be the same for all materials. However, the force versus time graph will be different for each material. The "best" material for a cell phone case is the one that results in the force being minimized and the time the collision takes place being maximized. It is the magnitude of the force that will damage the cell phone, so the best material is the one that minimizes the force acting on the cart during the collision.

Finally, the graphs students create will not appear linear like those shown in Figures 19.1 and 19.2. Instead, they will be curved. For students who are in calculus, this provides an opportunity to show contextual applications of integration. The integral of the force versus time graph will give the impulse. For those students who are not in calculus, they can still "qualitatively" answer the guiding question by making comparisons between graphs.

Timeline

The instructional time needed to complete this lab investigation is 200–280 minutes. Appendix 3 (p. 531) provides options for implementing this lab investigation over several class periods. Option C (230 minutes) should be used if students are unfamiliar with scientific writing, because this option provides extra instructional time for scaffolding the writing process. You can scaffold the writing process by modeling, providing examples, and providing hints as students write each section of the report. Option E can also be used if students are unfamiliar with any of the data collection and analysis tools. Option F (200 minutes) should be used if students are familiar with scientific writing and have developed the skills needed to write an investigation report on their own. In option F, students complete stage 6 (writing the investigation report) and stage 8 (revising the investigation report) as homework.

Materials and Preparation

The materials needed to implement this investigation are listed in Table 19.1 (p. 412). The items can be purchased from a science supply company such as Carolina, Flinn Scientific, PASCO, Vernier, or Ward's Science. The data collection and analysis software can be purchased from Vernier (Logger *Pro*) or PASCO (SPARKvue or Capstone). These companies also have apps that can be used on Apple- or Android-based tablets and cell phones. We recommend consulting with your school's information technology coordinator to determine the best option for your students.

TABLE 19.1

Materials list for Lab 19

Item	Quantity
Safety glasses or goggles	1 per student
Dynamics cart	1 per group
Dynamics track	1 per group
Force sensor	1 per group
Sensor interface	1 per group
Computer, tablet, or graphing calculator with data collection and analysis software	1 per group
Foam block	1 per group
Plastic block	1 per group
Rubber block	1 per group
Wooden block	1 per group
Metal block	1 per group
Investigation proposal A (optional)	1 per group
Whiteboard, 2' × 3'*	1 per group
Lab Handout	1 per student
Peer-review guide and teacher scoring rubric	1 per student
Checkout Questions	1 per student

* As an alternative, students can use computer and presentation software such as Microsoft PowerPoint or Apple Keynote to create their arguments.

Be sure to use a set routine for distributing and collecting the materials during the lab investigation. One option is to set up the materials for each group at each group's lab station before class begins. This option works well when there is a dedicated section of the classroom for lab work and the materials are large and difficult to move (such as a dynamics track). A second option is to have all the materials on a table or cart at a central location. You can then assign a member of each group to be the "materials manager." This individual is responsible for collecting all the materials his or her group needs from the table or cart during class and for returning all the materials at the end of the class. This option works well when the materials are small and easy to move (such as stopwatches, metersticks, or hanging masses). It also makes it easy to inventory the materials at the end of the class before students leave for the day.

Safety Precautions

Remind students to follow all normal lab safety rules. In addition, tell students to take the following safety precautions:

1. Wear sanitized safety glasses or goggles during lab setup, hands-on activity, and takedown.

2. Keep fingers and toes out of the way of moving objects.

3. Wash hands with soap and water after completing the lab.

Topics for the Explicit and Reflective Discussion

Reflecting on the Use of Core Ideas and Crosscutting Concepts During the Investigation

Teachers should begin the explicit and reflective discussion by asking students to discuss what they know about the core idea they used during the investigation. The following are some important concepts related to the core idea of forces and motion that students will need to identify a suitable material for a new protective cell phone case:

- The change in momentum of an object is a vector quantity in the direction of the net force exerted on the object.
- The change in momentum of an object occurs over a time interval.
- The force that one object exerts on a second object changes the momentum of the second object (in the absence of other forces on the second object).
- An *impulse* is the product of the average force acting on an object and the time interval during which the interaction occurred.

To help students reflect on what they know about forces and motion, we recommend showing them two or three images using presentation software that help illustrate these important ideas. You can then ask the students the following questions to encourage them to share how they are thinking about these important concepts:

1. What do we see going on in this image?

2. Does anyone have anything else to add?

3. What might be going on that we can't see?

4. What are some things that we are not sure about here?

You can then encourage students to think about how CCs played a role in their investigation. There are at least two CCs that students need to identify a suitable material for a new protective cell phone case: (a) Scale, Proportion, and Quantity; and (b) Structure and Function (see Appendix 2 [p. 527] for a brief description of these CCs). To help students reflect on what they know about these CCs, we recommend asking them the following questions:

1. Why is it important to think about issues of scale and quantity during an investigation?

2. What scales and quantities did you use during your investigation? Why was it useful?

3. The way an object is shaped or structured determines many of its properties and how it functions. Why is it useful to think about the relationship between structure and function during an investigation?

4. Why was it useful to examine the structure of a material when attempting to determine its ability to function as a protective cell phone case?

You can then encourage the students to think about how they used all these different concepts to help answer the guiding question and why it is important to use these ideas to help justify their evidence for their final arguments. Be sure to remind your students to explain why they included the evidence in their arguments and make the assumptions underlying their analysis and interpretation of the data explicit in order to provide an adequate justification of their evidence.

Reflecting on Ways to Design Better Investigations

It is important for students to reflect on the strengths and weaknesses of the investigation they designed during the explicit and reflective discussion. Students should therefore be encouraged to discuss ways to eliminate potential flaws, measurement errors, or sources of uncertainty in their investigations. To help students be more reflective about the design of their investigation and what they can do to make their investigations more rigorous in the future, you can ask them the following questions:

1. What were some of the strengths of the way you planned and carried out your investigation? In other words, what made it scientific?

2. What were some of the weaknesses of the way you planned and carried out your investigation? In other words, what made it less scientific?

3. What rules can we make, as a class, to ensure that our next investigation is more scientific?

Reflecting on the Nature of Scientific Knowledge and Scientific Inquiry

This investigation can be used to illustrate two important concepts related to the nature of scientific knowledge and the nature of scientific inquiry: (a) how the culture of science, societal needs, and current events influence the work of scientists; and (b) the nature and role of experiments in science (see Appendix 2 for a brief description of these two concepts). Be sure to review these concepts during and at the end of the explicit and reflective discussion. To help students think about these concepts in relation to what they did during the lab, you can ask them the following questions:

1. People view some types of research as being more important than other types of research because of cultural values and current events. Can you come up with some examples of how cultural values and current events have influenced the work of scientists?

2. Scientists share a set of values, norms, and commitments that shape what counts as knowing, how to represent or communicate information, and how to interact with other scientists. Can you work with your group to come up with a rule that you can use to decide if something is science or not science? Be ready to share in a few minutes.

3. I asked you to design and carry out an experiment as part of your investigation. Can you give me some examples of what experiments are used for in science?

4. Can you work with your group to come up with a rule that you can use to decide if an investigation is an experiment or not? Be ready to share in a few minutes.

You can also use presentation software or other techniques to encourage your students to think about these concepts. You can show examples of research projects that were influenced by cultural values and current events and ask students to think about what was going on at the time and why that research was viewed as being important for the greater good. You can also show images of different types of investigations (such as a physicist or an astronomer collecting data using a telescope as part of an observational or descriptive study, a person working in the library doing a literature review, a person working on a computer to analyze an existing data set, and an actual experiment) and ask students to indicate if they think each image represents an experiment and why or why not.

Be sure to remind your students that, to be proficient in science, it is important that they understand what counts as scientific knowledge and how that knowledge develops over time.

Hints for Implementing the Lab

- Allowing students to design their own procedures for collecting data gives students an opportunity to try, to fail, and to learn from their mistakes. However, you can scaffold students as they develop their procedure by having them fill out an investigation proposal. These proposals provide a way for you to offer students hints and suggestions without telling them how to do it. You can also check the proposals quickly during a class period. For this lab we suggest using Investigation Proposal A.

LAB 19

- Be sure to use blocks that are the same size.

- Encourage students to measure or calculate several different physical properties of the blocks. This information will enable them to relate the structure (physical properties) of the blocks to their function (decrease the force and increase the time of the collision).

- Learn how to use the dynamics cart and track, the force sensor, and the data collection and analysis software before the lab begins. It is important for you to know how to use the equipment so you can help students when technical issues arise.

- Allow students to become familiar with the dynamics cart and track, the force sensor, and the data collection and analysis software as part of the tool talk before they begin to design their investigation. Giving them 5–10 minutes to examine the available equipment and materials will let them see what they can and cannot do with them. This also gives students a chance to begin thinking about what variables to test and how to control for other variables.

- This lab provides students with an important opportunity to think about controlling variables in scientific investigations. Specifically, students must control the velocity and mass of the cart before impact. Students may also want to control the shape, mass, or size of the materials used during the collision. However, if students do not control these variables, this provides an important learning opportunity for the explicit and reflective discussion.

- Be sure to allow students to go back and re-collect data at the end of the argumentation session. Students often realize that they made numerous mistakes when they were collecting data as a result of their discussions during the argumentation session. The students, as a result, will want a chance to re-collect data, and the re-collection of data should be encouraged when time allows. This also offers an opportunity to discuss what scientists do when they realize a mistake is made inside the lab.

Connections to Standards

Table 19.2 highlights how the investigation can be used to address specific performance expectations from the *NGSS*; learning objectives from AP Physics 1; learning objectives from AP Physics C: Mechanics, *Common Core State Standards*, in English language arts (*CCSS ELA*); and *Common Core State Standards*, Mathematics (*CCSS Mathematics*).

TABLE 19.2

Lab 19 alignment with standards

***NGSS* performance expectations**	• HS-PS2-2: Use mathematical representations to support the claim that the total momentum of a system of objects is conserved when there is no net force on the system. • HS-PS2-3: Apply scientific and engineering ideas to design, evaluate, and refine a device that minimizes the force on a macroscopic object during a collision.
AP Physics 1 learning objectives	• 3.D.1.1: The student is able to justify the selection of data needed to determine the relationship between the direction of the force acting on an object and the change in momentum caused by that force. • 3.D.2.1: The student is able to justify the selection of routines for the calculation of the relationships between changes in momentum of an object, average force, impulse, and time of interaction. • 3.D.2.2: The student is able to predict the change in momentum of an object from the average force exerted on the object and the interval of time during which the force is exerted. • 3.D.2.3: The student is able to analyze data to characterize the change in momentum of an object from the average force exerted on the object and the interval of time during which the force is exerted. • 3.D.2.4: The student is able to design a plan for collecting data to investigate the relationship between changes in momentum and the average force exerted on an object over time.
AP Physics C: Mechanics learning objectives	• I.B.2.b: Students should understand how Newton's second law applies to an object subject to forces such as gravity, the pull of strings, or contact forces. • I.D.2.a: Relate mass, velocity, and linear momentum for a moving object, and calculate the total linear momentum of a system of objects. • I.D.2.b: Relate impulse to the change in linear momentum and the average force acting on an object. • I.D.2.c: State and apply the relations between linear momentum and center-of-mass motion for a system of particles. • I.D.2.d: Calculate the area under a force versus time graph and relate it to the change in momentum of an object. • I.D.2.e: Calculate the change in momentum of an object given a function $\mathbf{F}(t)$ for the net force acting on the object.
Literacy connections (*CCSS ELA*)	• *Reading:* Key ideas and details, craft and structure, integration of knowledge and ideas • *Writing:* Text types and purposes, production and distribution of writing, research to build and present knowledge, range of writing • *Speaking and listening:* Comprehension and collaboration, presentation of knowledge and ideas

continued

Table 19.2 (*continued*)

Mathematics connections (*CCSS Mathematics*)	• *Mathematical practices:* Reason abstractly and quantitatively, construct viable arguments and critique the reasoning of others, use appropriate tools strategically, attend to precision • *Number and quantity:* Reason quantitatively and use units to solve problems, represent and model with vector quantities, perform operations on vectors • *Algebra:* Interpret the structure of expressions, understand solving equations as a process of reasoning and explain the reasoning, solve equations and inequalities in one variable, represent and solve equations and inequalities graphically • *Functions:* Understand the concept of a function and use function notation, interpret functions that arise in applications in terms of the context, analyze functions using different representations, interpret expressions for functions in terms of the situation they model • *Statistics and probability:* Summarize, represent, and interpret data on two categorical and quantitative variables; understand and evaluate random processes underlying statistical experiments; make inferences and justify conclusions from sample surveys, experiments, and observational studies

Reference

Newton, I. 1687. *Philosophiae naturalis principia mathematica* [Mathematical principles of natural philosophy]. London: S. Pepys.

Lab Handout

Lab 19. Impulse and Materials: Which Material Is Most Likely to Provide the Best Protection for a Phone That Has Been Dropped?

Introduction

People throughout the United States use personal electronic devices. On any given day, a person might use a cell phone, smartwatch, laptop computer, or tablet in a number of different settings. Unfortunately, people often damage these devices by accidently dropping them or knocking them off a table. Most consumers, however, do not want to replace any of these devices on a regular basis because they tend to be expensive. This is one reason why so many people use a case to protect their different personal electronic devices. Many different companies sell cases that are designed to prevent phones, tablets, or laptops from breaking when they are dropped or knocked off a table.

When a person drops a cell phone, the force of gravity causes the phone to accelerate toward the ground at -9.8 m/s^2. Any object with a velocity has momentum, and as an object accelerates during free fall, its momentum increases. Upon reaching the ground, the ground exerts an upward, normal force that causes the momentum of the object to change. Figure L19.1 is an example of a momentum versus time graph for a cell phone that falls toward and then strikes the ground, bouncing upward once, and then coming to rest. Point A in the graph corresponds to the cell phone falling out of a person's hand. Between points B and C, the cell phone is hitting the ground, with the ground exerting a force on the cell phone, causing it to bounce back up. In other words, the segment of the graph between points B and C corresponds to the collision between the phone and the ground. At point D, the cell phone has reached the highest point of its bounce, and then at point E, the cell phone hits the ground and finally comes to rest at point F. The segment of the graph between E and F corresponds to the second collision between the phone and the ground.

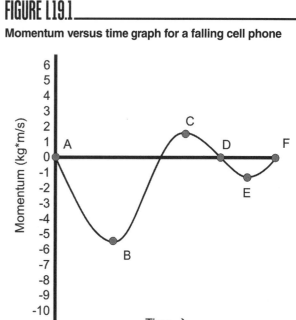

FIGURE L19.1

Momentum versus time graph for a falling cell phone

Cell phones tend to break easily when dropped because they are made out of lightweight materials, such as glass, aluminum, and plastic, which have a relatively low breaking point. The breaking point of a material is the maximum amount of force that it can absorb before it deforms. Companies make cell phones out of these materials because it allows the companies to make phones that are thin and light so they are easy to carry around. Unfortunately, a phone that is thin and light is also fragile.

Protective cases help alleviate this design trade-off. Protective phone cases tend to be made from a material that has two important characteristics. First, the material must be able to minimize the force that acts on the phone during a collision. This is important because the change of momentum associated with a phone in free fall suddenly hitting the ground is a vector that acts in the direction of the net force exerted on it. Second, the material must be able to maximize the time of the collision. This characteristic is important to consider because any change in momentum happens over a specific time interval. Your goal in this investigation is to use this information to identify a suitable material for a new protective cell phone case.

Your Task

Use what you know about momentum, impulses, the relationship between structure and function, and scale, proportional relationships, and vector quantities to design and carry out an investigation to compare how different materials change the momentum of an object during a collision.

The guiding question for this investigation is, *Which material is most likely to provide the best protection for a phone that has been dropped?*

Materials

You may use any of the following materials during your investigation:

- Safety glasses or goggles (required)
- Dynamics cart
- Dynamics track
- Force sensor
- Sensor interface
- Computer, tablet, or graphing calculator with data collection and analysis software
- Foam block
- Plastic block
- Rubber block
- Wooden block
- Metal block

Safety Precautions

Follow all normal lab safety rules. In addition, take the following safety precautions:

1. Wear sanitized safety glasses or goggles during lab setup, hands-on activity, and takedown.

2. Keep fingers and toes out of the way of moving objects.

3. Wash hands with soap and water after completing the lab.

Investigation Proposal Required?   ☐ Yes   ☐ No

Getting Started

To answer the guiding question, you will need to investigate how different materials change the momentum of an object during a collision. Figure L19.2 shows how you can attach a force sensor to a cart and then roll it down a frictionless track on an incline. The end of the force sensor should collide with the material you are testing at the end of the track. Starting the cart at the same point on the track for each test will allow you to control for the momentum of the cart prior to the collision with the block. Before you can begin to design your investigation using this equipment, however, you must determine what type of data you need to collect, how you will collect it, and how you will analyze it.

FIGURE L19.2 _____

One way to examine the change in momentum of the cart after a collision with a specific material

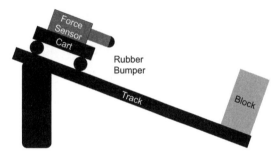

To determine *what type of data you need to collect*, think about the following questions:

- How might changes in the structure of a cell phone case affect the function of the case?
- What are the components of this system and how do they interact?
- How can you describe the components of the system quantitatively?
- How do you quantify a change in momentum?
- What measurements do you need to make?
- What will be the independent variable and the dependent variable for your experiment?

To determine *how you will collect the data*, think about the following questions:

- What other variables do you need to measure or control?
- What scale or scales should you use when you take your measurements?
- Which quantities are vectors, and which quantities are scalars?
- For any vector quantities, which directions are positive and which directions are negative?

- How will you make sure that your data are of high quality (i.e., how will you reduce error)?
- How will you keep track of and organize the data you collect?

To determine *how you will analyze the data,* think about the following questions:

- What type of calculations will you need to make?
- What types of comparisons will be useful?
- How could you use mathematics to document a difference between materials?
- What type of table or graph could you create to help make sense of your data?

Connections to the Nature of Scientific Knowledge and Scientific Inquiry

As you work through your investigation, you may want to consider

- how the culture of science, societal needs, and current events influence the work of scientists; and
- the nature and role of experiments in science.

Initial Argument

Once your group has finished collecting and analyzing your data, your group will need to develop an initial argument. Your initial argument needs to include a claim, evidence to support your claim, and a justification of the evidence. The *claim* is your group's answer to the guiding question. The *evidence* is an analysis and interpretation of your data. Finally, the *justification* of the evidence is why your group thinks the evidence matters. The justification of the evidence is important because scientists can use different kinds of evidence to support their claims. Your group will create your initial argument on a whiteboard. Your whiteboard should include all the information shown in Figure L19.3.

FIGURE L19.3 _____
Argument presentation on a whiteboard

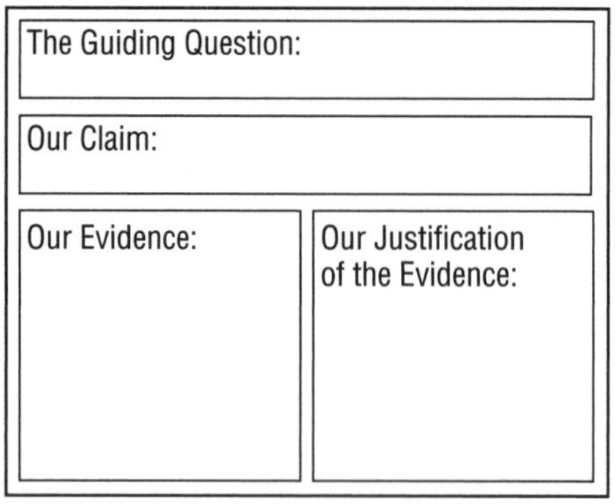

The Guiding Question:	
Our Claim:	
Our Evidence:	Our Justification of the Evidence:

Argumentation Session

The argumentation session allows all of the groups to share their arguments. One or two members of each group will stay at the lab station to share that group's argument, while the other members of the group go to the other lab stations to listen to and critique the other arguments. This is similar to what scientists do when they propose, support, evaluate, and refine new ideas during a poster session at a conference. If you are presenting your group's argument, your goal is to share your ideas and answer questions. You should also keep a record of the critiques and suggestions made by your classmates so you can use this feedback to make your initial argument stronger. You can keep track of specific critiques and suggestions for improvement that your classmates mention in the space below.

Critiques about our initial argument and suggestions for improvement:

If you are critiquing your classmates' arguments, your goal is to look for mistakes in their arguments and offer suggestions for improvement so these mistakes can be fixed. You should look for ways to make your initial argument stronger by looking for things that the other groups did well. You can keep track of interesting ideas that you see and hear during the argumentation in the space below. You can also use this space to keep track of any questions that you will need to discuss with your team.

Interesting ideas from other groups or questions to take back to my group:

LAB 19

Once the argumentation session is complete, you will have a chance to meet with your group and revise your initial argument. Your group might need to gather more data or design a way to test one or more alternative claims as part of this process. Remember, your goal at this stage of the investigation is to develop the best argument possible.

Report

Once you have completed your research, you will need to prepare an *investigation report* that consists of three sections. Each section should provide an answer to the following questions:

1. What question were you trying to answer and why?

2. What did you do to answer your question and why?

3. What is your argument?

Your report should answer these questions in two pages or less. This report must be typed, and any diagrams, figures, or tables should be embedded into the document. Be sure to write in a persuasive style; you are trying to convince others that your claim is acceptable or valid!

Checkout Questions

Lab 19. Impulse and Materials: Which Material Is Most Likely to Provide the Best Protection for a Phone That Has Been Dropped?

1. How might your lab results regarding force, impulse, and momentum inform engineers when they design cars? How can this information help keep people safe during a collision between two cars?

2. Scientists and engineers who study ways to transport people to other planets must account for a number of challenges in the design of spaceships. One problem is how to get a large spaceship to move with a fast enough velocity. In response, some have suggested using a solar sail, where a ship uses a specially designed sail to catch microscopic particles continuously emitted by the Sun. How do the results of your lab relate to solar sails? Why might scientists think solar sails are a good solution?

3. How an object is structured is related to its function.

    a. I agree with this statement.

    b. I disagree with this statement.

Explain your answer, using an example from your investigation about impulse and momentum.

4. The research done by a scientist is often influenced by current events or what is important in society.

    a. I agree with this statement.

    b. I disagree with this statement.

Explain your answer, using an example from your investigation about impulse and momentum.

5. Experiments are one type of research design used by scientists. Explain what an experiment is and what types of questions are best answered using an experiment, using examples from your investigation about impulse and momentum as well as from previous investigations in this class or your previous science classes.

6. Scientists often need to think about scales, proportional relationships, and vector quantities during an investigation. Explain why it is important for scientists to think about these things, using examples from your investigation about impulse and momentum as well as from previous investigations in this class or your previous science classes.

SECTION 7
Energy, Work, and Power

Introduction Labs

LAB 20

Teacher Notes

Lab 20. Kinetic and Potential Energy: How Can We Use the Work-Energy Theorem to Explain and Predict Behavior of a System That Consists of a Ball, a Ramp, and a Cup?

Purpose

The purpose of this lab is to *introduce* students to the disciplinary core idea (DCI) of Energy from the *NGSS* by having them use the work-energy theorem to explain and predict the behavior of a system that consists of a ball, a ramp, and a cup. This lab also gives students an opportunity to learn about the crosscutting concepts (CCs) of (a) Systems and System Models and (b) Energy and Matter: Flows, Cycles, and Conservation from the *NGSS*. In addition, this lab can be used to help students understand three big ideas from AP Physics: (a) objects and systems have properties such as mass and charge, (b) the interactions of an object with other objects can be described by forces, and (c) changes that occur as result of interactions are constrained by conservation laws. As part of the explicit and reflective discussion, students will also learn about (a) how scientific knowledge changes over time and (b) the difference between laws and theories in science.

Underlying Physics Concepts

The work-energy theorem states that work done by a force acting on an object over a displacement results in a change in the energy of the object. The change can be in terms of the type of energy or the amount of energy. In this lab, work is done by gravity to convert the energy of the ball from potential energy at the top of the ramp to kinetic energy at the bottom of the ramp. Work is then done by friction to dissipate the kinetic energy of the ball once it enters the cup as heat, eventually resulting in the ball and cup coming to rest.

To understand the work done by gravity, we first start with the law of conservation of energy. The law of conservation of energy states that in a closed system the total energy of the system is conserved. The energy of an object can be described based on its potential energy and kinetic energy. Potential energy is stored energy of position. Gravitational potential energy is stored energy due to an object's location in a gravitational field. Kinetic energy is the energy an object has due to its motion. The potential energy, PE, and kinetic energy, KE, of a ball on a ramp can be calculated using Equations 20.1 and 20.2, where m is the mass, g is the acceleration due to gravity, h is the height, and v is the velocity. In SI units, energy is measured in joules (J), mass is measured in kilograms (kg), height is measured in meters (m), and velocity is measured in meters per second (m/s). The acceleration due to gravity is constant and equal to 9.8 m/s².

$$(\text{Equation 20.1}) \quad PE = m\mathbf{g}\mathbf{h}$$

Kinetic and Potential Energy
How Can We Use the Work-Energy Theorem to Explain and Predict Behavior of a System
That Consists of a Ball, a Ramp, and a Cup?

(Equation 20.2) $KE = \frac{1}{2}m\mathbf{v}^2$

The total mechanical energy of an object is the sum total of its individual types of energy. This means that the total mechanical energy of an object (ME) is the sum of its potential and kinetic energy, shown in Equation 20.3.

(Equation 20.3) $ME = PE + KE$

There exist three unique points in the experiment where the kinetic and potential energies of the ball can be calculated, as shown in Figure 20.1. At point A, $PE_A = mg\mathbf{h}$ because the ball is in a gravitational field of height **h.** At point A, $KE_A = 0$ since the ball is at rest. Thus, at point A, all of the mechanical energy of the ball is in the form of potential energy. At point B, $PE_B = 0$ because the ball has a height of zero meters above the ground in the gravitational field, and $KE_B = \frac{1}{2}m\mathbf{v}^2$ because the ball has a non-zero velocity. At point B, all of the energy of the ball is in the form of kinetic energy.

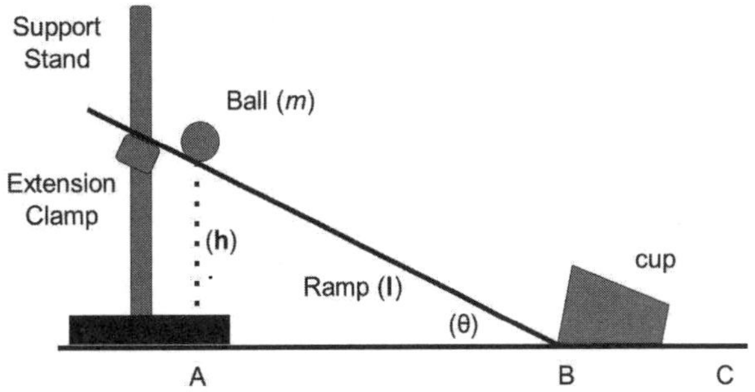

Note that because the total energy of a system is conserved, the total energy at point A must be equal to the total energy at point B. However, the type of energy of the ball has changed from point A to point B. Thus, work must have been done on the ball between these two points. As mentioned previously, work is a force acting on an object over a displacement. When the ball moves from point A to point B, the force acting on the ball is gravity. Thus, work is done by gravity to change the energy of the ball from potential energy to kinetic energy.

FIGURE 20.1

Energies at points A, B, and C for a ball of mass *m* rolling down a ramp of length l at an angle of inclination θ

Furthermore, as we have established, the kinetic energy at point A is equal to zero and the potential energy at point B is equal to zero. Thus, $PE_A = KE_B$ and therefore $mg\mathbf{h} = \frac{1}{2}m\mathbf{v}^2$ when the ball moves from point A to B. The mass value on both sides of the equation "divides to 1" (the more mathematically appropriate term for the commonly used "cancels out"), indicating that the mass of the ball does not affect the velocity of the ball at the bottom of the ramp.

The total energy of the ball at any point between point A and point B is constant. That is, the sum total of the potential energy and kinetic energy of the ball will not change as it rolls down the incline. Thus, the height and the velocity of the ball at point A will determine the total energy of the system. In most cases, the ball is released from rest. This means the total energy of the ball is a function of the height of point A above ground level. After the ball and cup have come to rest at point C, $PE_C = 0$ and $KE_C = 0$ since the ball and cup have a height and velocity of zero. In this instance, work done on the ball and cup results in a loss of energy.

The students can calculate the displacement of the ball and the cup before they come to rest after the collision at point B to verify KE_B. That is, the farther the cup moves (or the longer it takes for the cup to come to rest), the more energy the ball must have had at the bottom of the ramp. This is because the frictional force between the cup and the ground results in work being done on the cup to bring it to rest. The work-energy theorem states that work done on an object changes the mechanical energy of the object. This means work can be done to increase the mechanical energy, or, in the case of work done by friction, work can be done to decrease the mechanical energy of an object.

We can express this mathematically using the following approach. Work done by friction is directly proportional to the distance that it takes an object to come to rest—the longer the distance, the more work is done to bring the object to rest (assuming a constant force). Thus, the longer the distance it takes the cup to come to rest, the more work was done to dissipate the kinetic energy of the cup, and the more kinetic energy the cup must have had. Equation 20.4 shows the relationship between work and force, where W is work, $\mathbf{F}$ is force, and $\mathbf{x}$ is displacement. In SI units, work is measured in joules (J), force is measured in newtons (N), and displacement is measured in meters (m).

$$\textbf{(Equation 20.4)} \quad W = \mathbf{Fx}$$

Notice that work is measured in joules; this makes sense, because work leads to a change in the mechanical energy of an object, and energy is also measured in joules. Recognizing this fact, we can also relate the work done by friction to bring the cup to a stop to the kinetic energy of the ball at the bottom of the ramp by setting these two quantities equal to each other. This gives us Equation 20.5. In this equation, $\mathbf{F}_{fr}$ is the frictional force acting on the cup. Note that to calculate the force of friction acting on the cup, one must account for the mass of both the cup and the ball inside the cup. However, the value for the mass on the right side of the equation is only the mass of the ball prior to the ball entering the cup, because the cup does not have kinetic energy prior to the ball entering the cup.

$$\textbf{(Equation 20.5)} \quad \mathbf{F}_{fr}\mathbf{x} = \tfrac{1}{2}m\mathbf{v}^2$$

Because the kinetic energy at the bottom of the ramp is equal to the potential energy of the ball at the top of the ramp, we can also set the work done by friction to bring the cup to rest equal to the potential energy of the ball at the top of the ramp, shown in Equation 20.6. Again, recognize that the mass on the right side of the equation is only the mass of the ball, whereas to calculate the force of friction on the cup, one must account for both the mass of the ball and the mass of the cup.

$$\textbf{(Equation 20.6)} \quad F_{fr}x = mgh$$

To perform these calculations, we need to make several assumptions about the system. First, we must assume there is no friction between the ball and the ramp between points A and B in Figure 20.1. We also must assume that as the ball moves down the ramp (a) all of the potential energy transforms into translational kinetic energy and (b) none of the potential energy is transformed into rotational kinetic energy of the ball. Finally, we must assume that the collision between the ball and the cup at the bottom of the ramp is elastic, which means no kinetic energy is lost when the ball hits the back of the cup and they both begin to move. This approach makes sense, because we have identified the major factors in the motion of the ball. However, the empirical results will not match the predicted results that are derived from Equations 20.1–20.6. Because of this, we strongly recommend that students only build a conceptual model that identifies major relationships.

Timeline

The instructional time needed to complete this lab investigation is 170–230 minutes. Appendix 3 (p. 531) provides options for implementing this lab investigation over several class periods. Option C (230 minutes) should be used if students are unfamiliar with scientific writing, because this option provides extra instructional time for scaffolding the writing process. You can scaffold the writing process by modeling, providing examples, and providing hints as students write each section of the report. Option C can also be used if you are introducing students to the video analysis programs. Option D (170 minutes) should be used if students are familiar with scientific writing and have developed the skills needed to write an investigation report on their own. In option D, students complete stage 6 (writing the investigation report) and stage 8 (revising the investigation report) as homework.

Materials and Preparation

The materials needed to implement this investigation are listed in Table 20.1 (p. 436). The items can be purchased from a science supply company such as Carolina, Flinn Scientific, PASCO, Vernier, or Ward's Science. The 25 mm balls can be purchased as a set (often called a drilled ball set in science supply catalogs). These sets usually include two of each type of ball (brass, aluminum, steel, cork, wood, and copper).

The PVC pipe can be purchased at a home improvement store such as Home Depot or Lowe's. You can make six 1 m long ramps from one 10 ft. long piece of PVC pipe. To make the ramps, first cut the 10 ft. long PVC pipe into three equal-length pieces (each piece will be approximately 1 m in length) and then cut each 1 m piece lengthwise.

The use of video analysis is recommended for this lab. The video analysis software can be purchased from Vernier (Logger *Pro*) or PASCO (SPARKvue or Capstone). These companies also have apps that can be used on Apple- or Android-based tablets and cell phones. We recommend consulting with your school's information technology coordinator to determine the best option for your students.

TABLE 20.1

Materials list for Lab 20

Item	Quantity
Safety glasses or goggles	1 per student
Support stand	1 per group
Extension clamp	1 per group
Set of 25 mm balls (includes brass, aluminum, steel, cork, wood, and copper)	1 per group
PVC pipe (2 in. diameter, 1 m in length)	1 per group
Electronic or triple beam balance	1 per group
Plastic cup	1 per group
Meterstick	2 per group
Protractor	1 per group
Stopwatch	1 per student
Investigation Proposal A (optional)	1 per group
Whiteboard, 2' × 3'*	1 per group
Lab Handout	1 per student
Peer-review guide and teacher scoring rubric	1 per student
Checkout Questions	1 per student
Equipment for video analysis (optional)	
Video camera	1 per group
Computer or tablet with video analysis software	1 per group

* As an alternative, students can use computer and presentation software such as Microsoft PowerPoint or Apple Keynote to create their arguments.

Kinetic and Potential Energy
How Can We Use the Work-Energy Theorem to Explain and Predict Behavior of a System
That Consists of a Ball, a Ramp, and a Cup?

Be sure to use a set routine for distributing and collecting the materials during the lab investigation. One option is to set up the materials for each group at each group's lab station before class begins. This option works well when there is a dedicated section of the classroom for lab work and the materials are large and difficult to move (such as a dynamics track). A second option is to have all the materials on a table or cart at a central location. You can then assign a member of each group to be the "materials manager." This individual is responsible for collecting all the materials his or her group needs from the table or cart during class and for returning all the materials at the end of the class. This option works well when the materials are small and easy to move (such as stopwatches, metersticks, or hanging masses). It also makes it easy to inventory the materials at the end of the class before students leave for the day.

Safety Precautions

Remind students to follow all normal lab safety rules. In addition, tell students to take the following safety precautions:

1. Wear sanitized safety glasses or goggles during lab setup, hands-on activity, and takedown.

2. Keep fingers and toes out of the way of moving objects.

3. Wash hands with soap and water after completing the lab.

Topics for the Explicit and Reflective Discussion
Reflecting on the Use of Core Ideas and Crosscutting Concepts During the Investigation

Teachers should begin the explicit and reflective discussion by asking students to discuss what they know about the core idea they used during the investigation. The following are some important concepts related to the core idea of Energy that students will need to explain the behavior of a system that consist of a ball, a ramp, and a cup using the work-energy theorem:

- Energy is conserved in all systems under all circumstances.

- Conserved quantities are constant in an isolated or a closed system. An *open system* is one that exchanges any conserved quantity with its surroundings.

- The mechanical energy of a system includes the kinetic energy of the objects that make up the system and the potential energy of the configuration of the objects that make up the system.

- Changes in the potential energy of a system can result in changes to the kinetic energy of the system because energy is constant in a closed system.

- *Work* is defined as the product of a force applied on an object or system of objects over a displacement. Work done on an object or a system of objects results in a change in the type of energy (e.g., work done by gravity to change potential energy to kinetic energy) or a change in the amount of energy (e.g., work done by friction dissipates kinetic energy, thus bringing a moving object to rest).

To help students reflect on what they know about energy, we recommend showing them two or three images using presentation software that help illustrate these important ideas. You can then ask the students the following questions to encourage them to share how they are thinking about these important concepts:

1. What do we see going on in this image?

2. Does anyone have anything else to add?

3. What might be going on that we can't see?

4. What are some things that we are not sure about here?

You can then encourage students to think about how crosscutting concepts (CCs) played a role in their investigation. There are at least two CCs that students need to explain the behavior of a system that consist of a ball, a ramp, and a cup using the work-energy theorem: (a) Systems and System Models and (b) Energy and Matter: Flows, Cycles, and Conservation (see Appendix 2 [p. 527] for a brief description of these CCs). To help students reflect on what they know about these CCs, we recommend asking them the following questions:

1. Scientists often need to define the system under study and then make a model during an investigation. Why is developing a model of system so useful in science?

2. What were the boundaries and components of the system you studied during this investigation? What were the strengths and limitations of the model you developed?

3. Scientists often attempt to track how energy and matter move into, out of, and within systems as part of an investigation. Why is this useful?

4. How did you attempt to track how energy moves within the system you were studying? What did tracking energy allow you to do during your investigation?

You can then encourage the students to think about how they used all these different concepts to help answer the guiding question and why it is important to use these ideas to help justify their evidence for their final arguments. Be sure to remind your students to explain why they included the evidence in their arguments and make the assumptions underlying their analysis and interpretation of the data explicit in order to provide an adequate justification of their evidence.

Kinetic and Potential Energy

How Can We Use the Work-Energy Theorem to Explain and Predict Behavior of a System That Consists of a Ball, a Ramp, and a Cup?

Reflecting on Ways to Design Better Investigations

It is important for students to reflect on the strengths and weaknesses of the investigation they designed during the explicit and reflective discussion. Students should therefore be encouraged to discuss ways to eliminate potential flaws, measurement errors, or sources of uncertainty in their investigations. To help students be more reflective about the design of their investigation and what they can do to make their investigations more rigorous in the future, you can ask them the following questions:

1. What were some of the strengths of the way you planned and carried out your investigation? In other words, what made it scientific?

2. What were some of the weaknesses of the way you planned and carried out your investigation? In other words, what made it less scientific?

3. What rules can we make, as a class, to ensure that our next investigation is more scientific?

Reflecting on the Nature of Scientific Knowledge and Scientific Inquiry

This investigation can be used to illustrate two important concepts related to the nature of scientific knowledge and the nature of scientific inquiry: (a) how scientific knowledge changes over time and (b) the difference between laws and theories in science (see Appendix 2 for a brief description of these two concepts). Be sure to review these concepts during and at the end of the explicit and reflective discussion. To help students think about these concepts in relation to what they did during the lab, you can ask them the following questions:

1. Scientific knowledge can and does change over time. Can you tell me why it changes?

2. Can you work with your group to come up some examples of how scientific knowledge has changed over time? Be ready to share in a few minutes.

3. Is the work-energy theorem an example of a law or a theory? Why?

4. Can you work with your group to come up with a rule that you can use to decide if something is a law or a theory? Be ready to share in a few minutes.

You can also use presentation software or other techniques to encourage your students to think about these concepts. You can show examples of how our thinking about energy has changed over time and ask students to discuss what they think led to those changes. You can also show examples of either a law (such as $PE_{grav} = m\mathbf{g}h$) or a theory (such as *gravity is the curvature of four-dimensional space-time due to the presence of mass*) and ask students to indicate if they think it is law or a theory and why.

LAB 20

Be sure to remind your students that, to be proficient in science, it is important that they understand what counts as scientific knowledge and how that knowledge develops over time.

Hints for Implementing the Lab

- We recommend the use of video analysis in this lab, because there are numerous assumptions about the motion that are used to simplify the mathematics and isolate important variables but that contribute to a mismatch between the predicted and observed results. The use of video analysis will allow students to isolate different parts of the motion and analyze data for each part independently. For example, video analysis software will allow students to analyze the motion of the ball rolling down the ramp separate from the motion of the ball in the cup. This will increase the precision of their measurements.

- Allowing students to design their own procedures for collecting data gives students an opportunity to try, to fail, and to learn from their mistakes. However, you can scaffold students as they develop their procedure by having them fill out an investigation proposal. These proposals provide a way for you to offer students hints and suggestions without telling them how to do it. You can also check the proposals quickly during a class period. For this lab we suggest using Investigation Proposal A.

- To minimize friction as much as possible, use smooth PVC pipe for the ball to roll down and a smooth floor surface for the cup to move.

- You may want to use two-sided tape on the inside of the cup so when the ball comes into contact with the inside of the cup, it sticks and the recoiling and subsequent second collision with the cup is minimized.

- Allow the students to become familiar with the equipment and materials as part of the tool talk before they begin to design their investigation. Giving them 5–10 minutes to examine the equipment and materials will let them see what they can and cannot do with them. This also gives students a chance to begin thinking about what variables to test and how to control for other variables.

- You may want to have students use mathematics when they are developing their rules to explain and predict the motion of a ball at the bottom of a ramp and the distance a cup moves after the ball rolls down the ramp and enters the cup. This is a great opportunity for them to practice building a function that models a relationship between two quantities. However, it is not required for this lab.

- Be sure to allow students to go back and re-collect data at the end of the argumentation session. Students often realize that they made numerous mistakes when they were collecting data as a result of their discussions during the argumentation session. The students, as a result, will want a chance to re-collect data, and the re-collection of data should be encouraged when time allows. This also offers an opportunity to discuss what scientists do when they realize a mistake is made inside the lab.

If students use video analysis

- We suggest allowing students to familiarize themselves with the video analysis software before they finalize the procedure for the investigation, especially if they have not used such software previously. This gives students an opportunity to learn how to work with the software and to improve the quality of the video they take.

- Remind students to hold the video camera as still as possible. Any movement of the camera will introduce error into their analysis. If using actual camcorders, we recommend using a tripod to hold the camera steady. If students are using a camera on a cell phone or tablet, we recommend using a table to help steady the camera.

- Remind students to place a meterstick in the same field of view as the motion they are capturing with the video camera. Also, the meterstick should be approximately the same distance from the camera as the motion. Most video analysis software requires the user to define a scale in the video (this allows the software to establish distances and, subsequently, other variables dependent on distance and displacement).

Connections to Standards

Table 20.2 (p. 442) highlights how the investigation can be used to address learning objectives from AP Physics 1; learning objectives from AP Physics C: Mechanics, *Common Core State Standards,* in English language arts (*CCSS ELA*); and *Common Core State Standards,* Mathematics (*CCSS Mathematics*).

LAB 20

TABLE 20.2

Lab 20 alignment with standards

NGSS performance expectations	• None
AP Physics 1 learning objectives	• 3.A.1.1: The student is able to express the motion of an object using narrative, mathematical, and graphical representations. • 3.A.1.2: The student is able to design an experimental investigation of the motion of an object. • 3.A.1.3: The student is able to analyze experimental data describing the motion of an object and is able to express the results of the analysis using narrative, mathematical, and graphical representations. • 4.C.1.2: The student is able to predict changes in the total energy of a system due to changes in position and speed of objects or frictional interactions within the system. • 5.B.1.2: The student is able to translate between a representation of a single object, which can only have kinetic energy, and a system that includes the object, which may have both kinetic and potential energies. • 5.B.3.2: The student is able to make quantitative calculations of the internal potential energy of a system from a description or diagram of that system. • 5.B.3.3: The student is able to apply mathematical reasoning to create a description of the internal potential energy of a system from a description or diagram of the objects and interactions in that system.
AP Physics C: Mechanics learning objectives	• I.C.1.b(3): Apply the theorem to determine the change in an object's kinetic energy and speed that results from the application of specified forces, or to determine the force that is required in order to bring an object to rest in a specified distance. • I.C.2.b(5): Calculate the potential energy of one or more objects in a uniform gravitational field. • I.C.3.c: Students should be able to recognize and solve problems that call for application both of conservation of energy and Newton's Laws.
Literacy connections (CCSS ELA)	• *Reading:* Key ideas and details, craft and structure, integration of knowledge and ideas • *Writing:* Text types and purposes, production and distribution of writing, research to build and present knowledge, range of writing • *Speaking and listening:* Comprehension and collaboration, presentation of knowledge and ideas

Mathematics connections (*CCSS Mathematics*)	• *Mathematical practices:* Reason abstractly and quantitatively, construct viable arguments and critique the reasoning of others, use appropriate tools strategically, attend to precision
	• *Number and quantity:* Reason quantitatively and use units to solve problems, represent and model with vector quantities, perform operations on vectors
	• *Algebra:* Interpret the structure of expressions, create equations that describe numbers or relationships, understand solving equations as a process of reasoning and explain the reasoning, solve equations and inequalities in one variable, represent and solve equations and inequalities graphically
	• *Functions:* Understand the concept of a function and use function notation, interpret functions that arise in applications in terms of the context, analyze functions using different representations, build a function that models a relationship between two quantities, interpret expressions for functions in terms of the situation they model
	• *Statistics and probability:* Summarize, represent, and interpret data on two categorical and quantitative variables; understand and evaluate random processes underlying statistical experiments; make inferences and justify conclusions from sample surveys, experiments, and observational studies

Lab Handout

Lab 20. Kinetic and Potential Energy: How Can We Use the Work-Energy Theorem to Explain and Predict Behavior of a System That Consists of a Ball, a Ramp, and a Cup?

Introduction

You have learned how to use Newton's laws of motion to explain how objects move and to predict the motion of an object over time. Newton's laws of motion therefore function as a useful model that people can use to understand how the world works and to predict how different objects move after they interact with each other. Physicists, however, can use other models to explain and predict the motion of objects and to help understand natural phenomena. One such model views the motion of objects through a lens of work and energy. This model is called the *work-energy theorem*. To understand how work and energy can be used to explain and predict motion, it is important to understand the basic assumptions underlying the work-energy theorem.

The first basic assumption of the work-energy theorem is that an object can store energy as the result of its position. This stored energy is called *potential energy*. The second basic assumption of the work-energy theorem is that all objects in motion have energy because they are moving. The energy of motion is called *kinetic energy*. There are many forms of kinetic energy, including vibrational (the energy due to vibrational motion), rotational (the energy due to rotational motion), and translational (the energy due to motion from one location to another). The third assumption of the work-energy theorem is that doing work on an object results in a change in the *mechanical energy* of that object. Mechanical energy is the sum of the potential and kinetic energy of an object. A change in mechanical energy of an object can result from adding energy to an object, taking away energy from an object, or changing the type of energy an object has from one form to another. Work is done on an object when a force acts on an object over a displacement. Therefore, *work* is mathematically defined as the product of the force acting on an object times the displacement. The fourth, and final, basic assumption of the work-energy theorem is that energy cannot be created or destroyed—it just changes form as objects move or as it transfers from one object to another one.

In this investigation you will have an opportunity to explore the relationship between potential energy, kinetic energy, mechanical energy, work, and displacement in terms of the motion of objects. To explore the relationship between these various components of the work-energy theorem, you will study a simple system that consists of a ball, a ramp, and a cup. You goal is to develop a set of rules that you can use to explain and predict the motion of the ball at the bottom of the ramp and the distance the cup moves after the ball rolls down the ramp and enters the cup. To be valid or acceptable, your set of rules must be consistent with the basic assumptions underlying the work-energy theorem outlined in this section.

Your Task

Use what you know about the work-energy theorem and the importance of tracking matter and energy in a system to develop a conceptual model (i.e., a set of rules) that will allow you to explain and predict the behavior of a system that consists of a ball, a ramp, and a cup. To develop your model, you will need to design and carry out several experiments to determine how changes in several different components of the ball-ramp-cup system affect (1) the motion of the ball at the bottom of the ramp after it rolls down it and (2) the distance a cup moves when a ball is rolled down an incline and comes into contact with the cup. Once you have developed your rules, you will need to test them to determine if you can use them to make accurate predictions about the behavior of the ball-ramp-cup system.

The guiding question of this investigation is, *How can we use the work-energy theorem to explain and predict behavior of a system that consists of a ball, a ramp, and a cup?*

Materials

You may use any of the following materials during your investigation:

- Safety glasses or goggles (required)
- Support stand
- Extension clamp
- Set of 25 mm balls (includes brass, aluminum, steel, cork, wood, and copper)
- PVC pipe
- Electronic or triple beam balance
- Plastic cup
- 2 Metersticks
- Protractor
- Stopwatch

If you have access to the following equipment, you may also consider using a video camera and a computer or tablet with video analysis software.

Safety Precautions

Follow all normal lab safety rules. In addition, take the following safety precautions:

1. Wear sanitized safety glasses or goggles during lab setup, hands-on activity, and takedown.

2. Keep fingers and toes out of the way of moving objects.

3. Wash hands with soap and water after completing the lab.

LAB 20

Investigation Proposal Required? ☐ Yes ☐ No

Getting Started

Your first step in this investigation is to design and carry several experiments to determine how changes in several different components of the ball-ramp-cup system affect (1) the motion of the ball at the bottom of the ramp after it rolls down it and (2) the distance a cup moves when a ball is rolled down an incline and comes into contact with the cup. Figure L20.1 shows how you can set up the ball-ramp-cup system. The components of the system that you can change include the mass of the ball (**m**), the height of the ramp (**h**), the length of the ramp (**l**), and the angle of inclination (θ). Before you design or carry out your experiments, however, you must first determine what type of data you need to collect, how you will collect it, and how you will analyze it.

FIGURE L20.1 _____

The ball-ramp-cup system

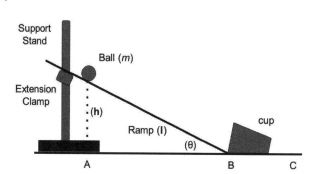

To determine *what type of data you need to collect*, think about the following questions:

- What are the boundaries and components of the system under study?
- How will you quantify the amount of potential, kinetic, and mechanical energy in the system?
- How can you track the potential, kinetic, and mechanical energy changes within this system?
- Which factors might control rates of change in the system under study?
- What will be the independent variable and the dependent variable for each experiment?

To determine *how you will collect your data*, think about the following questions:

- What other variables will you need to control in each experiment?
- What will be the reference point for measurement?

- What measurement scale or scales should you use to collect data?

- What equipment will you need to take your various measurements?

- How will you make sure that your data are of high quality (i.e., how will you reduce error)?

- How will you keep track of and organize the data you collect?

To determine *how you will analyze your data*, think about the following questions:

- What type of calculations will you need to make?

- How could you use mathematics to describe a relationship between variables?

- What types of comparisons will be useful?

- What type of table or graph could you create to help make sense of your data?

Once you have determined how changes in the different components of the ball-ramp-cup system affect (1) the motion of the ball at the bottom of the ramp after it rolls down it and (2) the distance a cup moves when a ball is rolled down an incline and comes into contact with the cup, your group will need to develop a set of rules that you can use to explain and predict the behavior of this system. Your set of rules must be consistent with the four basic assumptions underlying the work-energy theorem outlined in the "Introduction."

The last step in this investigation will be to test your model. To accomplish this goal, you can set components of the system to values that you did not test (e.g., if you tested the height of the ramp at 0.5 m and 0.75 m, then set the height to 0.6 m) to determine if you can use your rulesto make accurate predictions, then you will be able to generate the evidence you need to convince others that your rules are a valid way to explain and predict behavior of the ball and the cup in terms of the work-energy theorem of motion.

Connections to the Nature of Scientific Knowledge and Scientific Inquiry

As you work through your investigation, you may want to consider

- how scientific knowledge changes over time, and

- the difference between laws and theories in science.

Initial Argument

Once your group has finished collecting and analyzing your data, your group will need to develop an initial argument. Your initial argument must include a claim, evidence to support your claim, and a *justification* of the evidence. The *claim* is your group's answer to the guiding question. The *evidence* is an analysis and interpretation of your data. Finally, the *justification* of the evidence is why your group thinks the evidence matters. The justification of the evidence is important because scientists can use different kinds of evidence to support their claims. You group will create your initial argument on a whiteboard. Your whiteboard should include all the information shown in Figure L20.2.

Argumentation Session

The argumentation session allows all of the groups to share their arguments. One or two members of each group will stay at the lab station to share that group's argument, while the other members of the group go to the other lab stations to listen to and critique the other arguments. This is similar to what scientists do when they propose, support, evaluate, and refine new ideas during a poster session at a conference. If you are presenting your group's argument, your goal is to share your ideas and answer questions. You should also keep a record of the critiques and suggestions made by your classmates so you can use this feedback to make your initial argument stronger. You can keep track of specific critiques and suggestions for improvement that your classmates mention in the space below.

Critiques about our initial argument and suggestions for improvement:

FIGURE L20.2

Argument presentation on a whiteboard

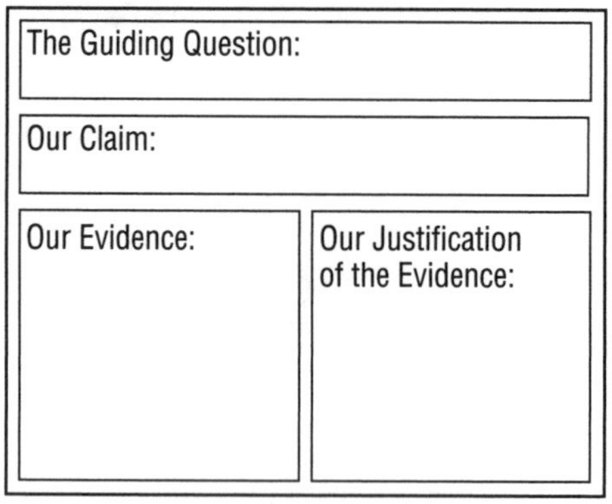

The Guiding Question:	
Our Claim:	
Our Evidence:	Our Justification of the Evidence:

If you are critiquing your classmates' arguments, your goal is to look for mistakes in their arguments and offer suggestions for improvement so these mistakes can be fixed. You should look for ways to make your initial argument stronger by looking for things that the other groups did well. You can keep track of interesting ideas that you see and hear during the argumentation in the space below. You can also use this space to keep track of any questions that you will need to discuss with your team.

Interesting ideas from other groups or questions to take back to my group:

Once the argumentation session is complete, you will have a chance to meet with your group and revise your initial argument. Your group might need to gather more data or design a way to test one or more alternative claims as part of this process. Remember, your goal at this stage of the investigation is to develop the best argument possible.

Report

Once you have completed your research, you will need to prepare an *investigation report* that consists of three sections. Each section should provide an answer to the following questions:

1. What question were you trying to answer and why?

2. What did you do to answer your question and why?

3. What is your argument?

Your report should answer these questions in two pages or less. This report must be typed, and any diagrams, figures, or tables should be embedded into the document. Be sure to write in a persuasive style; you are trying to convince others that your claim is acceptable or valid!

LAB 20

Lab 20. Kinetic and Potential Energy: How Can We Use the Work-Energy Theorem to Explain and Predict Behavior of a System That Consists of a Ball, a Ramp, and a Cup?

Use the figure below to answer questions 1 and 2. For the acceleration due to gravity, use the positive value for **g** (9.8 m/sec²).

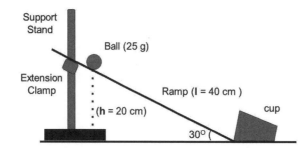

1. What is the potential energy of the ball at the moment it is released on the incline?

   How do you know?

2. What is the kinetic energy of the ball at the moment it strikes the cup?

   How do you know?

Kinetic and Potential Energy
How Can We Use the Work-Energy Theorem to Explain and Predict Behavior of a System
That Consists of a Ball, a Ramp, and a Cup?

3. Galileo hypothesized that free objects accelerate uniformly, or stated another way, that a falling object's velocity increases an equal amount in each equal time interval. Explain how the results of this ball and cup experiment could be used in support of this claim.

4. There is a difference between a law and a theory in science.

   a. I agree with this statement.
   b. I disagree with this statement.

   Explain your answer, using an example from your investigation about kinetic and potential energy.

5. Scientific knowledge, once proven to be true, does not change.

   a. I agree with this statement.
   b. I disagree with this statement.

   Explain your answer, using an example from your investigation about kinetic and potential energy.

6. Scientists often need to identify a system and then create a model of it as part of an investigation. Explain why it useful to create models of systems, using an example from your investigation about kinetic and potential energy and an example from previous investigations in this class or your previous science classes.

7. One of the important aims in science is to track how energy and matter move within a system and to determine if the energy and matter within the system are conserved. Explain why it is useful to track how energy and matter move within a system, using an example from your investigation about kinetic and potential energy and an example from previous investigations in this class or your previous science classes.

LAB 21

Lab 21. Conservation of Energy and Pendulums: How Does Placing a Nail in the Path of a Pendulum Affect the Height of a Pendulum Swing?

Purpose

The purpose of this lab is to *introduce* students to the disciplinary core idea (DCI) of Energy from the *NGSS* by giving them an opportunity to determine how placing a nail in the path of a pendulum affects the height of a pendulum swing. This lab also gives students an opportunity to learn about the crosscutting concepts (CCs) of (a) Patterns; (b) Cause and Effect: Mechanism and Explanation; and (c) Energy and Matter: Flows, Cycles, and Conservation from the *NGSS*. In addition, this lab can be used to help students understand three big ideas from AP Physics: (a) objects and systems have properties such as mass or charge, (b) the interactions of an object with other objects can be described by forces, and (c) changes that occur as result of interactions are constrained by conservation laws. As part of the explicit and reflective discussion, students will also learn about (a) the difference between observations and inferences in science and (b) how scientific knowledge can change over time.

Underlying Physics Concepts

The law of conservation of energy states that in a closed system the total energy of the system remains constant. The total energy of an object is the sum total of the different types of energy possessed by the object. This means that the total mechanical energy (*ME*) of an object is the sum of its potential energy (PE) and kinetic energy (KE), as shown in Equation 21.1.

$$\text{(Equation 21.1)} \quad ME = PE + KE$$

The potential energy and kinetic energy of the pendulum can be found in Equations 21.2 and 21.3, where m is the mass of the pendulum bob, $\mathbf{g}$ is the acceleration due to gravity and is a constant 9.8 m/s², $\mathbf{h}$ is the height above the equilibrium point, and $\mathbf{v}$ is the velocity of the pendulum at any given point. In SI units, mass is measured in kilograms (kg), height is measured in meters (m), and velocity is measured in meters per second (m/s).

$$\text{(Equation 21.2)} \quad PE = m\mathbf{g}\mathbf{h}$$

$$\text{(Equation 21.3)} \quad KE = \tfrac{1}{2} m\mathbf{v}^2$$

As the pendulum swings, there are three important points, shown in Figure 21.1. Point A is the release point of the pendulum and is some height above point B. Because the pendulum mass does have a height relative to point B, the pendulum has potential energy. At point A, the velocity of the mass on the end of the pendulum is zero. Pursuant to Equation 21.3, if the velocity of an object is zero, the kinetic energy of that object is also equal to zero.

Equation 21.1 suggests that the total mechanical energy of the system is equal to the sum of the potential energy and kinetic energy. At point A, this means $ME = PE_A + KE_A$. We have established that the kinetic energy of the pendulum is zero at point A, so the total mechanical energy of the pendulum is equal to the potential energy at point A.

FIGURE 21.1

A pendulum

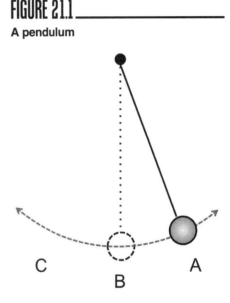

Point B is referred to as the equilibrium point, because at point B there is no net force acting on the pendulum mass. For a pendulum, point B is also defined as "ground level" for the height of the mass. In other words, $h = 0$ at point B. Pursuant to Equation 21.2, this means that the potential energy at point B is equal to zero. The velocity of the pendulum is maximum at point B, and as such, the kinetic energy is also a maximum. Furthermore, if the potential energy at point B is equal to zero, then the kinetic energy at point B is equal to the potential energy at point A.

As the pendulum passes through the equilibrium point, it will continue to swing, reach a maximum height at point C, and then swing back toward the equilibrium point. At point C, the velocity of the pendulum mass is equal to zero. We can show this conceptually as follows: Before reaching point C, the mass had a negative, non-zero velocity to the left. After it begins to swing back toward equilibrium, the pendulum has a positive, non-zero velocity to the right. Because the velocity is a continuous function with respect to time, when it accelerates from a negative to a positive velocity, there must be a point where the velocity is equal to zero.

If the velocity is equal to zero, then the total mechanical energy of the pendulum at point C is equal to the potential energy at point C. As we have already established, the total mechanical energy of the system is equal to the potential energy of the pendulum at point A. Thus, the potential energy at point C is equal to the potential energy at point A, shown mathematically in Equation 21.4.

$$\textbf{(Equation 21.4)} \quad mgh_A = mgh_C$$

Because the mass of the ball does not change as the pendulum swings, and the acceleration due to gravity is constant, the height at point C must also be equal to the height at point A.

LAB 21

In this lab, students will place a nail in the path of the pendulum to determine how the nail affects the height of the swing (shown in Figure 21.2). Despite placing the nail in the path of the pendulum, the height of the mass above the pendulum swing remains the same, as shown in Equation 21.4 (with the one caveat that the initial release point must be lower than the height of the nail). However, the angle the swing makes with respect to the equilibrium point does change. In Figure 21.2, the measure of angle ac is always greater than the measure of angle ab (as long as the nail is above the release point). Finally, the discussion thus far has ignored the role or

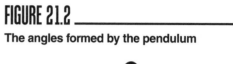
FIGURE 21.2 _____
The angles formed by the pendulum

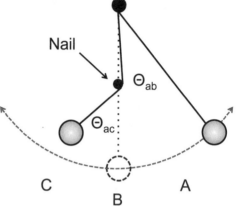

air resistance on the motion of the pendulum bob as it swings, because this simplifies the mathematical analysis of the motion. Although the air resistance on the pendulum in motion will eventually bring the pendulum to rest, through one swing it should be that the height after contact with the nail is equal to that of the release height.

Timeline

The instructional time needed to complete this lab investigation is 170–230 minutes. Appendix 3 (p. 531) provides options for implementing this lab investigation over several class periods. Option C (230 minutes) should be used if students are unfamiliar with scientific writing, because this option provides extra instructional time for scaffolding the writing process. You can scaffold the writing process by modeling, providing examples, and providing hints as students write each section of the report. Option C can also be used if you are introducing students to the video analysis programs. Option D (170 minutes) should be used if students are familiar with scientific writing and have developed the skills needed to write an investigation report on their own. In option D, students complete stage 6 (writing the investigation report) and stage 8 (revising the investigation report) as homework.

Materials and Preparation

The materials needed to implement this investigation are listed in Table 21.1. The items can be purchased from a science supply company such as Carolina, Flinn Scientific, PASCO, Vernier, or Ward's Science. The video analysis software can be purchased from Vernier (Logger *Pro*) or PASCO (SPARKvue or Capstone). These companies also have apps that can be used on Apple- or Android-based tablets and cell phones. We recommend consulting with your school's information technology coordinator to determine the best option for your students.

TABLE 21.1

Materials list for Lab 21

Item	Quantity
Safety glasses or goggles	1 per student
Video camera	1 per group
Computer or tablet with video analysis software	1 per group
Support stand	1 per group
Clamps	2 per group
Nail (3-inch framing)	1 per group
Hanging mass set	1 per group
String	As needed
Meterstick	1 per group
Investigation Proposal A (optional)	1 per group
Whiteboard, 2' × 3'*	1 per group
Lab Handout	1 per student
Peer-review guide and teacher scoring rubric	1 per student
Checkout Questions	1 per student

* As an alternative, students can use computer and presentation software such as Microsoft PowerPoint or Apple Keynote to create their arguments.

Be sure to use a set routine for distributing and collecting the materials during the lab investigation. One option is to set up the materials for each group at each group's lab station before class begins. This option works well when there is a dedicated section of the classroom for lab work and the materials are large and difficult to move (such as a dynamics track). A second option is to have all the materials on a table or cart at a central location. You can then assign a member of each group to be the "materials manager." This individual is responsible for collecting all the materials his or her group needs from the table or cart during class and for returning all the materials at the end of the class. This option works well when the materials are small and easy to move (such as stopwatches, metersticks, or hanging masses). It also makes it easy to inventory the materials at the end of the class before students leave for the day.

Safety Precautions

Remind students to follow all normal lab safety rules. In addition, tell students to take the following safety precautions:

LAB 21

1. Wear sanitized safety glasses or goggles during lab setup, hands-on activity, and takedown.

2. Handle nails with care. Nail ends can be sharp and can puncture or scrape skin.

3. Keep fingers and toes out of the way of moving objects.

4. Wash hands with soap and water after completing the lab.

Topics for the Explicit and Reflective Discussion

Reflecting on the Use of Core Ideas and Crosscutting Concepts During the Investigation

Teachers should begin the explicit and reflective discussion by asking students to discuss what they know about the core idea they used during the investigation. The following are some important concepts related to the core idea of Energy that students will need to determine how placing a nail in the path of a pendulum affects the height of a pendulum swing:

- Energy is conserved in all systems under all circumstances.
- Conserved quantities are constant in an isolated or a closed system. An *open system* is one that exchanges any conserved quantity with its surroundings.
- The mechanical energy of a system includes the kinetic energy of the objects that make up the system and the potential energy of the configuration of the objects that make up the system.
- Changes in the potential energy of a system can result in changes to the kinetic energy system because energy is constant in a closed system.

To help students reflect on what they know about energy, we recommend showing them two or three images using presentation software that help illustrate these important ideas. You can then ask the students the following questions to encourage them to share how they are thinking about these important concepts:

1. What do we see going on in this image?

2. Does anyone have anything else to add?

3. What might be going on that we can't see?

4. What are some things that we are not sure about here?

You can then encourage students to think about how CCs played a role in their investigation. There are at least three CCs that students need to determine how placing a nail in the path of a pendulum affects the height of a pendulum swing: (a) Patterns; (b) Cause and Effect: Mechanism and Explanation; and (c) Energy and Matter: Flows, Cycles, and Conservation (see Appendix 2 [p. 527] for a brief description of these CCs). To help students reflect on what they know about these CCs, we recommend asking them the following questions:

1. Why is it important to look for patterns during an investigation?

2. What patterns did you identify during your investigation? What did the identification of these patterns allow you to do?

3. Why is it important to identify causal relationships in science?

4. What did you have to do during your investigation to determine what causes a change in the height of a pendulum swing? Why was that useful to do?

5. Scientists often attempt to track how energy and matter move into, out of, and within systems as part of an investigation. Why is this useful?

6. How did you attempt to track how energy moves within the system you were studying? What did tracking energy allow you to do during your investigation?

You can then encourage the students to think about how they used all these different concepts to help answer the guiding question and why it is important to use these ideas to help justify their evidence for their final arguments. Be sure to remind your students to explain why they included the evidence in their arguments and make the assumptions underlying their analysis and interpretation of the data explicit in order to provide an adequate justification of their evidence.

Reflecting on Ways to Design Better Investigations

It is important for students to reflect on the strengths and weaknesses of the investigation they designed during the explicit and reflective discussion. Students should therefore be encouraged to discuss ways to eliminate potential flaws, measurement errors, or sources of uncertainty in their investigations. To help students be more reflective about the design of their investigation and what they can do to make their investigations more rigorous in the future, you can ask them the following questions:

1. What were some of the strengths of the way you planned and carried out your investigation? In other words, what made it scientific?

2. What were some of the weaknesses of the way you planned and carried out your investigation? In other words, what made it less scientific?

3. What rules can we make, as a class, to ensure that our next investigation is more scientific?

Reflecting on the Nature of Scientific Knowledge and Scientific Inquiry

This investigation can be used to illustrate two important concepts related to the nature of scientific knowledge and the nature of scientific inquiry: (a) the difference between observations and inferences in science and (b) how scientific knowledge changes over time (see Appendix 2 for a brief description of these two concepts). Be sure to review these concepts during and at the end of the explicit and reflective discussion. To help students think about these concepts in relation to what they did during the lab, you can ask them the following questions:

1. You had to make observations and inferences during your investigation. Can you give me some examples of these observations and inferences?

2. Can you work with your group to come up with a rule that you can use to decide if a piece of information is an observation or an inference? Be ready to share in a few minutes.

3. Scientific knowledge can and does change over time. Can you tell me why it changes?

4. Can you work with your group to come up some examples of how scientific knowledge has changed over time? Be ready to share in a few minutes.

You can also use presentation software or other techniques to encourage your students to think about these concepts. You can show examples of information from the investigation that are either observations or inferences and ask students to classify each example and explain their thinking. You can also show examples of how our thinking about energy has changed over time and ask students to discuss what they think led to those changes.

Be sure to remind your students that, to be proficient in science, it is important that they understand what counts as scientific knowledge and how that knowledge develops over time.

Hints for Implementing the Lab

- Allowing students to design their own procedures for collecting data gives students an opportunity to try, to fail, and to learn from their mistakes. However, you can scaffold students as they develop their procedure by having them fill out an investigation proposal. These proposals provide a way for you to offer students hints and suggestions without telling them how to do it. You can also check the proposals quickly during a class period. For this lab we suggest using Investigation Proposal A.

- For optimal results, make the support stand and associated setup as rigid as possible. A pendulum length of 1 m or more is also useful.

- If the release point of the pendulum is above the height of the nail, the pendulum will swing and wrap around the nail. For best results, remind students to release the pendulum from a point below the height of the nail.

- Allow the students to become familiar with the equipment and materials, including the video analysis software, as part of the tool talk before they begin to design their investigation. Giving them 5–10 minutes to examine the equipment and materials will let them see what they can and cannot do with them. This also gives students a chance to begin thinking about what variables to test and how to control for other variables.

- Remind students to hold the video camera as still as possible. Any movement of the camera will introduce error into their analysis. If using actual camcorders, we recommend using a tripod to hold the camera steady. If students are using a camera on a cell phone or tablet, we recommend using a table to help steady the camera.

- Remind students to place a meterstick in the same field of view as the motion they are capturing with the video camera. Also, the meterstick should be approximately the same distance from the camera as the motion. Most video analysis software requires the user to define a scale in the video (this allows the software to establish distances and, subsequently, other variables dependent on distance and displacement).

- Be sure to allow students to go back and re-collect data at the end of the argumentation session. Students often realize that they made numerous mistakes when they were collecting data as a result of their discussions during the argumentation session. The students, as a result, will want a chance to re-collect data, and the re-collection of data should be encouraged when time allows. This also offers an opportunity to discuss what scientists do when they realize a mistake is made inside the lab.

Connections to Standards

Table 21.2 (p. 462) highlights how the investigation can be used to address learning objectives from AP Physics 1; learning objectives from AP Physics C: Mechanics, *Common Core State Standards*, in English language arts (*CCSS ELA*); and *Common Core State Standards*, Mathematics (*CCSS Mathematics*).

TABLE 21.2_____

Lab 21 alignment with standards

***NGSS* performance expectations**	• None
AP Physics 1 learning objectives	• 3.A.1.1: The student is able to express the motion of an object using narrative, mathematical, and graphical representations. • 3.A.1.2: The student is able to design an experimental investigation of the motion of an object. • 3.A.1.3: The student is able to analyze experimental data describing the motion of an object and is able to express the results of the analysis using narrative, mathematical, and graphical representations. • 4.C.1.2: The student is able to predict changes in the total energy of a system due to changes in position and speed of objects or frictional interactions within the system. • 5.B.1.2: The student is able to translate between a representation of a single object, which can only have kinetic energy, and a system that includes the object, which may have both kinetic and potential energies. • 5.B.3.2: The student is able to make quantitative calculations of the internal potential energy of a system from a description or diagram of that system. • 5.B.3.3: The student is able to apply mathematical reasoning to create a description of the internal potential energy of a system from a description or diagram of the objects and interactions in that system.
AP Physics C: Mechanics learning objectives	• I.C.3.c: Students should be able to recognize and solve problems that call for application both of conservation of energy and Newton's laws. • I.C.3.a(3): Analyze situations in which an object's mechanical energy is changed by friction or by a specified externally applied force.
Literacy connections (*CCSS ELA*)	• *Reading:* Key ideas and details, craft and structure, integration of knowledge and ideas • *Writing:* Text types and purposes, production and distribution of writing, research to build and present knowledge, range of writing • *Speaking and listening:* Comprehension and collaboration, presentation of knowledge and ideas

National Science Teachers Association

Mathematics connections (*CCSS Mathematics*)	• *Mathematical practices:* Reason abstractly and quantitatively, construct viable arguments and critique the reasoning of others, use appropriate tools strategically, attend to precision • *Number and quantity:* Reason quantitatively and use units to solve problems, represent and model with vector quantities, perform operations on vectors • *Algebra:* Interpret the structure of expressions, understand solving equations as a process of reasoning and explain the reasoning, solve equations and inequalities in one variable, represent and solve equations and inequalities graphically. • *Functions:* Understand the concept of a function and use function notation, interpret functions that arise in applications in terms of the context, analyze functions using different representations, interpret expressions for functions in terms of the situation they model • *Statistics and probability:* Summarize, represent, and interpret data on two categorical and quantitative variables; understand and evaluate random processes underlying statistical experiments; make inferences and justify conclusions from sample surveys, experiments, and observational studies

LAB 21

Lab 21. Conservation of Energy and Pendulums: How Does Placing a Nail in the Path of a Pendulum Affect the Height of a Pendulum Swing?

Introduction

Two of the most influential thinkers in history were Aristotle in the 4th century BC and Galileo in the 16th–17th centuries. Aristotle took a philosophical approach to understanding the natural world, whereas Galileo preferred a more empirical one. Thus, they developed some very different explanations about the motion of objects. For example, Aristotle claimed that heavier bodies fall faster than lighter ones in the same medium, whereas Galileo claimed that in a vacuum all bodies fall with the same speed.

Aristotle and Galileo also did not agree about the nature of energy. Historians of science attribute the first use of the word *energy* to Aristotle (in classical Greek, *energeia*). Historians debate what exactly Aristotle meant by the word *energeia*, with the most common translation being "activity or operation." Historians do agree, however, that Aristotle used the concept of energeia not only to analyze the motion of objects but also to understand ethics and psychology. According to Aristotle, a person's desire to be happy was a type of energeia. Galileo, on the other hand, provided a more modern view of energy and suggested that energy was a property of objects. Thus, he decoupled the term *energy* from the study of ethics or psychology.

Since Galileo's time, other physicists have contributed to our understanding of forces, motion, and energy. Isaac Newton, in the 17th and early 18th centuries, provided the mathematical foundation for the study of force and motion. In the 19th century, Thomas Young was the first to define *energy* in the formal sense, and Gaspard-Gustave de Coriolis expanded on Young's definition of energy by introducing the concept of kinetic energy. Also in the 19th century, William Rankine was the first to propose the idea of potential energy, and James Prescott Joule and William Thomson (better known as Lord Kelvin) developed the law of conservation of energy, which states that the total energy in a closed system remains constant. We now use these ideas about energy to help understand and explain a wide range of natural phenomena.

In this investigation, you will have an opportunity to use these important ideas about energy to explore the motion of a pendulum. The motion of a pendulum has been studied for hundreds of years. Galileo, for example, used the motion of a pendulum to help quantify his ideas about falling objects. Other scientists have also used the pendulum to investigate forces and energy. One of Galileo's most informative and important investigations about the behavior of a pendulum, however, was when he placed a nail in the path of a swinging pendulum as shown in Figure L21.1. This simple modification to a basic pendulum allowed Galileo to explore the energy of the pendulum bob as it swings back and forth.

Your Task

Use what you know about energy, cause-and-effect relationships in science, and the importance of identifying and explaining patterns in nature to design and carry out an investigation to determine how the height of a pendulum swing changes after coming into contact with a nail in its path.

The guiding question of this investigation is, *How does placing a nail in the path of a pendulum affect the height of a pendulum swing?*

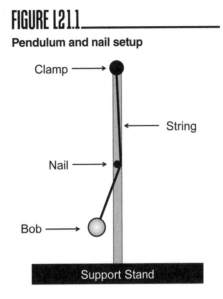

FIGURE L21.1

Pendulum and nail setup

Materials

You may use any of the following materials during your investigation:

- Safety glasses or goggles (required)
- Video camera
- Computer or tablet with video analysis software
- Support stand Clamps Nail
- Hanging mass set
- String
- Meterstick

Safety Precautions

Follow all normal lab safety rules. In addition, take the following safety precautions:

1. Wear sanitized safety glasses or goggles during lab setup, hands-on activity, and takedown.

2. Handle nails with care. Nail ends can be sharp and can puncture or scrape skin.

3. Keep fingers and toes out of the way of moving objects.

4. Wash hands with soap and water after completing the lab.

Investigation Proposal Required?   ☐ Yes   ☐ No

Getting Started

To answer the guiding question, you will need to design and carry out an investigation to determine how the height of a pendulum swing changes after the string comes into contact with a nail as the bob swings back and forth. Be sure to focus only on the height of a pendulum swing during the first few oscillations, because air resistance will cause the pendulum to eventually slow down and come to rest. It is also important to examine pendulums with different lengths of string and different bob masses during your investigation. With these issues in mind, you can now decide what type of data you need to collect, how you will collect it, and how you will analyze it.

To determine *what type of data you need to collect,* think about the following questions:

- What are the boundaries and components of the system you are studying?
- How can you describe the components of the system quantitatively?
- How could you keep track of changes in this system quantitatively?
- What could cause a change in the height of the pendulum swing?
- How will you determine the height of pendulum swing?
- Is it useful to track how energy flows into, out of, or within this system?
- What else will you need to measure during the investigation?

To determine *how you will collect the data,* think about the following questions:

- What type of research design needs to be used to establish a cause-and-effect relationship?
- How could you track the flow of energy within this system?
- How long will you need to observe the pendulum swinging back and forth?
- What types of comparisons will be useful?
- How will you vary the length of the string and the mass of the bob?
- What will be the reference point for your measurements?
- What equipment will you need to use in order to make the measurements you need?
- How will you make sure that your data are of high quality (i.e., how will you reduce error)?
- How will you keep track of and organize the data you collect?

To determine *how you will analyze the data,* think about the following questions:

- What types of patterns might you look for as you analyze your data?
- How could you use mathematics to determine if there is a difference between conditions?
- How precise is your video (frames per second)?
- What type of table or graph could you create to help make sense of your data?

Connections to the Nature of Scientific Knowledge and Scientific Inquiry

As you work through your investigation, you may want to consider

- the difference between observations and inferences in science, and
- how scientific knowledge changes over time.

Initial Argument

Once your group has finished collecting and analyzing your data, your group will need to develop an initial argument. Your argument must include a claim, evidence to support your claim, and a justification of the evidence. The *claim* is your group's answer to the guiding question. The *evidence* is an analysis and interpretation of your data. Finally, the *justification* of the evidence is why your group thinks the evidence matters. The justification of the evidence is important because scientists can use different kinds of evidence to support their claims. Your group will create your initial argument on a whiteboard. Your whiteboard should include all the information shown in Figure L21.2.

FIGURE L21.2

Argument presentation on a whiteboard

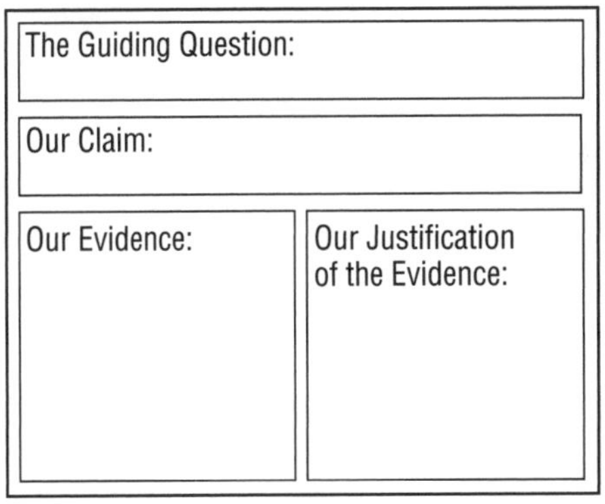

The Guiding Question:	
Our Claim:	
Our Evidence:	Our Justification of the Evidence:

Argumentation Session

The argumentation session allows all of the groups to share their arguments. One or two members of each group will stay at the lab station to share that group's argument, while the other members of the group go to the other lab stations to listen to and critique the other arguments. This is similar to what scientists do when they propose, support, evaluate, and refine new ideas during a poster session at a conference. If you are presenting your group's argument, your goal is to share your ideas and answer questions. You should also keep a record of the critiques and suggestions made by your classmates so you can use this feedback to make your initial argument stronger. You can keep track of specific critiques and suggestions for improvement that your classmates mention in the space below.

Critiques about our initial argument and suggestions for improvement:

If you are critiquing your classmates' arguments, your goal is to look for mistakes in their arguments and offer suggestions for improvement so these mistakes can be fixed. You should look for ways to make your initial argument stronger by looking for things that the other groups did well. You can keep track of interesting ideas that you see and hear during the argumentation in the space below. You can also use this space to keep track of any questions that you will need to discuss with your team.

Interesting ideas from other groups or questions to take back to my group:

Once the argumentation session is complete, you will have a chance to meet with your group and revise your initial argument. Your group might need to gather more data or design a way to test one or more alternative claims as part of this process. Remember, your goal at this stage of the investigation is to develop the best argument possible.

Report

Once you have completed your research, you will need to prepare an *investigation report* that consists of three sections. Each section should provide an answer to the following questions:

1. What question were you trying to answer and why?

2. What did you do to answer your question and why?

3. What is your argument?

Your report should answer these questions in two pages or less. This report must be typed, and any diagrams, figures, or tables should be embedded into the document. Be sure to write in a persuasive style; you are trying to convince others that your claim is acceptable or valid!

Checkout Questions

Lab 21. Conservation of Energy and Pendulums: How Does Placing a Nail in the Path of a Pendulum Affect the Height of a Pendulum Swing?

1. Pictured at right is a pendulum. Let **h** = 0 represent the height when the bob is at equilibrium and acceleration due to gravity is 9.8 m/s² With these facts, calculate the kinetic energy of the bob when it is at the equilibrium position, assuming the pendulum is released from the point where **h** = 10 cm.

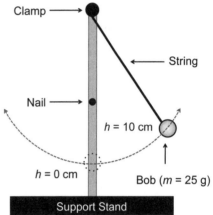

2. The length of the pendulum has an effect on the height of the bob after contacting and sweeping through the nail (assuming the initial height is not higher than the nail).

   a. I agree with this statement.
   b. I disagree with this statement.

   Explain your answer, using the findings from your investigation about placing a nail in the path of a pendulum.

3. There is no difference between observations and inferences in science.

   a. I agree with this statement.
   b. I disagree with this statement.

   Explain your answer, using an example from your investigation about placing a nail in the path of a pendulum.

4. Scientific knowledge can change over time.

   a. I agree with this statement.
   b. I disagree with this statement.

   Explain your answer, using an example from your investigation about placing a nail in the path of a pendulum.

5. Scientists often look for or attempt to identify patterns in nature. Explain why this is a useful practice, using an example from your investigation about placing a nail in the path of a pendulum.

6. In science, understanding cause-and-effect relationships is an important goal. Sometimes, scientists hypothesize a causal relationship that is not supported by the data. Does this mean there was something wrong with the investigation? Explain your answer, using an example from your investigation about placing a nail in the path of a pendulum.

7. In a pendulum, energy is transferred from potential to kinetic energy. Discuss the relationship between the potential and kinetic energy of the pendulum as it swings back and forth. Is energy conserved? Why is it important to keep track of the energy as it is transferred from potential to kinetic energy?

Teacher Notes

Lab 22. Conservation of Energy and Wind Turbines: How Can We Maximize the Amount of Electrical Energy That Will Be Generated by a Wind Turbine Based on the Design of Its Blades?

Purpose

The purpose of this lab is to *introduce* students to the disciplinary core idea (DCI) of Energy from the *NGSS* by giving them an opportunity to determine how to maximize the amount of electrical energy produced by a wind turbine. This lab also gives students an opportunity to learn about the crosscutting concepts (CCs) of (a) Systems and System Models; (b) Energy and Matter: Flows, Cycles, and Conservation; and (c) Structure and Function from the *NGSS*. In addition, this lab can be used to help students understand four big ideas from AP Physics: (a) objects and systems have properties such as mass or charge, (b) the interactions of an object with other objects can be described by forces, (c) interactions between systems can result in changes in those systems, and (d) changes that occur as result of interactions are constrained by conservation laws. As part of the explicit and reflective discussion, students will also learn about (a) how scientific knowledge changes over time and (b) how the culture of science, societal needs, and current events influence the work of scientists.

Underlying Physics Concepts

Wind turbines work by converting the translational kinetic energy of wind particles to the rotational kinetic energy of the wind turbine. The rotational kinetic energy of the wind turbine is then converted to electrical energy by an induction generator. An induction generator works by manually turning a rotor that is attached to a wire. The wire rotates in a magnetic field, thereby producing an alternating current. In this way, the kinetic energy of wind particles can be converted into electrical energy.

It is important to note that the energy conversions are not 100% efficient. That is, energy is not completely converted from one form to the other. Some energy is lost as heat due to friction when the rotor is spinning. Furthermore, the collisions between the wind particles and the turbine blades are not elastic, so kinetic energy is not conserved during the collision (the total energy is conserved, but some kinetic energy may be lost as heat).

When analyzing a collision, it is always best to start with momentum, because the momentum of the system is always conserved (momentum is not lost to the surroundings). We can therefore ask, How is the translational momentum of the wind transferred to angular momentum of the turbine? Assuming the strength of the current generated

Conservation of Energy and Wind Turbines

How Can We Maximize the Amount of Electrical Energy That Will Be Generated by a Wind Turbine Based on the Design of Its Blades?

is a function of the angular velocity of the rotor, then it is also possible to answer this question based on the induced angular momentum of the turbine's blades. Furthermore, if we assume that the wind blows at a constant speed and the density of the air remains constant (valid assumptions when using a fan to produce the wind in a controlled lab setting), then the translational momentum of the wind remains constant. This means that the angular momentum of the wind turbine is a function of the design of the turbine. For this lab, the relationship of the design of the turbine to the induced angular momentum of the turbine is overshadowed by the relationship between the design of the turbine and the electrical energy produced by the turbine. That being said, the angular momentum of the turbine mediates the relationship between the kinetic energy of the wind and the electrical energy produced by the turbine.

In this lab, students must investigate various factors that affect the induced angular momentum of the turbine and the resultant production of electrical energy. In the Lab Handout we identify several possible factors that affect the electrical energy produced by the wind turbine. These factors include the angle of the blade relative to the wind, the shape of the blade, and the number of blades on the turbine. This is not meant to be an exhaustive list, and students may choose to investigate other potential variables. In general, the angle of the blade with respect to the wind that will produce a maximum energy output is approximately 20°. It is important to note, however, that other factors, such as the wind speed, will affect the exact angle that produces maximum energy output.

Timeline

The instructional time needed to complete this lab investigation is 220–280 minutes. Appendix 3 (p. 531) provides options for implementing this lab investigation over several class periods. Option A (280 minutes) should be used if students are unfamiliar with scientific writing, because this option provides extra instructional time for scaffolding the writing process. You can scaffold the writing process by modeling, providing examples, and providing hints as students write each section of the report. Option B (220 minutes) should be used if students are familiar with scientific writing and have developed the skills needed to write an investigation report on their own. In option B, students complete stage 6 (writing the investigation report) and stage 8 (revising the investigation report) as homework.

Materials and Preparation

The materials needed to implement this investigation are listed in Table 22.1 (p. 474). The items can be purchased from a science supply company such as Carolina, Flinn Scientific, PASCO, Vernier, or Ward's Science. We recommend the wind turbine kits from Vernier. You can purchase the KidWind Advanced Wind Experiment Kit (KW-AWXC) or just the KidWind Basic Turbine Building Parts Kit (KW-BTPART). The Advanced Experiment Kit Classroom Pack includes three turbines, extra hubs, and blade consumables for approximately 24 students. The Basic Turbine Building Parts Kit includes a hub, a wind turbine generator, and 25 dowels. You can purchase one kit for each group and then build the turbine base for each group using PVC pipe. Figure 22.1 shows a station built from PVC pipe from two different angles. Instructions for building the PVC stand can be found at *www1.eere.energy.gov/education/pdfs/wind_basicpvcwindturbine.pdf.*

TABLE 22.1

Materials list for Lab 22

Item	Quantity
Safety glasses or goggles	1 per student
Wind turbine kit with adjustable blades	1 per group
Fan to generate wind	2–4 per class
Multimeter or galvanometer	1 per group
Electric wire	As needed
Lightbulbs (small)	3 per group
Ruler	1 per group
Protractor	1 per group
Investigation Proposal C (optional)	3 per group
Whiteboard, 2' × 3'*	1 per group
Lab Handout	1 per student
Peer-review guide and teacher scoring rubric	1 per student
Checkout Questions	1 per student

* As an alternative, students can use computer and presentation software such as Microsoft PowerPoint or Apple Keynote to create their arguments.

Be sure to use a set routine for distributing and collecting the materials during the lab investigation. One option is to set up the materials for each group at each group's lab station before class begins. This option works well when there is a dedicated section of the classroom for lab work and the materials are large and difficult to move (such as a dynamics track). A second option is to have all the materials on a table or cart at a central location. You can then assign a member of each group to be the "materials manager." This individual is responsible for collecting all the materials his or her group needs from the table or cart during class and for returning all the materials at the end of the class. This option works well when the materials are small and easy to move (such as stopwatches, metersticks, or hanging masses). It also makes it easy to inventory the materials at the end of the class before students leave for the day.

FIGURE 22.1 _____

Wind turbine setup: (a) front view and (b) side view

(a)    (b)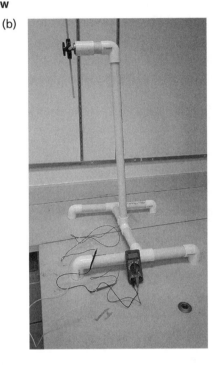

Safety Precautions

Remind students to follow all normal lab safety rules. In addition, tell students to take the following safety precautions:

1. Wear sanitized safety glasses or goggles during lab setup, hands-on activity, and takedown.

2. Handle blades with care. They can puncture skin.

3. Electric wire may get hot. Use caution when touching exposed parts of wires.

4. Handle lightbulbs with care. They can get hot and burn skin; also, they are fragile and can break easily, causing a sharp hazard that can cut or puncture skin.

5. Keep fingers and toes out of the way of moving objects.

6. Wash hands with soap and water after completing the lab.

LAB 22

Topics for the Explicit and Reflective Discussion

Reflecting on the Use of Core Ideas and Crosscutting Concepts During the Investigation

Teachers should begin the explicit and reflective discussion by asking students to discuss what they know about the core idea they used during the investigation. The following are some important concepts related to the core idea of Energy that students will need to determine how to maximize the amount of electrical energy produced by a wind turbine:

- Energy, linear momentum, and angular momentum are conserved in all systems under all circumstances.

- Conserved quantities are constant in an isolated or a closed system. An open system is one that exchanges any conserved quantity with its surroundings.

- Kinetic energy after an elastic collision is the same as the kinetic energy before the collision in an isolated system.

To help students reflect on what they know about energy, we recommend showing them two or three images using presentation software that help illustrate these important ideas. You can then ask the students the following questions to encourage them to share how they are thinking about these important concepts:

1. What do we see going on in this image?

2. Does anyone have anything else to add?

3. What might be going on that we can't see?

4. What are some things that we are not sure about here?

You can then encourage students to think about how CCs played a role in their investigation. There are at least three CCs that students need to use to determine how to maximize the amount of electrical energy produced by a wind turbine: (a) Systems and System Models; (b) Energy and Matter: Flows, Cycles, and Conservation; and (c) Structure and Function (see Appendix 2 [p. 527] for a brief description of these CCs). To help students reflect on what they know about these CCs, we recommend asking them the following questions:

1. Scientists often need to define the system under study and then make a model during an investigation. Why is developing a model of a system so useful in science?

2. What were the boundaries and components of the system you studied during this investigation? What were the strengths and limitations of the model you developed?

3. Scientists often attempt to track how energy and matter move into, out of, and within systems as part of an investigation. Why is this useful?

Conservation of Energy and Wind Turbines

How Can We Maximize the Amount of Electrical Energy That Will Be Generated by a Wind Turbine Based on the Design of Its Blades?

4. How did you attempt to track how energy moves within the system you were studying? What did tracking energy allow you to do during your investigation?

5. The way an object is shaped or structured determines many of its properties and how it functions. Why is it useful to think about the relationship between structure and function during an investigation?

6. Why was it useful to examine the structure of a wind turbine when attempting to determine how much electrical energy it produces?

You can then encourage the students to think about how they used all these different concepts to help answer the guiding question and why it is important to use these ideas to help justify their evidence for their final arguments. Be sure to remind your students to explain why they included the evidence in their arguments and make the assumptions underlying their analysis and interpretation of the data explicit in order to provide an adequate justification of their evidence.

Reflecting on Ways to Design Better Investigations

It is important for students to reflect on the strengths and weaknesses of the investigation they designed during the explicit and reflective discussion. Students should therefore be encouraged to discuss ways to eliminate potential flaws, measurement errors, or sources of uncertainty in their investigations. To help students be more reflective about the design of their investigation and what they can do to make their investigations more rigorous in the future, you can ask them the following questions:

1. What were some of the strengths of the way you planned and carried out your investigation? In other words, what made it scientific?

2. What were some of the weaknesses of the way you planned and carried out your investigation? In other words, what made it less scientific?

3. What rules can we make, as a class, to ensure that our next investigation is more scientific?

Reflecting on the Nature of Scientific Knowledge and Scientific Inquiry

This investigation can be used to illustrate two important concepts related to the nature of scientific knowledge and the nature of scientific inquiry: (a) how scientific knowledge changes over time and (b) how the culture of science, societal needs, and current events influence the work of scientists (see Appendix 2 for a brief description of these two concepts). Be sure to review these concepts during and at the end of the explicit and reflective discussion. To help students think about these concepts in relation to what they did during the lab, you can ask them the following questions:

1. Scientific knowledge can and does change over time. Can you tell me why it changes?

2. Can you work with your group to come up some example of how scientific knowledge has changed over time? Be ready to share in a few minutes.

3. People view some types of research as being more important than other types of research because of cultural values and current events. Can you come up with some examples of how cultural values and current events have influenced the work of scientists?

4. Scientists share a set of values, norms, and commitments that shape what counts as knowing, how to represent or communicate information, and how to interact with other scientists. Can you work with your group to come up with a rule that you can use to decide if something is science or not science? Be ready to share in a few minutes.

You can also use presentation software or other techniques to encourage your students to think about these concepts. You can show examples of how our thinking about energy has changed over time and ask students to discuss what they think led to those changes. You can also show examples of research projects that were influenced by cultural values and current events and ask students to think about what was going on at the time and why that research was viewed as being important for the greater good.

Be sure to remind your students that, to be proficient in science, it is important that they understand what counts as scientific knowledge and how that knowledge develops over time.

Hints for Implementing the Lab

- Allowing students to design their own procedures for collecting data gives students an opportunity to try, to fail, and to learn from their mistakes. However, you can scaffold students as they develop their procedure by having them fill out an investigation proposal. These proposals provide a way for you to offer students hints and suggestions without telling them how to do it. You can also check the proposals quickly during a class period. We recommend having students fill out an investigation proposal for each experiment. For this lab we suggest using Investigation Proposal C.

- Learn how to use the wind turbine generator before the lab begins. It is important for you to know how to use the equipment so you can help students when technical issues arise.

- Allow students to become familiar with the wind turbine kit as part of the tool talk before they begin to design their investigation. Giving them 5–10 minutes to examine the equipment will let them see what they can and cannot do with it. This

Conservation of Energy and Wind Turbines

How Can We Maximize the Amount of Electrical Energy That Will Be Generated by a Wind Turbine Based on the Design of Its Blades?

also gives students a chance to begin thinking about what variables to test and how to control for other variables.

- We recommend setting up two to four "testing stations" throughout your classrooms. Students can then arrange the angle of their turbine blades on the rotor and just place the rotor on the generator to test their design. They can then remove the rotor (the black piece in Figure 22.1) from the generator while they adjust the pitch, thereby allowing another group to collect data. This will reduce the cost of equipment.

- Be sure to allow students to go back and re-collect data at the end of the argumentation session. Students often realize that they made numerous mistakes when they were collecting data as a result of their discussions during the argumentation session. The students, as a result, will want a chance to re-collect data, and the re-collection of data should be encouraged when time allows. This also offers an opportunity to discuss what scientists do when they realize a mistake is made inside the lab.

Connections to Standards

Table 22.2 (p. 480) highlights how the investigation can be used to address specific performance expectations from the *NGSS;* learning objectives from AP Physics 1; learning objectives from AP Physics C: Mechanics, *Common Core State Standards,* in English language arts (*CCSS ELA*); and *Common Core State Standards,* Mathematics (*CCSS Mathematics*).

LAB 22

TABLE 22.2

Lab 22 alignment with standards

***NGSS* performance expectations**	• HS-PS3-3: Design, build, and refine a device that works within given constraints to convert one form of energy into another form of energy.
AP Physics 1 learning objectives	• 5.D.1.4: The student is able to design an experimental test of an application of the principle of the conservation of linear momentum, predict an outcome of the experiment using the principle, analyze data generated by that experiment whose uncertainties are expressed numerically, and evaluate the match between the prediction and the outcome.
AP Physics C: Mechanics learning objectives	• I.C.3.a(2): Describe and identify situations in which mechanical energy is converted to other forms of energy. • I.C.3.b(1): Identify situations in which mechanical energy is or is not conserved.
Literacy connections (*CCSS ELA*)	• *Reading:* Key ideas and details, craft and structure, integration of knowledge and ideas • *Writing:* Text types and purposes, production and distribution of writing, research to build and present knowledge, range of writing • *Speaking and listening:* Comprehension and collaboration, presentation of knowledge and ideas
Mathematics connections (*CCSS Mathematics*)	• *Mathematical practices:* Make sense of problems and persevere in solving them, reason abstractly and quantitatively, construct viable arguments and critique the reasoning of others, model with mathematics, use appropriate tools strategically, attend to precision, look for and make use of structure, look for and express regularity in repeated reasoning • *Number and quantity:* Reason quantitatively and use units to solve problems, represent and model with vector quantities, perform operations on vectors • *Algebra:* Interpret the structure of expressions, create equations that describe numbers or relationships, understand solving equations as a process of reasoning and explain the reasoning, solve equations and inequalities in one variable, represent and solve equations and inequalities graphically • *Functions:* Understand the concept of a function and use function notation, interpret functions that arise in applications in terms of the context, analyze functions using different representations, build a function that models a relationship between two quantities, construct and compare linear and exponential models and solve problems, interpret expressions for functions in terms of the situation they model • *Statistics and probability:* Summarize, represent, and interpret data on two categorical and quantitative variables; interpret linear models; understand and evaluate random processes underlying statistical experiments; make inferences and justify conclusions from sample surveys, experiments, and observational studies

Conservation of Energy and Wind Turbines

How Can We Maximize the Amount of Electrical Energy That Will Be Generated by a Wind Turbine Based on the Design of Its Blades?

Lab Handout

Lab 22. Conservation of Energy and Wind Turbines: How Can We Maximize the Amount of Electrical Energy That Will Be Generated by a Wind Turbine Based on the Design of Its Blades?

Introduction

The United States relies heavily on fossil fuels, such as oil, coal, and natural gas, for energy. Americans burn fossil fuels to power their cars, to heat and light their homes, and to run their consumer electronics and household appliances. Unfortunately, when the supply of fossil fuels declines, it can lead to an energy crisis. In 1973, for example, the Organization of the Petroleum Exporting Companies (OPEC) declared an oil embargo and drastically reduced the amount of oil they sent to the United States. This six-month embargo caused widespread gasoline shortages and significant price increases. A second oil crisis in the United States occurred in 1979 as a result of a decrease in the output of oil from the Middle East along with price increases. This decrease in output resulted in gasoline shortages and long lines at gasoline stations throughout the United States (see Figure L22.1). These events led to a political push for policies that would make America more energy independent. A country that is energy independent does not need to import oil or other forms of energy from other countries to meet the energy demands of its citizens.

More recently, evidence that the burning of fossil fuels contributes to climate change has exacerbated the need to develop new energy sources. To combat the effects of climate change, scientists have been working on ways to harness and use renewable energy sources. Renewable energy sources are forms of energy that are continuously replenished. Solar, tidal, and geothermal energy are exam-

FIGURE L22.1

A long line at a gas station in Maryland as a result of the 1979 oil crisis

ples of renewable energy sources. Furthermore, all renewable energy sources are clean, which means they do not contribute to climate change. These sources of energy, however, are often difficult to access without the development of new technologies.

One of the more promising renewable energy sources is wind. Wind is just the movement of a large number of air particles from one area to another. In the United States, wind energy is a very attractive option because the Great Plains states of Texas, Oklahoma, Kansas, Nebraska, and the Dakotas have vast open spaces where wind blows for extended periods of time. Wind energy takes advantage of the law of conservation of energy to convert the kinetic energy of air particles (gases such nitrogen [N_2], oxygen [O_2], water vapor, and others) into electrical energy. To achieve this conversion, several huge wind turbines are built to create a wind farm (see Figure L22.2). When air particles collide with the blades on the turbine, they transfer some momentum to the blades. This momentum transfer causes the turbine to spin, resulting in an increase in the kinetic energy of the turbine blades. The turbine then turns a generator, producing an electric current from a transfer of kinetic energy to electric energy. Wires carry the current to other locations, for use in homes and businesses.

FIGURE L22.2

A wind farm

When designing a wind turbine, scientists and engineers must account for a variety of factors related to how and where wind blows. As mentioned earlier, wind farms are best placed in the Great Plains, yet only a small percentage of people live in this geographic location, so scientists and engineers must also think about how to move the energy generated to more populous areas. Another factor is the height of the turbine. Wind tends to blow faster at higher altitudes, meaning the height of the turbine affects the amount of energy produced. Other potential factors that may influence the amount of electrical energy produced include the angle of the blades relative to the wind, the number of blades on the turbine, and the shape and mass of the turbine blades.

Your Task

Use what you know about the conservation of energy, systems, models, and structure and function to test the effect of changes to the design of wind turbine blades on the amount of electrical energy produced by the wind turbine. Your goal is to determine how the angle, number, and shape of the blades that are attached to the wind turbine affect the amount of electrical energy that the wind turbine is able to produce.

The guiding question of this investigation is, *How can we maximize the amount of electrical energy that will be generated by a wind turbine based on the design of its blades?*

Conservation of Energy and Wind Turbines

How Can We Maximize the Amount of Electrical Energy That Will Be Generated by a Wind Turbine Based on the Design of Its Blades?

Materials

You may use any of the following materials during your investigation:

- Safety glasses or goggles (required)
- Wind turbine kit with adjustable blades
- Fan to generate wind
- Multimeter or galvanometer
- Electric wires
- Lightbulbs
- Ruler
- Protractor

Safety Precautions

Follow all normal lab safety rules. In addition, take the following safety precautions:

1. Wear sanitized safety glasses or goggles during lab setup, hands-on activity, and takedown.

2. Handle blades with care. They can puncture skin.

3. Electric wire may get hot. Use caution when touching exposed parts of wires.

4. Handle lightbulbs with care. They can get hot and burn skin; also, they are fragile and can break easily, causing a sharp hazard that can cut or puncture skin.

5. Keep fingers and toes out of the way of moving objects.

6. Wash hands with soap and water after completing the lab.

Investigation Proposal Required?   ☐ Yes   ☐ No

Getting Started

You will need to design and carry out at least three different experiments to determine how to maximize the amount of electrical energy that is generated by a wind turbine based on the design of its blades. You will need to conduct three different experiments because you will need to be able to answer the following questions before you can develop an answer to the guiding question for this lab:

1. How does changing the number of blades affect the energy output of the turbine?

2. How does changing the shape of the blades affect the energy output of the turbine?

3. How does changing the angle of the blades affect the energy output of the turbine?

It will be important for you to determine what type of data you need to collect, how you will collect it, and how you will analyze it for each experiment, because each experiment is slightly different.

LAB 22

To determine *what type of data you need to collect,* think about the following questions:

- What are the boundaries and components of the system you are studying?
- How do the components of the system interact with each other?
- How can you describe the components of the system quantitatively?
- How could you keep track of changes in this system quantitatively?
- How might the structure of a wind turbine blade relate to its function?
- How might changes to the structure of a wind turbine blade affect how it functions?
- Is it useful to track how energy flows into, out of, or within this system?
- What will be the independent variable and the dependent variable for each experiment?

To determine *how you will collect the data,* think about the following questions:

- How will you vary the independent variable during each experiment?
- What will you do to hold the other variables constant during each experiment?
- When will you need to take measurements or observations during each experiment?
- What scale or scales should you use when you take your measurements?

To determine *how you will analyze the data,* think about the following questions:

- What types of calculations will you need to make?
- What types of comparisons will you need to make?
- How could you use mathematics to determine if there is a difference between the groups?
- What type of table or graph could you create to help make sense of your data?

Connections to the Nature of Scientific Knowledge and Scientific Inquiry

As you work through your investigation, you may want to consider

- how scientific knowledge changes over time, and
- how the culture of science, societal needs, and current events influence the work of scientists.

Initial Argument

Once your group has finished collecting and analyzing your data, your group will need to develop an initial argument. Your initial argument needs to include a claim, evidence to support your claim, and a justification of the evidence. The *claim* is your group's answer to the guiding question. The *evidence* is an analysis and interpretation of your data. Finally, the

Conservation of Energy and Wind Turbines

How Can We Maximize the Amount of Electrical Energy That Will Be Generated by a Wind Turbine Based on the Design of Its Blades?

justification of the evidence is why your group thinks the evidence matters. The justification of the evidence is important because scientists can use different kinds of evidence to support their claims. Your group will create your initial argument on a whiteboard. Your whiteboard should include all the information shown in Figure L22.3.

Argumentation Session

The argumentation session allows all of the groups to share their arguments. One or two members of each group will stay at the lab station to share that group's argument, while the other members of the group go to the other lab stations to listen to and critique the other arguments. This is similar to what scientists do when they propose, support, evaluate, and refine new ideas during a poster session at a conference. If you are presenting your group's argument, your goal is to share your ideas and answer questions. You should also keep a record of the critiques and suggestions made by your classmates so you can use this feedback to make your initial argument stronger. You can keep track of specific critiques and suggestions for improvement that your classmates mention in the space below.

Critiques about our initial argument and suggestions for improvement:

FIGURE L22.3

Argument presentation on a whiteboard

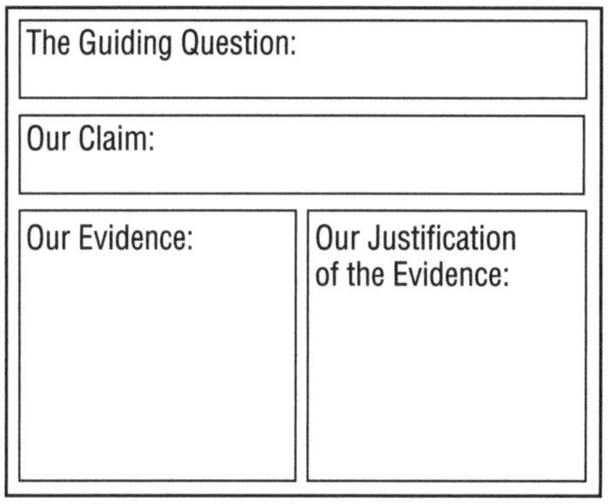

The Guiding Question:	
Our Claim:	
Our Evidence:	Our Justification of the Evidence:

If you are critiquing your classmates' arguments, your goal is to look for mistakes in their arguments and offer suggestions for improvement so these mistakes can be fixed. You should look for ways to make your initial argument stronger by looking for things that the other groups did well. You can keep track of interesting ideas that you see and hear during the argumentation in the space below. You can also use this space to keep track of any questions that you will need to discuss with your team.

LAB 22

Interesting ideas from other groups or questions to take back to my group:

Once the argumentation session is complete, you will have a chance to meet with your group and revise your initial argument. Your group might need to gather more data or design a way to test one or more alternative claims as part of this process. Remember, your goal at this stage of the investigation is to develop the best argument possible.

Report

Once you have completed your research, you will need to prepare an *investigation report* that consists of three sections. Each section should provide an answer to the following questions:

1. What question were you trying to answer and why?

2. What did you do to answer your question and why?

3. What is your argument?

Your report should answer these questions in two pages or less. This report must be typed, and any diagrams, figures, or tables should be embedded into the document. Be sure to write in a persuasive style; you are trying to convince others that your claim is acceptable or valid!

Checkout Questions

Lab 22. Conservation of Energy and Wind Turbines: How Can We Maximize the Amount of Electrical Energy That Will Be Generated by a Wind Turbine Based on the Design of Its Blades?

1. The electrical energy produced by a wind turbine originates as solar energy. Describe the processes that transfer solar energy from the Sun into electrical energy in the wires produced by the turbine.

2. Using concepts of (1) conservation of momentum, (2) conservation of energy, (3) the definition of momentum as a vector quantity, and (4) the definition of energy as a scalar quantity, explain why there is a maximum value for the energy output for one of the variables you tested.

3. People view some research as being more important than other research because of current events or what is important in society.

    a. I agree with this statement.
    b. I disagree with this statement.

Explain your answer, using an example from your investigation about wind turbines.

4. Scientific knowledge, once it has been proven true, does not change.

    a. I agree with this statement.
    b. I disagree with this statement.

Explain your answer, using an example from your investigation about wind turbines.

Conservation of Energy and Wind Turbines

How Can We Maximize the Amount of Electrical Energy That Will Be Generated by a Wind Turbine Based on the Design of Its Blades?

5. How something is structured can affect that object's function.

   a. I agree with this statement.

   b. I disagree with this statement.

   Explain your answer, using an example from your investigation about wind turbines.

6. Scientists often need to track how matter moves into and within a system. Explain why this is important, using an example from your investigation about wind turbines.

7. Scientists often need to define a system under study and then create a model during an investigation. Explain why models of systems are useful in science, using an example from your investigation about wind turbines.

Application Lab

LAB 23

Lab 23. Power: Which Toy Car Has the Engine With the Greatest Horsepower?

Purpose

The purpose of this lab is for students to *apply* what they know about the disciplinary core idea (DCI) of Energy from the *NGSS* to measure the horsepower of a toy car. This lab also gives students an opportunity to learn about the crosscutting concepts (CCs) of (a) Systems and System Models and (b) Energy and Matter: Flows, Cycles, and Conservation from the *NGSS*. In addition, this lab can be used to help students understand four big ideas from AP Physics: (a) objects and systems have properties such as mass and charge, (b) the interactions of an object with other objects can be described by forces, (c) interactions between systems can result in changes in those systems, and (d) changes that occur as result of interactions are constrained by conservation laws. As part of the explicit and reflective discussion, students will also learn about (a) the difference between observations and inferences in science and (b) the role of imagination and creativity in science.

Underlying Physics Concepts

Power is defined as the rate at which work is done or the rate at which energy is produced or used. The work done on an object is the magnitude of a constant force applied to the object times the displacement of that object in the direction of the force. We can mathematically represent power and work through Equations 23.1 and 23.2, respectively. In these equations, P is power, W is work, t is time, $\mathbf{F}$ is force, and $\mathbf{d}$ is displacement. Also, notice that a capital P is used for power, because lowercase $\mathbf{p}$ is used for momentum. In SI units, power is measured in watts (W), work is measured in joules (J), time is measured in seconds (s), force is measured in newtons (N), and displacement is measured in meters (m).

$$\textbf{(Equation 23.1)} \quad P = W/t$$

$$\textbf{(Equation 23.2)} \quad W = \mathbf{Fd}$$

We also know that force is the mass of an object times its acceleration, which gives us Equation 23.3, where m is mass and $\mathbf{a}$ is acceleration. In SI units, mass is measured in kilograms (kg) and acceleration is measured in meters per second squared (m/s^2).

$$\textbf{(Equation 23.3)} \quad W = m\mathbf{ad}$$

We can then combine Equations 23.1 and 23.3 to get Equation 23.4.

(Equation 23.4) $P = mad/t$

What is important to note about all of these equations is the relationship between work and power. Work causes a transformation in the amount or type of energy that an object has. For example, when an object falls toward the ground, work is done by gravity to transfer the energy of the object from potential energy to kinetic energy. Yet, the total energy of the object remains unchanged. At the same time, we can do work on a stationary object to give it kinetic energy by pushing it or pulling it. In this case, work changes the energy of an object. Power is the change in energy of an object divided by the time it takes to induce that change in energy. Thus, power is a measure of the rate at which work is done on an object. Put another way, power is a measure of the rate of change of the energy of an object due to work done on that object.

Finally, if the work done to an object is to change the potential energy of the object, then the power of the object doing the work can be calculated using the change in potential energy. Figure 23.1 shows a person lifting a box of mass m with a pulley system. The person does work on the box by pulling it up a displacement of Δ**h.** To calculate the power of the person when pulling the box, we need to calculate the change in energy of the box and divide that by the total time it took to lift the box. Because the person is doing work to change the potential energy of the box, the work done is equal to the change in gravitational potential energy when the box is lifted Δ**h.** Mathematically, the power output of the person is expressed in Equation 23.5, where **g** is the acceleration due to gravity and is a constant equal to 9.8 m/s².

FIGURE 23.1 _____

Lifting an object

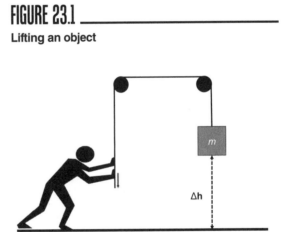

(Equation 23.5) $P = mg\Delta h/t$

The easiest way to measure the horsepower of a toy car during this investigation is to suspend it in the air and attach a mass to each wheel. Then, measure the time it takes for each wheel to lift the mass 1 meter off the ground. The power of the engine is the total amount of potential energy added to the system divided by the amount of time during which the energy was added. In other words, add the potential energy for each mass while the wheels spin and divide the total potential energy by the time it takes to lift the masses. Figure 23.2 shows how this method can be used to measure the horsepower of a toy car engine. Notice that each wheel has a mass attached to it.

LAB 23

FIGURE 23.2

A possible way to measure the horsepower of a toy car.

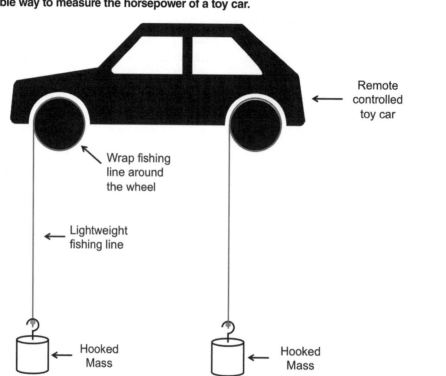

Remote
controlled
toy car

Wrap fishing
line around
the wheel

Lightweight
fishing line

Hooked
Mass

Hooked
Mass

A second way to measure the horsepower of a toy car during this investigation is to determine the amount of time it takes the car to pull a mass behind it. The students can then calculate the sum of the forces on the mass, and assuming they calculate the coefficient of friction between the mass and the floor, they can then calculate the force applied by the car's engine to pull the mass. Because the car pulled the mass a specific distance in a specific amount of time, we know the power of the car's engine is the force applied by the car times the displacement of the mass divided by the time it takes the car to pull the mass. Mathematically, this is expressed in Equation 23.6.

$$\textbf{(Equation 23.6)} \quad P = \mathbf{F}_{car}\mathbf{d}_{mass}/t$$

A third possible way to measure the horsepower of a car is to find the amount of time it takes the car to reach its maximum velocity when starting from rest. Students can then use equations of motion to calculate the average acceleration of the car. The average force applied by the engine is the average acceleration of the car times the mass of the car. In this case, power can be calculated via Equation 23.7.

(Equation 23.7) $\quad P = \mathbf{m}_{car}\mathbf{a}_{car}\mathbf{d}/t$

The first method has the least amount of error inherent in the design. It is fine if students use the other two methods, because it provides an opportunity for students to think about experimental design and error.

Timeline

The instructional time needed to complete this lab investigation is 200–280 minutes. Appendix 3 (p. 531) provides options for implementing this lab investigation over several class periods. Option E (280 minutes) should be used if students are unfamiliar with scientific writing, because this option provides extra instructional time for scaffolding the writing process. You can scaffold the writing process by modeling, providing examples, and providing hints as students write each section of the report. Option E can also be used if you are introducing students to the video analysis programs. Option F (200 minutes) should be used if students are familiar with scientific writing and have developed the skills needed to write an investigation report on their own. In option F, students complete stage 6 (writing the investigation report) and stage 8 (revising the investigation report) as homework.

Materials and Preparation

The materials needed to implement this investigation are listed in Table 23.1. The items can be purchased from a science supply company such as Carolina, Flinn Scientific, PASCO, Vernier, or Ward's Science. The remote control or toy car can be purchased from Wal-Mart or online at Amazon.com. Video analysis software can be purchased from Vernier (Logger *Pro*) or PASCO (SPARKvue or Capstone). These companies also have apps that can be used on Apple- or Android-based tablets and cell phones. We recommend consulting with your school's information technology coordinator to determine the best option for your students.

TABLE 23.1

Materials list for Lab 23

Item	Quantity
Consumables	
Tape	1 roll per group
String or fishing line	1 roll per group
Equipment and other materials	
Safety glasses or goggles	1 per student
Remote control car type or brand A	2 per class

continued

Table 23.1 (*continued*)

Remote control car type or brand B	2 per class
Remote control car type or brand C	2 per class
Remote control car type or brand D	2 per class
Hanging mass set	1 per group
Electronic or triple beam balance	1 per group
Stopwatches	2 per group
Ruler	1 per student
Metersticks	4 per group
Investigation Proposal A (optional)	1 per group
Whiteboard, 2' × 3'*	1 per group
Lab Handout	1 per student
Peer-review guide and teacher scoring rubric	1 per student
Checkout Questions	1 per student
Equipment for video analysis (optional)	
Video camera	1 per group
Computer or tablet with video analysis software	1 per group

* As an alternative, students can use computer and presentation software such as Microsoft PowerPoint or Apple Keynote to create their arguments.

Be sure to use a set routine for distributing and collecting the materials during the lab investigation. One option is to set up the materials for each group at each group's lab station before class begins. This option works well when there is a dedicated section of the classroom for lab work and the materials are large and difficult to move (such as a dynamics track). A second option is to have all the materials on a table or cart at a central location. You can then assign a member of each group to be the "materials manager." This individual is responsible for collecting all the materials his or her group needs from the table or cart during class and for returning all the materials at the end of the class. This option works well when the materials are small and easy to move (such as stopwatches, metersticks, or hanging masses). It also makes it easy to inventory the materials at the end of the class before students leave for the day.

Safety Precautions

Remind students to follow all normal lab safety rules. In addition, tell students to take the following safety precautions:

1. Wear sanitized safety glasses or goggles during lab setup, hands-on activity, and takedown.

2. Keep fingers and toes out of the way of moving objects.

3. Wash hands with soap and water after completing the lab.

Topics for the Explicit and Reflective Discussion

Reflecting on the Use of Core Ideas and Crosscutting Concepts During the Investigation

Teachers should begin the explicit and reflective discussion by asking students to discuss what they know about the core idea they used during the investigation. The following are some important concepts related to the core idea of Energy that students will need to determine the horsepower of a toy car:

- The change in an object's kinetic energy depends on the force exerted on the object and on the displacement of the object during the time interval that the force is exerted.

- Mechanical energy, which is the sum of kinetic and potential energy, is transferred into or out of a system when an external force is exerted on a system such that a component of the force is parallel to its displacement. The process through which mechanical energy is transferred into or out of a system is called *work*.

- If the force is constant during a given displacement, then the work done is the product of the displacement and the component of the force parallel or antiparallel to the displacement.

To help students reflect on what they know about energy, we recommend showing them two or three images using presentation software that help illustrate these important ideas. You can then ask the students the following questions to encourage them to share how they are thinking about these important concepts:

1. What do we see going on in this image?

2. Does anyone have anything else to add?

3. What might be going on that we can't see?

4. What are some things that we are not sure about here?

You can then encourage students to think about how CCs played a role in their investigation. There are at least two CCs that students need to use to determine the horsepower of a toy car: (a) Systems and System Models and (b) Energy and Matter: Flows, Cycles, and Conservation (see Appendix 2 [p. 527] for a brief description of these CCs). To help students reflect on what they know about these CCs, we recommend asking them the following questions:

1. Scientists often need to define the system under study and then make a model during an investigation. Why is developing a model of system so useful in science?

2. What were the boundaries and components of the system you studied during this investigation?

3. Scientists often attempt to track how energy and matter move into, out of, and within systems as part of an investigation. Why is this useful?

4. How did you attempt to track how energy moves within the system you were studying? What did tracking energy allow you to do during your investigation?

You can then encourage the students to think about how they used all these different concepts to help answer the guiding question and why it is important to use these ideas to help justify their evidence for their final arguments. Be sure to remind your students to explain why they included the evidence in their arguments and make the assumptions underlying their analysis and interpretation of the data explicit in order to provide an adequate justification of their evidence.

Reflecting on Ways to Design Better Investigations

It is important for students to reflect on the strengths and weaknesses of the investigation they designed during the explicit and reflective discussion. Students should therefore be encouraged to discuss ways to eliminate potential flaws, measurement errors, or sources of uncertainty in their investigations. To help students be more reflective about the design of their investigation and what they can do to make their investigations more rigorous in the future, you can ask them the following questions:

1. What were some of the strengths of the way you planned and carried out your investigation? In other words, what made it scientific?

2. What were some of the weaknesses of the way you planned and carried out your investigation? In other words, what made it less scientific?

3. What rules can we make, as a class, to ensure that our next investigation is more scientific?

Reflecting on the Nature of Scientific Knowledge and Scientific Inquiry

This investigation can be used to illustrate two important concepts related to the nature of scientific knowledge and the nature of scientific inquiry: (a) the difference between observations and inferences in science and (b) the role of imagination and creativity in science (see Appendix 2 for a brief description of these two concepts). Be sure to review these concepts during and at the end of the explicit and reflective discussion. To help students think about these concepts in relation to what they did during the lab, you can ask them the following questions:

1. You had to make observations and inferences during your investigation. Can you give me some examples of these observations and inferences?

2. Can you work with your group to come up with a rule that you can use to decide if a piece of information is an observation or an inference? Be ready to share in a few minutes.

3. Some people think that there is no room for imagination or creativity in science. What do you think?

4. Can you work with your group to come up with different ways that you needed to use your imagination or be creative during this investigation? Be ready to share in a few minutes.

You can also use presentation software or other techniques to encourage your students to think about these concepts. You can show examples of information from the investigation that are either observations or inferences and ask students to classify each example and explain their thinking. You can also show students an image of the following quote by E. O. Wilson from *Letters to a Young Scientist* (2013) and ask them what they think he meant by it:

> The ideal scientist thinks like a poet and only later works like a bookkeeper. Keep in mind that innovators in both literature and science are basically dreamers and storytellers. In the early stages of the creation of both literature and science, everything in the mind is a story. There is an imagined ending, and usually an imagined beginning, and a selection of bits and pieces that might fit in between. In works of literature and science alike, any part can be changed, causing a ripple among the other parts, some of which are discarded and new ones added. (p. 74)

Be sure to remind your students that, to be proficient in science, it is important that they understand what counts as scientific knowledge and how that knowledge develops over time.

Hints for Implementing the Lab

- Allowing students to design their own procedures for collecting data gives students an opportunity to try, to fail, and to learn from their mistakes. However, you can scaffold students as they develop their procedure by having them fill out an investigation proposal. These proposals provide a way for you to offer students hints and suggestions without telling them how to do it. You can also check the proposals quickly during a class period. For this lab we suggest using Investigation Proposal A.

- We recommend that you try each potential method (setup) on your own before students collect data. This will allow you to become familiar with the range of masses that can be added to the car. Too much mass added to the car can cause the engine to burn out.

- Allow the students to become familiar with the equipment and materials as part of the tool talk before they begin to design their investigation. Giving them 5–10 minutes to examine the available equipment and materials will let them see what they can and cannot do with them. This also gives students a chance to begin thinking about what variables to test and how to control for other variables.

- Do not give students masses that are large enough to lead to the engine burning out.

- We recommend using lightweight but high-strength fishing wire for this investigation. Lightweight wire will add negligible mass to each wheel but will be strong enough to support the tension.

- Students can suspend the car on two metersticks taped to a table with the ends hanging off the table, or they can suspend the car on a ring stand. If students do use metersticks taped to the table, make sure the metersticks are secure and able to support the mass of the toy car.

- We recommend using remote control toy cars with large tires (such as the one shown in Figure 23.2 [p. 494]) for this lab. You only need to have four different types or brands of cars for the students to test. That way, you can purchase two cars of each type or brand (2 cars × 4 different types or brands = total of 8 cars). Each group can develop their method for measuring horsepower using one car and then test the other three types or brands of toy car. This will allow each group to test the same four types or brands of cars without having to purchase four different cars for each group to test.

- Be sure to allow students to go back and re-collect data at the end of the argumentation session. Students often realize that they made numerous mistakes when they were collecting data as a result of their discussions during the argumentation session. The students, as a result, will want a chance to re-collect data, and the re-collection of data should be encouraged when time allows. This also offers an opportunity to discuss what scientists do when they realize a mistake is made inside the lab.

If students use video analysis

- We suggest allowing students to familiarize themselves with the video analysis software before they finalize the procedure for the investigation, especially if they have not used such software previously. This gives students an opportunity to learn how to work with the software and to improve the quality of the video they take.

- Remind students to hold the video camera as still as possible. Any movement of the camera will introduce error into their analysis. If using actual camcorders, we recommend using a tripod to hold the camera steady. If students are using a camera on a cell phone or tablet, we recommend using a table to help steady the camera.

- Remind students to place a meterstick in the same field of view as the motion they are capturing with the video camera. Also, the meterstick should be approximately the same distance from the camera as the motion. Most video analysis software requires the user to define a scale in the video (this allows the software to establish distances and, subsequently, other variables dependent on distance and displacement).

Connections to Standards

Table 23.2 highlights how the investigation can be used to address learning objectives from AP Physics 1; learning objectives from AP Physics C: Mechanics, *Common Core State Standards*, in English language arts (*CCSS ELA*); and *Common Core State Standards*, Mathematics (*CCSS Mathematics*).

TABLE 23.2

Lab 23 alignment with standards

NGSS performance expectations	• None
AP Physics 1 learning objectives	• 4.C.2.2: The student is able to apply the concepts of conservation of energy and the work-energy theorem to determine qualitatively and/or quantitatively that work done on a two-object system in linear motion will change the kinetic energy of the center of mass of the system, the potential energy of the systems, and/or the internal energy of the system. • 5.B.1.2: The student is able to translate between a representation of a single object, which can only have kinetic energy, and a system that includes the object, which may have both kinetic and potential energies. • 5.B.5.1: The student is able to design an experiment and analyze data to examine how a force exerted on an object or system does work on the object or system as it moves through a distance. • 5.B.5.4: The student is able to make claims about the interaction between a system and its environment in which the environment exerts a force on the system, thus doing work on the system and changing the energy of the system (kinetic energy plus potential energy). • 5.B.5.5: The student is able to predict and calculate the energy transfer to (i.e., the work done on) an object or system from information about a force exerted on the object or system through a distance.

continued

LAB 23

Table 23.2 (*continued*)

AP Physics C: Mechanics learning objectives	• I.C.1.a: Students should understand the definition of work, including when it is positive, negative, or zero. • I.C.1.a(1): Calculate the work done by a specified constant force on an object that undergoes a specified displacement. • I.C.2.b(2): Calculate a potential energy function associated with a specified one-dimensional force $F(x)$. • I.C.3.a(1): State and apply the relation between the work performed on an object by nonconservative forces and the change in an object's mechanical energy. • I.C.3.b(1): Identify situations in which mechanical energy is or is not conserved. • I.C.4.b: Calculate the work performed by a force that supplies constant power, or the average power supplied by a force that performs a specified amount of work.
Literacy connections (*CCSS ELA*)	• *Reading*: Key ideas and details, craft and structure, integration of knowledge and ideas • *Writing*: Text types and purposes, production and distribution of writing, research to build and present knowledge, range of writing • *Speaking and listening:* Comprehension and collaboration, presentation of knowledge and ideas
Mathematics connections (*CCSS Mathematics*)	• *Mathematical practices:* Reason abstractly and quantitatively, construct viable arguments and critique the reasoning of others, use appropriate tools strategically, attend to precision • *Number and quantity:* Reason quantitatively and use units to solve problems, represent and model with vector quantities, perform operations on vectors • *Algebra:* Interpret the structure of expressions, understand solving equations as a process of reasoning and explain the reasoning, solve equations and inequalities in one variable, represent and solve equations and inequalities graphically • *Functions:* Understand the concept of a function and use function notation, interpret functions that arise in applications in terms of the context, analyze functions using different representations, interpret expressions for functions in terms of the situation they model • *Statistics and probability:* Summarize, represent, and interpret data on two categorical and quantitative variables; understand and evaluate random processes underlying statistical experiments; make inferences and justify conclusions from sample surveys, experiments, and observational studies

Reference

Wilson, E. O. 2013. *Letters to a young scientist.* New York: Liveright Publishing.

Lab Handout

Lab 23. Power: Which Toy Car Has the Engine With the Greatest Horsepower?

Introduction

With the onset of the Industrial Revolution, many people wanted a way to compare the capabilities of new machines with existing ways of doing things. For example, when James Watt started building and selling steam engines in the late 18th century, one of his customers asked Watt to compare the power of the engine with the power of a horse. The customer, in other words, wanted to know if the new steam engine could do the same things a horse could do. Watt used the term *horsepower* as a convenient way of comparing the power of his steam engine with the power of a horse. Figure L23.1 shows a painting

by Carl Rakeman of a race in 1830 between a horse-drawn cart and the first steam locomotive built in the United States, which was called Tom Thumb. This race was an important event leading to the rise of the railroad industry in America because it demonstrated that a steam-powered locomotive could outperform a horse. Since then, *horsepower* has been given a formal definition as a unit of measure, with 1 horsepower (hp) equal to 746 watts (W).

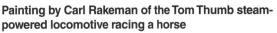

FIGURE L23.1

Painting by Carl Rakeman of the Tom Thumb steam-powered locomotive racing a horse

Note: A full-color, high-resolution version of this image is available on the book's Extras page at *www.nsta.org/ adi-physics1.*

Although most countries use the watt or kilowatt as the standard unit of power, in the United States horsepower is still quite common. Most devices that are powered by an engine or motor are advertised based upon their horsepower; these include household devices such as blenders, lawnmowers, and electric garage door openers. The most common use for horsepower is to describe the power of a car engine. Most everyday cars have an engine of approximately 150 hp, but the fastest sports cars can have engines that are up to 600 hp. And race cars can have engines that are over 800 hp.

An independent group must verify the advertised horsepower of a new engine before a company can sell that engine or a vehicle with that engine in it. An independent group can use several different methods to measure the horsepower of an engine that they did not build; these methods are based on the conservation of energy and the relationship between work and power. The two most common methods are (1) to measure how long it takes the engine to lift or pull a mass a given distance and (2) to measure the amount of time it takes the car to reach its maximum velocity when starting from rest and then use the work-energy theorem to determine the maximum horsepower of the engine. In this investigation, you will have an opportunity to develop your own method for measuring the horsepower of a toy car's engine using the work-energy theorem.

Your Task

Use what you know about the conservation of energy, the relationship between work and power, systems and system models, and the importance of tracking the flow of energy in a system to develop a method for measuring the horsepower of a toy car engine. You will then test the engines of several different toy cars using your method to determine which one has the greatest horsepower.

The guiding question of this investigation is, *Which toy car has the engine with the greatest horsepower?*

Materials

You may use any of the following materials during your investigation:

Consumables
- Tape
- String or fishing line

Equipment
- Safety glasses or goggles (required)
- 4 Different types or brands of remote control cars
- Hanging mass set
- Electronic or triple beam balance
- Stopwatches
- Ruler
- Metersticks

If you have access to the following equipment, you may also consider using a video camera and a computer or tablet with video analysis software.

Safety Precautions

Follow all normal lab safety rules. In addition, take the following safety precautions:

1. Wear sanitized safety glasses or goggles during lab setup, hands-on-activity, and takedown.

2. Keep fingers and toes out of the way of moving objects.

3. Wash hands with soap and water after completing the lab.

Investigation Proposal Required? ☐ Yes ☐ No

Getting Started

To answer the guiding question, you will need to create a method that you can use to determine the horsepower of the engine found inside a toy car. This means you will need to determine the maximum power that the engine is able to produce. *Power* is defined as the rate at which work is done on an object. The equation for power is $P = W/t$, where W is work and t is time. The term *work* is used to describe any situation when a force acts on an object over a displacement of that object. For a force to qualify as having done work on an object, there must be a displacement and the force must be either parallel to the displacement (e.g., a force to the right with a displacement to the right) or antiparallel to the displacement (e.g., a force to the right with a displacement to the left). The equation for work is $W = \mathbf{F}\mathbf{d}$, where $\mathbf{F}$ is force and $\mathbf{d}$ is displacement. Finally, and perhaps most important in terms of your goal for this investigation, energy cannot be created or destroyed—it just changes form as objects move or as it transfers from one object to another one. You will need to use these fundamental ideas as the foundation for the method you will develop, but there may also be other ideas that you need to use.

To determine *what type of data you need to collect,* think about the following questions:

- What are the boundaries and components of the system under study?
- How can you describe the components of the system quantitatively?
- How could you keep track of changes in this system quantitatively?
- Is it useful to track how energy flows into, out of, or within this system?
- How might the structure of the toy car affect its function?
- How might the structure of the test of horsepower affect its function?

To determine *how you will collect the data,* think about the following questions:

- What types of equipment will you need to use and how will you use it?
- How will you track the flow of energy into, out of, and within the system under study?
- What scale or scales should you use when you take your measurements?
- How will you make sure that your data are of high quality (i.e., how will you reduce error)?
- How will you keep track of the data you collect?

To determine *how you will analyze the data,* think about the following questions:

- What type of calculations will you need to make?
- What types of comparisons will you need to make?

- What type of table or graph could you create to help make sense of your data?

Connections to the Nature of Scientific Knowledge and Scientific Inquiry

As you work through your investigation, you may want to consider

- the difference between observations and inferences in science, and
- the role of imagination and creativity in science.

Initial Argument

Once your group has finished collecting and analyzing your data, your group will need to develop an initial argument. Your argument must include a claim, evidence to support your claim, and a justification of the evidence. The *claim* is your group's answer to the guiding question. The *evidence* is an analysis and interpretation of your data. Finally, the *justification* of the evidence is why your group thinks the evidence matters. The justification of the evidence is important because scientists can use different kinds of evidence to support their claims. Your group will create your initial argument on a whiteboard. Your whiteboard should include all the information shown in Figure L23.2.

FIGURE L23.2

Argument presentation on a whiteboard

The Guiding Question:	
Our Claim:	
Our Evidence:	Our Justification of the Evidence:

Argumentation Session

The argumentation session allows all of the groups to share their arguments. One or two members of each group will stay at the lab station to share that group's argument, while the other members of the group go to the other lab stations to listen to and critique the other arguments. This is similar to what scientists do when they propose, support, evaluate, and refine new ideas during a poster session at a conference. If you are presenting your group's argument, your goal is to share your ideas and answer questions. You should also keep a record of the critiques and suggestions made by your classmates so you can use this feedback to make your initial argument stronger. You can keep track of specific critiques and suggestions for improvement that your classmates mention in the space below.

Critiques about our initial argument and suggestions for improvement:

If you are critiquing your classmates' arguments, your goal is to look for mistakes in their arguments and offer suggestions for improvement so these mistakes can be fixed. You should look for ways to make your initial argument stronger by looking for things that the other groups did well. You can keep track of interesting ideas that you see and hear during the argumentation in the space below. You can also use this space to keep track of any questions that you will need to discuss with your team.

Interesting ideas from other groups or questions to take back to my group:

Once the argumentation session is complete, you will have a chance to meet with your group and revise your initial argument. Your group might need to gather more data or design a way to test one or more alternative claims as part of this process. Remember, your goal at this stage of the investigation is to develop the best argument possible.

Report

Once you have completed your research, you will need to prepare an *investigation report* that consists of three sections. Each section should provide an answer to the following questions:

1. What question were you trying to answer and why?

2. What did you do to answer your question and why?

3. What is your argument?

Your report should answer these questions in two pages or less. This report must be typed, and any diagrams, figures, or tables should be embedded into the document. Be sure to write in a persuasive style; you are trying to convince others that your claim is acceptable or valid!

LAB 23

Checkout Questions

Lab 23. Power: Which Toy Car Has the Engine With the Greatest Horsepower?

1. How do the law of conservation of energy and the work-energy theorem help engineers determine the horsepower of a motor?

2. In the figure at right, a person uses a pulley to lift a box off the ground. Assuming that the person lifts the box of mass m with a constant velocity **v** and lifts the box a height of Δh in t seconds, create a mathematical expression for the rate at which the person does work on the box.

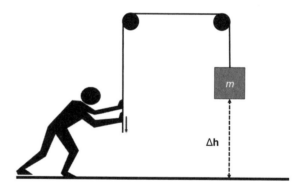

3. Scientists need to be creative and have a good imagination.

    a. I agree with this statement.

    b. I disagree with this statement.

Explain your answer, using an example from your investigation about the horse-power of a toy car.

4. A scientist must first make an observation before he or she can make an inference.

    a. I agree with this statement.

    b. I disagree with this statement.

Explain your answer, using an example from your investigation about the horse-power of a toy car.

5. Scientists often need to define a system under study during an investigation. Explain why it is useful to define a system under study during an investigation, using an example from your investigation about the horsepower of a toy car.

6. Scientists often need to track how matter moves into and within a system. Explain why this is important, using an example from your investigation about the horsepower of a toy car.

SECTION 8
Appendixes

Standards Matrix A: Alignment of the Argument-Driven Inquiry Lab Investigations With the Scientific Practices, Crosscutting Concepts, and Core Ideas in *A Framework for K–12 Science Education* (NRC 2012)

Aspect of the NRC *Framework*	Lab 1. Acceleration and Velocity	Lab 2. Acceleration and Gravity	Lab 3. Projectile Motion	Lab 4. The Coriolis Effect	Lab 5. Force, Mass, and Acceleration	Lab 6. Forces on a Pulley	Lab 7. Forces on an Incline	Lab 8. Friction	Lab 9. Falling Objects and Air Resistance	Lab 10. Rotational Motion	Lab 11. Circular Motion	Lab 12. Torque and Rotation	Lab 13. Simple Harmonic Motion and Pendulums	Lab 14. Simple Harmonic Motion and Springs	Lab 15. Simple Harmonic Motion and Rubber Bands	Lab 16. Linear Momentum and Collisions	Lab 17. Impulse and Momentum	Lab 18. Elastic and Inelastic Collisions	Lab 19. Impulse and Materials	Lab 20. Kinetic and Potential Energy	Lab 21. Conservation of Energy and Pendulums	Lab 22. Conservation of Energy and Wind Turbines	Lab 23. Power
Scientific practices																							
Asking Questions and Defining Problems	■	■	■	■	■	■	■	■	■	■	■	■	■	■	■	■	■	■	■	■	■	■	■
Developing and Using Models			■		■		■	■					■	■		■				■		■	■
Planning and Carrying Out Investigations	■	■	■	■	■	■	■	■	■	■	■	■	■	■	■	■	■	■	■	■	■	■	■
Analyzing and Interpreting Data	■	■	■	■	■	■	■	■	■	■	■	■	■	■	■	■	■	■	■	■	■	■	■
Using Mathematics and Computational Thinking	■	■	■	■	■	■	■	■	■	■	■	■	■	■	■	■	■	■	■	■	■	■	■
Constructing Explanations and Designing Solutions	■	■	■	■	■	■	■	■	■	■	■	■	■	■	■	■	■	■	■	■	■	■	■
Engaging in Argument From Evidence	■	■	■	■	■	■	■	■	■	■	■	■	■	■	■	■	■	■	■	■	■	■	■

Key: ■ = strong alignment; □ = moderate alignment.

Aspect of the NRC *Framework*	Lab Investigation																						
	Lab 1. Acceleration and Velocity	Lab 2. Acceleration and Gravity	Lab 3. Projectile Motion	Lab 4. The Coriolis Effect	Lab 5. Force, Mass, and Acceleration	Lab 6. Forces on a Pulley	Lab 7. Forces on an Incline	Lab 8. Friction	Lab 9. Falling Objects and Air Resistance	Lab 10. Rotational Motion	Lab 11. Circular Motion	Lab 12. Torque and Rotation	Lab 13. Simple Harmonic Motion and Pendulums	Lab 14. Simple Harmonic Motion and Springs	Lab 15. Simple Harmonic Motion and Rubber Bands	Lab 16. Linear Momentum and Collisions	Lab 17. Impulse and Momentum	Lab 18. Elastic and Inelastic Collisions	Lab 19. Impulse and Materials	Lab 20. Kinetic and Potential Energy	Lab 21. Conservation of Energy and Pendulums	Lab 22. Conservation of Energy and Wind Turbines	Lab 23. Power
Obtaining, Evaluating, and Communicating Information	■	■	■	■	■	■	■	■	■	■	■	■	■	■	■	■	■	■	■	■	■	■	■
Crosscutting concepts																							
Patterns	■	■	■				■			■			■	■							■		
Cause and Effect: Mechanism and Explanation			■		■								■								■		
Scale, Proportion, and Quantity	■			■	■							■				■	■		■				
Systems and System Models				■		■		■	■			■	■		■	■		■		■		■	■
Energy and Matter: Flows, Cycles, and Conservation																■	■	■		■	■	■	■
Structure and Function						■	■	■	■	■			■						■			■	
Stability and Change		■					■						■	■									
Core ideas																							
PS2: Motion and Stability: Forces and Interactions	■	■	■	■	■	■	■	■	■	■	■	■	■	■	■	■	■	■	■				
PS3: Energy																				■	■	■	■

Key: ■ = strong alignment; □ = moderate alignment.

Standards Matrix B: Alignment of the Argument-Driven Inquiry Lab Investigations With the Nature of Scientific Knowledge (NOSK) and the Nature of Scientific Inquiry (NOSI) Concepts*

NOSK and NOSI concepts	Lab 1. Acceleration and Velocity	Lab 2. Acceleration and Gravity	Lab 3. Projectile Motion	Lab 4. The Coriolis Effect	Lab 5. Force, Mass, and Acceleration	Lab 6. Forces on a Pulley	Lab 7. Forces on an Incline	Lab 8. Friction	Lab 9. Falling Objects and Air Resistance	Lab 10. Rotational Motion	Lab 11. Circular Motion	Lab 12. Torque and Rotation	Lab 13. Simple Harmonic Motion and Pendulums	Lab 14. Simple Harmonic Motion and Springs	Lab 15. Simple Harmonic Motion and Rubber Bands	Lab 16. Linear Momentum and Collisions	Lab 17. Impulse and Momentum	Lab 18. Elastic and Inelastic Collisions	Lab 19. Impulse and Materials	Lab 20. Kinetic and Potential Energy	Lab 21. Conservation of Energy and Pendulums	Lab 22. Conservation of Energy and Wind Turbines	Lab 23. Power
NOSK																							
The difference between observations and inferences in science	■		■			■	■			■							■				■		■
How scientific knowledge changes over time					■															■	■	■	
The difference between laws and theories in science		■			■						■						■	■		■			
The difference between data and evidence in science		■		■						■	■	■					■	■					
NOSI																							
How the culture of science, societal needs, and current events influence the work of scientists						■		■	■							■	■			■		■	
How scientists use different methods to answer different types of questions							■	■						■									
The role of imagination and creativity in science				■					■	■		■		■	■								■
The nature and role of experiments in science	■		■										■						■				

Key: ■ = strong alignment; ☐ = moderate alignment.

*The NOSK/NOSI concepts listed in this matrix are based on the work of Abd-El-Khalick and Lederman 2000; Akerson, Abd-El-Khalick, and Lederman 2000; Lederman et al. 2002, 2014; NGSS Lead States 2013 (see Appendix H, "Understanding the Scientific Enterprise: The Nature of Science in the *Next Generation Science Standards*"); and Schwartz, Lederman, and Crawford 2004.

Standards Matrix C: Alignment of the Argument-Driven Inquiry Lab Investigations With the *NGSS* (NGSS Lead States 2013) Performance Expectations for High School Physical Science

NGSS performance expectation	Lab 1. Acceleration and Velocity	Lab 2. Acceleration and Gravity	Lab 3. Projectile Motion	Lab 4. The Coriolis Effect	Lab 5. Force, Mass, and Acceleration	Lab 6. Forces on a Pulley	Lab 7. Forces on an Incline	Lab 8. Friction	Lab 9. Falling Objects and Air Resistance	Lab 10. Rotational Motion	Lab 11. Circular Motion	Lab 12. Torque and Rotation	Lab 13. Simple Harmonic Motion and Pendulums	Lab 14. Simple Harmonic Motion and Springs	Lab 15. Simple Harmonic Motion and Rubber Bands	Lab 16. Linear Momentum and Collisions	Lab 17. Impulse and Momentum	Lab 18. Elastic and Inelastic Collisions	Lab 19. Impulse and Materials	Lab 20. Kinetic and Potential Energy	Lab 21. Conservation of Energy and Pendulums	Lab 22. Conservation of Energy and Wind Turbines	Lab 23. Power
HS-PS2-1: Analyze data to support the claim that Newton's second law of motion describes the mathematical relationship among the net force on a macroscopic object, its mass, and its acceleration.	□	□	□		■	■	■	■	■														
HS-PS2-2: Use mathematical representations to support the claim that the total momentum of a system of objects is conserved when there is no net force on the system.																■	■	■	■				
HS-PS2-3: Apply scientific and engineering ideas to design, evaluate, and refine a device that minimizes the force on a macroscopic object during a collision.																	□	□	■				
HS-PS2-4: Use mathematical representations of Newton's Law of Gravitation and Coulomb's Law to describe and predict the gravitational and electrostatic forces between objects.		□	□			□																	

Key: ■ = strong alignment; □ = moderate alignment.

	Lab Investigation																						
NGSS performance expectation	Lab 1. Acceleration and Velocity	Lab 2. Acceleration and Gravity	Lab 3. Projectile Motion	Lab 4. The Coriolis Effect	Lab 5. Force, Mass, and Acceleration	Lab 6. Forces on a Pulley	Lab 7. Forces on an Incline	Lab 8. Friction	Lab 9. Falling Objects and Air Resistance	Lab 10. Rotational Motion	Lab 11. Circular Motion	Lab 12. Torque and Rotation	Lab 13. Simple Harmonic Motion and Pendulums	Lab 14. Simple Harmonic Motion and Springs	Lab 15. Simple Harmonic Motion and Rubber Bands	Lab 16. Linear Momentum and Collisions	Lab 17. Impulse and Momentum	Lab 18. Elastic and Inelastic Collisions	Lab 19. Impulse and Materials	Lab 20. Kinetic and Potential Energy	Lab 21. Conservation of Energy and Pendulums	Lab 22. Conservation of Energy and Wind Turbines	Lab 23. Power
HS-PS3-1: Create a computational model to calculate the change in the energy of one component in a system when the change in energy of the other component(s) and energy flows in and out of the system are known.																					□	□	
HS-PS3-3: Design build, and refine a device that works within given constraints to convert one form of energy into another form of energy.																						■	

Key: ■ = strong alignment; □ = moderate alignment.

Standards Matrix D: Alignment of the Argument-Driven Inquiry Lab Investigations With the Science Practices and Big Ideas in AP Physics I

AP Physics I science practices and big ideas	Lab 1. Acceleration and Velocity	Lab 2. Acceleration and Gravity	Lab 3. Projectile Motion	Lab 4. The Coriolis Effect	Lab 5. Force, Mass, and Acceleration	Lab 6. Forces on a Pulley	Lab 7. Forces on an Incline	Lab 8. Friction	Lab 9. Falling Objects and Air Resistance	Lab 10. Rotational Motion	Lab 11. Circular Motion	Lab 12. Torque and Rotation	Lab 13. Simple Harmonic Motion and Pendulums	Lab 14. Simple Harmonic Motion and Springs	Lab 15. Simple Harmonic Motion and Rubber Bands	Lab 16. Linear Momentum and Collisions	Lab 17. Impulse and Momentum	Lab 18. Elastic and Inelastic Collisions	Lab 19. Impulse and Materials	Lab 20. Kinetic and Potential Energy	Lab 21. Conservation of Energy and Pendulums	Lab 22. Conservation of Energy and Wind Turbines	Lab 23. Power
Science practices																							
Use representations and models to communicate scientific phenomena and solve scientific problems.	■	■	■	■	■	■	■	■	■	■	■	■	■	■	■	■	■	■	■	■	■	■	■
Use mathematics appropriately.	■	■	■	■	■	■	■	■	■	■	■	■	■	■	■	■	■	■	■	■	■	■	■
Engage in scientific questioning to extend thinking or to guide investigations.	■	■	■	■	■	■	■	■	■	■	■	■	■	■	■	■	■	■	■	■	■	■	■
Plan and implement data collection strategies in relation to a particular scientific question.	■	■	■	■	■	■	■	■	■	■	■	■	■	■	■	■	■	■	■	■	■	■	■
Perform data analysis and evaluation of evidence.	■	■	■	■	■	■	■	■	■	■	■	■	■	■	■	■	■	■	■	■	■	■	■
Work with scientific explanations and theories.	■	■	■	■	■	■	■	■	■	■	■	■	■	■	■	■	■	■	■	■	■	■	■
Connect and relate knowledge across various scales, concepts, and representations in and across domains.	■	■	■	■	■	■	■	■	■	■	■	■	■	□	■	■	■	■	■	■	■	■	■

Key: ■ = strong alignment; □ = moderate alignment.

AP Physics I science practices and big ideas	Lab Investigation																						
	Lab 1. Acceleration and Velocity	Lab 2. Acceleration and Gravity	Lab 3. Projectile Motion	Lab 4. The Coriolis Effect	Lab 5. Force, Mass, and Acceleration	Lab 6. Forces on a Pulley	Lab 7. Forces on an Incline	Lab 8. Friction	Lab 9. Falling Objects and Air Resistance	Lab 10. Rotational Motion	Lab 11. Circular Motion	Lab 12. Torque and Rotation	Lab 13. Simple Harmonic Motion and Pendulums	Lab 14. Simple Harmonic Motion and Springs	Lab 15. Simple Harmonic Motion and Rubber Bands	Lab 16. Linear Momentum and Collisions	Lab 17. Impulse and Momentum	Lab 18. Elastic and Inelastic Collisions	Lab 19. Impulse and Materials	Lab 20. Kinetic and Potential Energy	Lab 21. Conservation of Energy and Pendulums	Lab 22. Conservation of Energy and Wind Turbines	Lab 23. Power
Big ideas																							
Objects and systems have properties such as mass and charge. Systems may have internal structure.					■		■	■												■	■	■	■
Fields existing in space can be used to explain interactions.		■	■			■	■		■				■		■								
The interactions of an object with other objects can be described by forces.	■	■	■	■	■	■	■	■	■	■	■	■	■	■	■	■	■	■	■	■	■	■	■
Interactions between systems can result in changes in those systems.	■	■		■	■			■	■	■	■	■		■		■	■	■	■			■	■
Changes that occur as a result of interactions are constrained by conservation laws.																■	■	■	■	■	■	■	■

Key: ■ = strong alignment; ☐ = moderate alignment.

Standards Matrix E: Alignment of the Argument-Driven Inquiry Lab Investigations With the Course Content and Science Practices and Skills in AP Physics C: Mechanics

AP Physics C: Mechanics content areas and laboratory objectives	Lab 1. Acceleration and Velocity	Lab 2. Acceleration and Gravity	Lab 3. Projectile Motion	Lab 4. The Coriolis Effect	Lab 5. Force, Mass, and Acceleration	Lab 6. Forces on a Pulley	Lab 7. Forces on an Incline	Lab 8. Friction	Lab 9. Falling Objects and Air Resistance	Lab 10. Rotational Motion	Lab 11. Circular Motion	Lab 12. Torque and Rotation	Lab 13. Simple Harmonic Motion and Pendulums	Lab 14. Simple Harmonic Motion and Springs	Lab 15. Simple Harmonic Motion and Rubber Bands	Lab 16. Linear Momentum and Collisions	Lab 17. Impulse and Momentum	Lab 18. Elastic and Inelastic Collisions	Lab 19. Impulse and Materials	Lab 20. Kinetic and Potential Energy	Lab 21. Conservation of Energy and Pendulums	Lab 22. Conservation of Energy and Wind Turbines	Lab 23. Power
Content areas																							
Kinematics	■	■	■	■																			
Newton's laws of motion					■	■	■	■	■														
Work, energy, and power																				■	■	■	■
Systems of particles and linear momentum																■	■	■	■				
Circular motion and rotation										■	■	■											
Oscillations and gravitation													■	■	■								
Laboratory objectives																							
Design experiments	■	■	■	■	■	■	■	■	■	■	■	■	■	■	■	■	■	■	■	■	■	■	■
Observe and measure real phenomena	■	■	■	■	■	■	■	■	■	■	■	■	■	■	■	■	■	■	■	■	■	■	■
Analyze data	■	■	■	■	■	■	■	■	■	■	■	■	■	■	■	■	■	■	■	■	■	■	■
Analyze error	■	■	■	■	■	■	■	■	■	■	■	■	■	■	■	■	■	■	■	■	■	■	■
Communicate results	■	■	■	■	■	■	■	■	■	■	■	■	■	■	■	■	■	■	■	■	■	■	■

Key: ■ = strong alignment; □ = moderate alignment

Standards Matrix F: Alignment of the Argument-Driven Inquiry Lab Investigations With the *Common Core State Standards* for English Language Arts (*CCSS ELA*; NGAC and CCSSO 2010)

CCSS ELA	Lab Investigation																						
	Lab 1. Acceleration and Velocity	Lab 2. Acceleration and Gravity	Lab 3. Projectile Motion	Lab 4. The Coriolis Effect	Lab 5. Force, Mass, and Acceleration	Lab 6. Forces on a Pulley	Lab 7. Forces on an Incline	Lab 8. Friction	Lab 9. Falling Objects and Air Resistance	Lab 10. Rotational Motion	Lab 11. Circular Motion	Lab 12. Torque and Rotation	Lab 13. Simple Harmonic Motion and Pendulums	Lab 14. Simple Harmonic Motion and Springs	Lab 15. Simple Harmonic Motion and Rubber Bands	Lab 16. Linear Momentum and Collisions	Lab 17. Impulse and Momentum	Lab 18. Elastic and Inelastic Collisions	Lab 19. Impulse and Materials	Lab 20. Kinetic and Potential Energy	Lab 21. Conservation of Energy and Pendulums	Lab 22. Conservation of Energy and Wind Turbines	Lab 23. Power
Reading																							
Key ideas and details	■	■	■	■	■	■	■	■	■	■	■	■	■	■	■	■	■	■	■	■	■	■	■
Craft and structure	■	■	■	■	■	■	■	■	■	■	■	■	■	■	■	■	■	■	■	■	■	■	■
Integration of knowledge and ideas	■	■	■	■	■	■	■	■	■	■	■	■	■	■	■	■	■	■	■	■	■	■	■
Writing																							
Text types and purposes	■	■	■	■	■	■	■	■	■	■	■	■	■	■	■	■	■	■	■	■	■	■	■
Production and distribution of writing	■	■	■	■	■	■	■	■	■	■	■	■	■	■	■	■	■	■	■	■	■	■	■
Research to build and present knowledge	■	■	■	■	■	■	■	■	■	■	■	■	■	■	■	■	■	■	■	■	■	■	■
Range of writing	■	■	■	■	■	■	■	■	■	■	■	■	■	■	■	■	■	■	■	■	■	■	■
Speaking and listening																							
Comprehension and collaboration	■	■	■	■	■	■	■	■	■	■	■	■	■	■	■	■	■	■	■	■	■	■	■
Presentation of knowledge and ideas	■	■	■	■	■	■	■	■	■	■	■	■	■	■	■	■	■	■	■	■	■	■	■

Key: ■ = strong alignment; □ = moderate alignment.

Standards Matrix G: Alignment of the Argument-Driven Inquiry Lab Investigations With the *Common Core State Standards* for Mathematics (*CCSS Mathematics;* NGAC and CCSSO 2010)

CCSS Mathematics	Lab 1. Acceleration and Velocity	Lab 2. Acceleration and Gravity	Lab 3. Projectile Motion	Lab 4. The Coriolis Effect	Lab 5. Force, Mass, and Acceleration	Lab 6. Forces on a Pulley	Lab 7. Forces on an Incline	Lab 8. Friction	Lab 9. Falling Objects and Air Resistance	Lab 10. Rotational Motion	Lab 11. Circular Motion	Lab 12. Torque and Rotation	Lab 13. Simple Harmonic Motion and Pendulums	Lab 14. Simple Harmonic Motion and Springs	Lab 15. Simple Harmonic Motion and Rubber Bands	Lab 16. Linear Momentum and Collisions	Lab 17. Impulse and Momentum	Lab 18. Elastic and Inelastic Collisions	Lab 19. Impulse and Materials	Lab 20. Kinetic and Potential Energy	Lab 21. Conservation of Energy and Pendulums	Lab 22. Conservation of Energy and Wind Turbines	Lab 23. Power
Mathematical practices																							
Make sense of problems and persevere in solving them	■	■	■		■		■		■				■	■								■	
Reason abstractly and quantitatively	■	■	■	■	■	■	■	■	■	■	■	■	■	■	■	■	■	■	■	■	■	■	■
Construct viable arguments and critique the reasoning of others	■	■	■	■	■	■	■	■	■	■	■	■	■	■	■	■	■	■	■	■	■	■	■
Model with mathematics	■		■		■		■		■				■	■								■	
Use appropriate tools strategically	■	■	■	■	■	■	■	■	■	■	■	■	■	■	■	■	■	■	■	■	■	■	■
Attend to precision	■	■	■	■	■	■	■	■	■	■	■	■	■	■	■	■	■	■	■	■	■	■	■
Look for and make use of structure		■		■		■			■				■	■								■	
Look for and express regularity in repeated reasoning	■	■	■		■		■		■				■	■								■	
Number and quantity																							
Extend the properties of exponents to rational exponents																							
Use properties of rational and irrational numbers																							
Reason quantitatively and use units to solve problems	■	■	■	■	■	■	■	■	■	■	■	■	■	■	■	■	■	■	■	■	■	■	■
Perform arithmetic operations with complex numbers																							

Key: ■ = strong alignment; □ = moderate alignment.

CCSS Mathematics	Lab 1. Acceleration and Velocity	Lab 2. Acceleration and Gravity	Lab 3. Projectile Motion	Lab 4. The Coriolis Effect	Lab 5. Force, Mass, and Acceleration	Lab 6. Forces on a Pulley	Lab 7. Forces on an Incline	Lab 8. Friction	Lab 9. Falling Objects and Air Resistance	Lab 10. Rotational Motion	Lab 11. Circular Motion	Lab 12. Torque and Rotation	Lab 13. Simple Harmonic Motion and Pendulums	Lab 14. Simple Harmonic Motion and Springs	Lab 15. Simple Harmonic Motion and Rubber Bands	Lab 16. Linear Momentum and Collisions	Lab 17. Impulse and Momentum	Lab 18. Elastic and Inelastic Collisions	Lab 19. Impulse and Materials	Lab 20. Kinetic and Potential Energy	Lab 21. Conservation of Energy and Pendulums	Lab 22. Conservation of Energy and Wind Turbines	Lab 23. Power
Number and quantity (continued)																							
Represent complex numbers and their operations on the complex plane																							
Use complex numbers in polynomial identities and equations																							
Represent and model with vector quantities	■	■	■	■	■	■	■	■	■	■	■	■	■	■	■	■	■	■	■	■	■	■	■
Perform operations on vectors	■	■	■	■	■	■	■	■	■	■	■	■	■	■	■	■	■	■	■	■	■	■	■
Perform operations on matrices and use matrices in applications																							
Algebra																							
Interpret the structure of expressions	■	■	■		■	■	■	■	■	■	■	■	■	■	■	■	■	■	■	■	■	■	■
Write expressions in equivalent forms to solve problems					■		■																
Perform arithmetic operations on polynomials																							
Understand the relationship between zeros and factors of polynomials																							
Use polynomial identities to solve problems																							
Rewrite rational functions																							
Create equations that describe numbers or relationships			■		■		■	■	■					■						■		■	

Key: ■ = strong alignment; □ = moderate alignment.

CCSS Mathematics	Lab 1. Acceleration and Velocity	Lab 2. Acceleration and Gravity	Lab 3. Projectile Motion	Lab 4. The Coriolis Effect	Lab 5. Force, Mass, and Acceleration	Lab 6. Forces on a Pulley	Lab 7. Forces on an Incline	Lab 8. Friction	Lab 9. Falling Objects and Air Resistance	Lab 10. Rotational Motion	Lab 11. Circular Motion	Lab 12. Torque and Rotation	Lab 13. Simple Harmonic Motion and Pendulums	Lab 14. Simple Harmonic Motion and Springs	Lab 15. Simple Harmonic Motion and Rubber Bands	Lab 16. Linear Momentum and Collisions	Lab 17. Impulse and Momentum	Lab 18. Elastic and Inelastic Collisions	Lab 19. Impulse and Materials	Lab 20. Kinetic and Potential Energy	Lab 21. Conservation of Energy and Pendulums	Lab 22. Conservation of Energy and Wind Turbines	Lab 23. Power
Algebra *(continued)*																							
Understand solving equations as a process of reasoning and explain the reasoning	■	■	■		■	■	■	■	■	■	■	■	■	■	■	■	■	■	■	■	■	■	■
Solve equations and inequalities in one variable	■	■	■		■		■	■	■	■	■	■	■	■	■	■	■	■	■	■	■	■	■
Solve systems of equations																							
Represent and solve equations and inequalities graphically	■	■	■		■	■	■	■	■	■			■	■		■	■	■	■	■	■	■	■
Functions																							
Understand the concept of a function and use function notation	■	■	■	■	■	■	■	■	■		■	■	■	■	■	■	■	■	■	■	■	■	■
Interpret functions that arise in applications in terms of the context	■	■	■	■	■	■	■	■	■		■	■	■	■	■	■	■	■	■	■	■	■	■
Analyze functions using different representations	■	■	■		■		■		■				■	■	■	■	■	■	■	■	■	■	■
Build a function that models a relationship between two quantities			■		■		■		■					■							■	■	
Build new functions from existing functions																							
Construct and compare linear and exponential models and solve problems	■		■		■		■		■					■								■	
Interpret expressions for functions in terms of the situation they model	■	■	■		■	■	■	■	■	■		■	■	■	■	■	■	■	■	■	■	■	■

Key: ■ = strong alignment; □ = moderate alignment.

CCSS Mathematics	Lab 1. Acceleration and Velocity	Lab 2. Acceleration and Gravity	Lab 3. Projectile Motion	Lab 4. The Coriolis Effect	Lab 5. Force, Mass, and Acceleration	Lab 6. Forces on a Pulley	Lab 7. Forces on an Incline	Lab 8. Friction	Lab 9. Falling Objects and Air Resistance	Lab 10. Rotational Motion	Lab 11. Circular Motion	Lab 12. Torque and Rotation	Lab 13. Simple Harmonic Motion and Pendulums	Lab 14. Simple Harmonic Motion and Springs	Lab 15. Simple Harmonic Motion and Rubber Bands	Lab 16. Linear Momentum and Collisions	Lab 17. Impulse and Momentum	Lab 18. Elastic and Inelastic Collisions	Lab 19. Impulse and Materials	Lab 20. Kinetic and Potential Energy	Lab 21. Conservation of Energy and Pendulums	Lab 22. Conservation of Energy and Wind Turbines	Lab 23. Power
Functions *(continued)*																							
Extend the domain of trigonometric functions using the unit circle														■									
Model periodic phenomena with trigonometric functions														■									
Prove and apply trigonometric identities														■									
Statistics and probability																							
Summarize, represent, and interpret data on a single count or measurement variable																							
Summarize, represent, and interpret data on two categorical and quantitative variables	■	■	■		■	■	■		■		■	■	■	■	■	■	■	■	■	■	■	■	■
Interpret linear models	■		■		■	■	■		■					■								■	
Understand and evaluate random processes underlying statistical experiments		■	■		■	■	■	■	■	■	■	■	■	■	■	■	■	■	■	■	■	■	■
Make inferences and justify conclusions from sample surveys, experiments, and observational studies	■	■	■		■	■	■		■	■	■	■	■	■	■	■	■	■	■	■	■	■	■
Understand independence and conditional probability and use them to interpret data																							

Key: ■ = strong alignment; □ = moderate alignment.

CCSS Mathematics	Lab Investigation																						
	Lab 1. Acceleration and Velocity	Lab 2. Acceleration and Gravity	Lab 3. Projectile Motion	Lab 4. The Coriolis Effect	Lab 5. Force, Mass, and Acceleration	Lab 6. Forces on a Pulley	Lab 7. Forces on an Incline	Lab 8. Friction	Lab 9. Falling Objects and Air Resistance	Lab 10. Rotational Motion	Lab 11. Circular Motion	Lab 12. Torque and Rotation	Lab 13. Simple Harmonic Motion and Pendulums	Lab 14. Simple Harmonic Motion and Springs	Lab 15. Simple Harmonic Motion and Rubber Bands	Lab 16. Linear Momentum and Collisions	Lab 17. Impulse and Momentum	Lab 18. Elastic and Inelastic Collisions	Lab 19. Impulse and Materials	Lab 20. Kinetic and Potential Energy	Lab 21. Conservation of Energy and Pendulums	Lab 22. Conservation of Energy and Wind Turbines	Lab 23. Power
Statistics and probability *(continued)*																							
Use the rules of probability to compute probabilities of compound events in a uniform probability model																							
Calculate expected values and use them to solve problems																							
Use probability to evaluate outcomes of decisions																							

Key: ■ = strong alignment; □ = moderate alignment.

References

Abd-El-Khalick, F., and N. G. Lederman. 2000. Improving science teachers' conceptions of nature of science: A critical review of the literature. *International Journal of Science Education* 22: 665–701.

Akerson, V., F. Abd-El-Khalick, and N. Lederman. 2000. Influence of a reflective explicit activity-based approach on elementary teachers' conception of nature of science. *Journal of Research in Science Teaching* 37 (4): 295–317.

Lederman, N. G., F. Abd-El-Khalick, R. L. Bell, and R. S. Schwartz. 2002. Views of nature of science questionnaire: Toward a valid and meaningful assessment of learners' conceptions of nature of science. *Journal of Research in Science Teaching* 39 (6): 497–521.

Lederman, J., N. Lederman, S. Bartos, S. Bartels, A. Meyer, and R. Schwartz. 2014. Meaningful assessment of learners' understanding about scientific inquiry: The Views About Scientific Inquiry (VASI) questionnaire. *Journal of Research in Science Teaching* 51 (1): 65–83.

National Governors Association Center for Best Practices and Council of Chief State School Officers (NGAC and CCSSO). 2010. *Common core state standards.* Washington, DC: NGAC and CCSSO.

NGSS Lead States. 2013. *Next Generation Science Standards: For states, by states.* Washington, DC: National Academies Press. *www.nextgenscience.org/next-generation-science-standards.*

National Research Council (NRC). 2012. *A framework for K–12 science education: Practices, crosscutting concepts, and core ideas.* Washington, DC: National Academies Press.

Schwartz, R. S., N. Lederman, and B. Crawford. 2004. Developing views of nature of science in an authentic context: An explicit approach to bridging the gap between nature of science and scientific inquiry. *Science Education* 88: 610–645.

APPENDIX 2

OVERVIEW OF THE *NGSS* CROSSCUTTING CONCEPTS

Patterns

Scientists look for patterns in nature and attempt to understand the underlying cause of these patterns. Scientists, for example, often collect data and then look for patterns to identify a relationship between two variables, a trend over time, or a difference between groups.

Cause and Effect: Mechanism and Explanation

Natural phenomena have causes, and uncovering causal relationships (e.g., how changes in x affect y) is a major activity of science.

Scale, Proportion, and Quantity

Students need to understand that it is critical for scientists to be able to recognize what is relevant at different sizes, times, and scales. Scientists must also be able to recognize proportional relationships between categories, groups, or quantities. In physics, quantity takes on additional importance as physicists differentiate between vector quantities (quantities with magnitude and direction) and scalar quantities (quantities with only magnitude).

Systems and System Models

Students need to understand that defining a system under study and making a model of it are tools for developing a better understanding of natural phenomena in science.

Energy and Matter: Flows, Cycles, and Conservation

Students should realize that in science it is important to track how energy and matter move into, out of, and within systems.

Structure and Function

In nature, the way an object or a material is structured or shaped determines how it functions and places limits on what it can and cannot do.

Stability and Change

It is critical to understand what makes a system stable or unstable and what controls rates of change in system.

OVERVIEW OF NATURE OF SCIENTIFIC KNOWLEDGE AND SCIENTIFIC INQUIRY CONCEPTS

Nature of Scientific Knowledge Concepts

The difference between observations and inferences in science

An *observation* is a descriptive statement about a natural phenomenon, whereas an *inference* is an interpretation of an observation. Students should also understand that current scientific knowledge and the perspectives of individual scientists guide both observations and inferences. Thus, different scientists can have different but equally valid interpretations of the same observations due to differences in their perspectives and background knowledge.

How scientific knowledge changes over time

A person can have confidence in the validity of scientific knowledge but must also accept that scientific knowledge may be abandoned or modified in light of new evidence or because existing evidence has been reconceptualized by scientists. There are many examples in the history of science of both *evolutionary changes* (i.e., the slow or gradual refinement of ideas) and *revolutionary changes* (i.e., the rapid abandonment of a well-established idea) in scientific knowledge.

The difference between laws and theories in science

A *scientific law* describes the behavior of a natural phenomenon or a generalized relationship under certain conditions; a *scientific theory* is a well-substantiated explanation of some aspect of the natural world. Theories do not become laws even with additional evidence; they explain laws. However, not all scientific laws have an accompanying explanatory theory. It is also important for students to understand that scientists do not discover laws or theories; the scientific community develops them over time.

The difference between data and evidence in science

Data are measurements, observations, and findings from other studies that are collected as part of an investigation. *Evidence,* in contrast, is analyzed data and an interpretation of the analysis.

Nature of Scientific Inquiry Concepts

How the culture of science, societal needs, and current events influence the work of scientists

Scientists share a set of values, norms, and commitments that shape what counts as knowing, how to represent or communicate information, and how to interact with other scientists. The culture of science affects who gets to do science, what scientists choose to investigate, how investigations are conducted, how research findings are interpreted, and what people see as implications. People also view some research as being more important than other research because of cultural values and current events.

How scientists use different methods to answer different types of questions

Examples of methods include experiments, systematic observations of a phenomenon, literature reviews, and analysis of existing data sets; the choice of method depends on the objectives of the research. There is no universal step-by step scientific method that all scientists follow; rather, different scientific disciplines (e.g., chemistry vs. physics) and fields within a discipline (e.g., organic vs. physical chemistry) use different types of methods, use different core theories, and rely on different standards to develop scientific knowledge.

The role of imagination and creativity in science

Students should learn that developing explanations for, or models of, natural phenomena and then figuring out how these explanations or models can be put to the test of reality is as creative as writing poetry, composing music, or designing video games. Scientists must also use their imagination and creativity to figure out new ways to test ideas and collect or analyze data.

The nature and role of experiments in science

Scientists use experiments to test the validity of a hypothesis (i.e., a tentative explanation) for an observed phenomenon. Experiments include a test and the formulation of predictions (expected results) if the test is conducted and the hypothesis is valid. The experiment is then carried out and the predictions are compared with the actual results of the experiment. If the predictions match the actual results, then the hypothesis is supported. If the actual results do not match the predicted results, then the hypothesis is not supported. A signature feature of an experiment is the control of variables to help eliminate alternative explanations for the results.

APPENDIX 3
Timeline Options for Implementing ADI Lab Investigations

Option A: 6 days (280 minutes), no homework

Day	Stage	Time
1	1: Introduce the task and the guiding question	20 minutes
	2: Design a method	30 minutes
2	2: Collect data	50 minutes
3	3: Develop an initial argument	20 minutes
	4: Argumentation session (and revise initial argument)	30 minutes
4	5: Explicit and reflective discussion	20 minutes
	6: Write investigation report (draft)	30 minutes
5	7: Double-blind peer review	50 minutes
6	8: Revise and submit the investigation report	30 minutes

Option B: 5 days (220 minutes), writing done as homework

Day	Stage	Time
1	1: Introduce the task and the guiding question	20 minutes
	2: Design a method	30 minutes
2	2: Collect data	50 minutes
3	3: Develop an initial argument	20 minutes
	4: Argumentation session (and revise initial argument)	30 minutes
4	5: Explicit and reflective discussion	20 minutes
	6: Write investigation report (draft)	Homework
5	7: Double-blind peer review	50 minutes
	8: Revise and submit the investigation report	Homework

Option C: 5 days (230 minutes), no homework

Day	Stage	Time
1	1: Introduce the task and the guiding question	20 minutes
	2: Design a method and collect data	30 minutes
2	3: Develop an initial argument	20 minutes
	4: Argumentation session (and revise initial argument)	30 minutes
3	5: Explicit and reflective discussion	20 minutes
	6: Write investigation report (draft)	30 minutes
4	7: Double-blind peer review	50 minutes
5	8: Revise and submit the investigation report	30 minutes

Option D: 4 days (170 minutes), writing done as homework

Day	Stage	Time
1	1: Introduce the task and the guiding question	20 minutes
	2: Design a method and collect data	30 minutes
2	3: Develop an initial argument	20 minutes
	4: Argumentation session (and revise initial argument)	30 minutes
3	5: Explicit and reflective discussion	20 minutes
	6: Write investigation report (draft)	Homework
4	7: Double-blind peer review	50 minutes
	8: Revise and submit the investigation report	Homework

Option E: 6 days (280 minutes), no homework

Day	Stage	Time
1	1: Introduce the task and the guiding question	20 minutes
	2: Design a method	30 minutes
2	2: Collect data	30 minutes
	3: Develop an initial argument	20 minutes
3	4: Argumentation session (and revise initial argument)	30 minutes
	5: Explicit and reflective discussion	20 minutes
4	6: Write investigation report (draft)	50 minutes
5	7: Double-blind peer review	50 minutes
6	8: Revise and submit the investigation report	30 minutes

Option F: 4 days (200 minutes), writing done as homework

Day	Stage	Time
1	1: Introduce the task and the guiding question	20 minutes
	2: Design a method	30 minutes
2	2: Collect data	30 minutes
	3: Develop an initial argument	20 minutes
3	4: Argumentation session (and revise initial argument)	30 minutes
	5: Explicit and reflective discussion	20 minutes
	6: Write investigation report (draft)	Homework
4	7: Double-blind peer review	50 minutes
	8: Revise and submit the investigation report	Homework

APPENDIX 4
Investigation Proposal Options

This appendix presents six investigation proposals (long and short versions of three different types of proposals) that may be used in most labs. Investigation Proposal A is appropriate for descriptive studies, whereas Investigation Proposal B and Investigation Proposal C are appropriate for comparative or experimental studies. The development of these proposals was supported by the Institute of Education Sciences, U.S. Department of Education, through grant R305A100909 to Florida State University.

The format of investigation proposals B and C is modeled after a hypothetical deductive-reasoning guide described in *Exploring the Living World* (Lawson 1995) and modified from an investigation guide described in an article by Maguire, Myerowitz, and Sampson (2010).

References

Lawson, A. E. 1995. Exploring the living world: A laboratory manual for biology. McGraw-Hill College.

Maguire, L., L. Myerowitz, and V. Sampson. 2010. Diffusion and osmosis in cells: A guided inquiry activity. *The Science Teacher* 77 (8): 55–60.

Investigation Proposal A: Descriptive Study

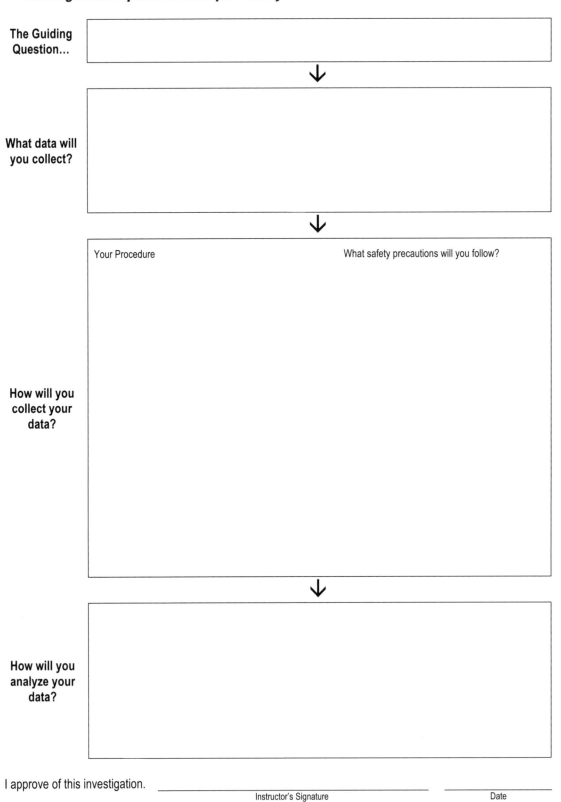

The Guiding
Question...

What data will
you collect?

How will you
collect your
data?

Your Procedure   What safety precautions will you follow?

How will you
analyze your
data?

I approve of this investigation. _____ _____
                                          Instructor's Signature   Date

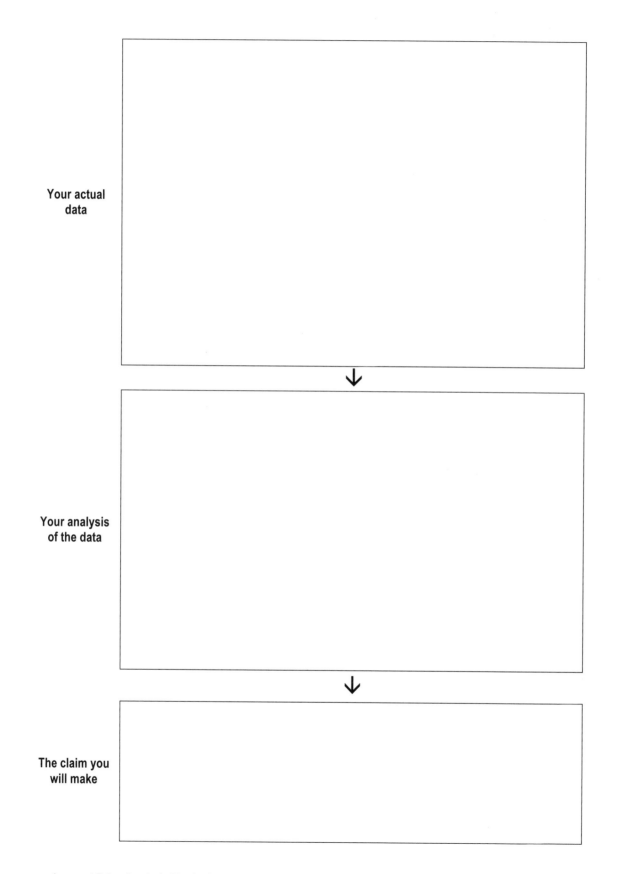

Your actual data

Your analysis of the data

The claim you will make

Investigation Proposal A: Descriptive Study (Short Form)

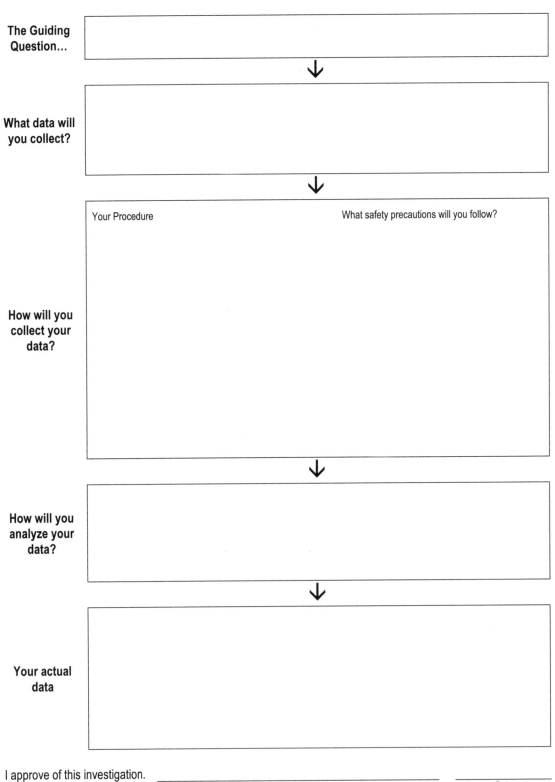

The Guiding Question…

↓

What data will you collect?

↓

Your Procedure   What safety precautions will you follow?

How will you collect your data?

↓

How will you analyze your data?

↓

Your actual data

I approve of this investigation. _____   _____

Instructor's Signature   Date

Investigation Proposal B: Comparative or Experimental Study

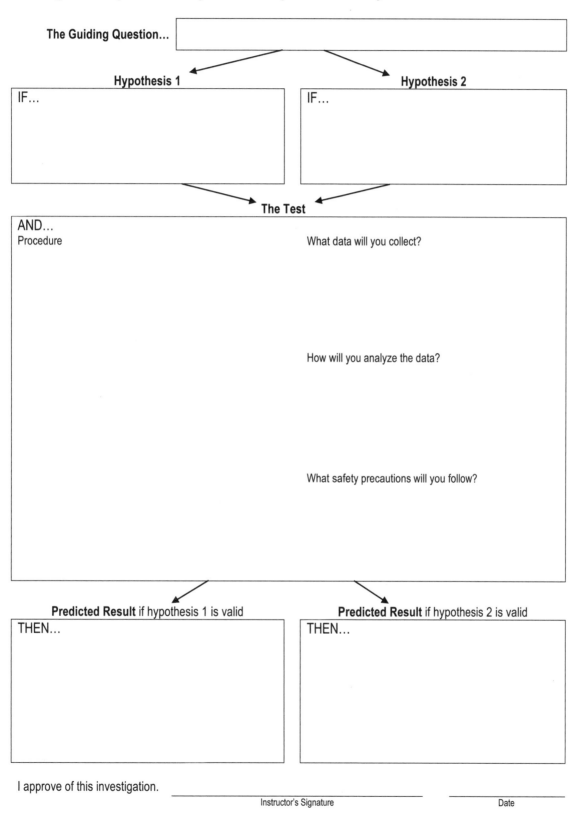

The Guiding Question...

Hypothesis 1

Hypothesis 2

IF...

IF...

The Test

AND...
Procedure

What data will you collect?

How will you analyze the data?

What safety precautions will you follow?

Predicted Result if hypothesis 1 is valid

Predicted Result if hypothesis 2 is valid

THEN...

THEN...

I approve of this investigation. _____   _____
Instructor's Signature   Date

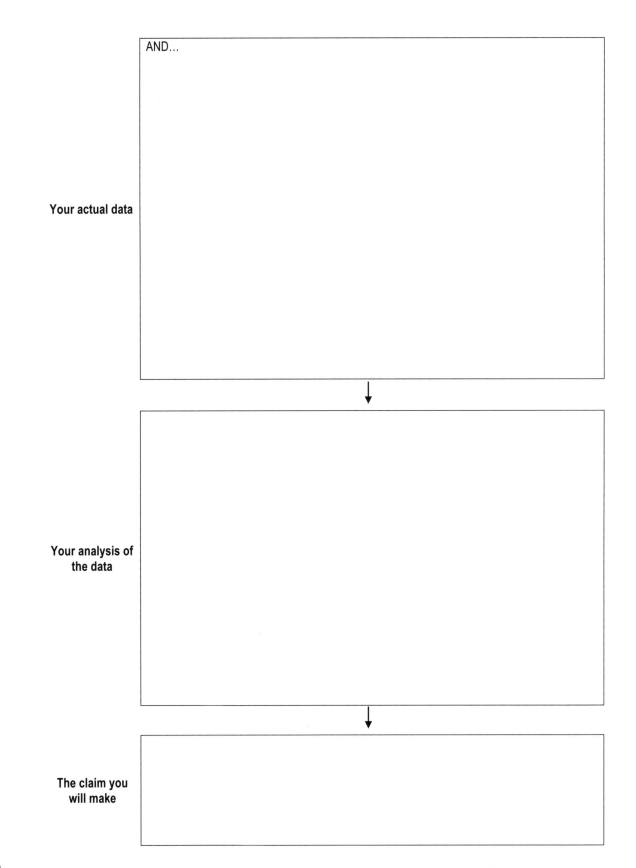

AND...

Your actual data

Your analysis of the data

The claim you will make

Investigation Proposal B: Comparative or Experimental Study (Short Form)

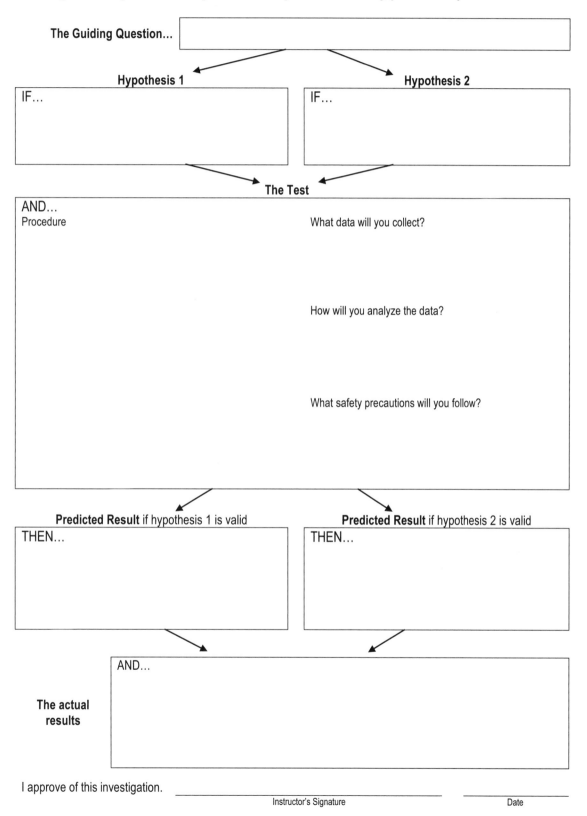

The Guiding Question...

Hypothesis 1   Hypothesis 2

IF...   IF...

The Test

AND...
Procedure

What data will you collect?

How will you analyze the data?

What safety precautions will you follow?

Predicted Result if hypothesis 1 is valid   **Predicted Result** if hypothesis 2 is valid

THEN...   THEN...

AND...

The actual results

I approve of this investigation. _____ _____

Instructor's Signature   Date

Investigation Proposal C: Comparative or Experimental Study

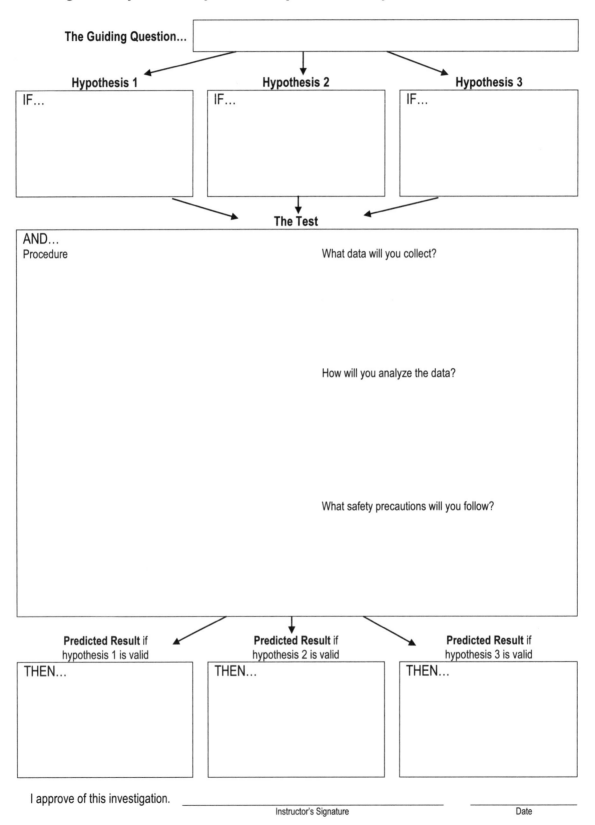

The Guiding Question...

Hypothesis 1
IF...

Hypothesis 2
IF...

Hypothesis 3
IF...

The Test

AND...
Procedure

What data will you collect?

How will you analyze the data?

What safety precautions will you follow?

Predicted Result if
hypothesis 1 is valid
THEN...

Predicted Result if
hypothesis 2 is valid
THEN...

Predicted Result if
hypothesis 3 is valid
THEN...

I approve of this investigation. _____ _____
Instructor's Signature   Date

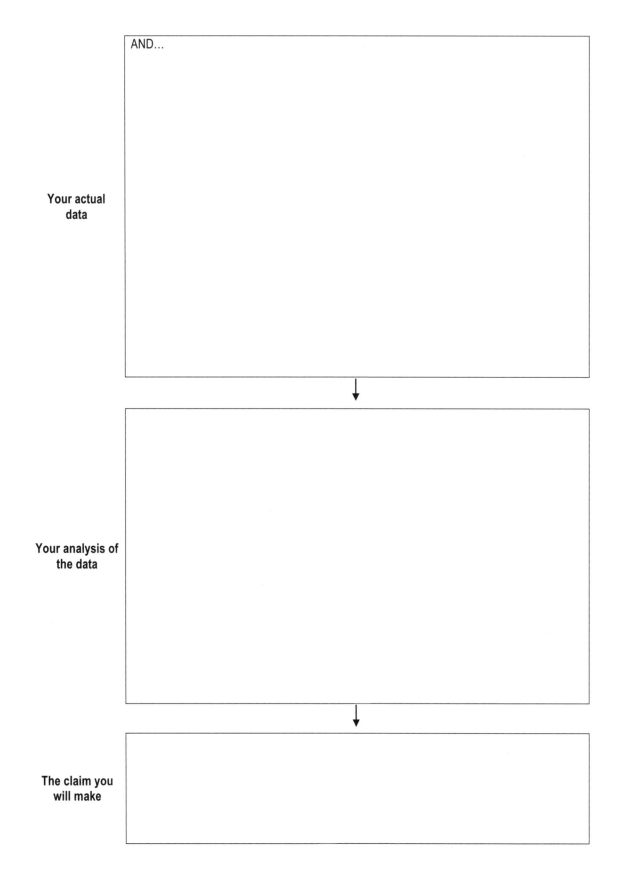

AND...

Your actual data

Your analysis of the data

The claim you will make

Investigation Proposal C: Comparative or Experimental Study (Short Form)

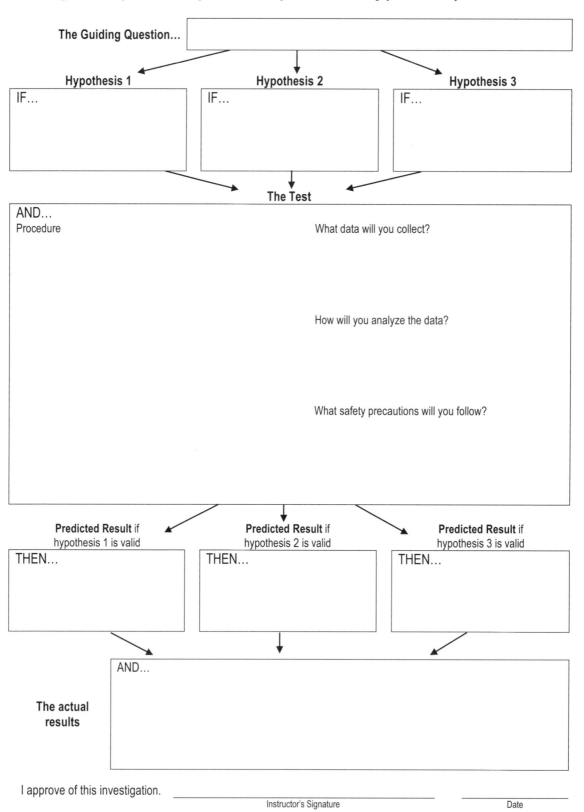

The Guiding Question...

Hypothesis 1   Hypothesis 2   Hypothesis 3

IF...   IF...   IF...

The Test

AND...
Procedure

What data will you collect?

How will you analyze the data?

What safety precautions will you follow?

Predicted Result if hypothesis 1 is valid   **Predicted Result** if hypothesis 2 is valid   **Predicted Result** if hypothesis 3 is valid

THEN...   THEN...   THEN...

AND...

The actual results

I approve of this investigation. _____   _____
Instructor's Signature   Date

APPENDIX 5
Investigation Report Peer-Review Guide: High School Version

Report By: _____
ID Number

Author: Did the reviewers do a good job? 1 2 3 4 5
Rate the overall quality of the peer review

Reviewed By: _____ _____ _____ _____
ID Number   ID Number   ID Number   ID Number

Section 1: Introduction and Guiding Question	Reviewer Rating			Teacher Score
1. Did the author provide enough **background information**?	☐ No	☐ Partially	☐ Yes	0 1 2
2. Is the background information **accurate**?	☐ No	☐ Partially	☐ Yes	0 1 2
3. Did the author **describe the goal** of the study?	☐ No	☐ Partially	☐ Yes	0 1 2
4. Did the author make the **guiding question** explicit and explain how the guiding question is related to the background information?	☐ No	☐ Partially	☐ Yes	0 1 2

Reviewers: If your group made any "No" or "Partially" marks in this section, please **explain how the author could improve** this part of his or her report.	**Author:** What revisions did you make in your report? Is there anything you decided to keep the same even though the reviewers suggested otherwise? Be sure to explain why.

Section 2: Method	Reviewer Rating			Teacher Score
1. Did the author describe **the procedure** he or she used to gather data and then explain why he or she used this procedure?	☐ No	☐ Partially	☐ Yes	0 1 2
2. Did the author explain **what data** were collected (or used) during the investigation and why they were collected (or used)?	☐ No	☐ Partially	☐ Yes	0 1 2
3. Did the author describe **how he or she analyzed the data** and explain why the analysis helped him or her answer the guiding question?	☐ No	☐ Partially	☐ Yes	0 1 2

Section 2: Method *(continued)*	Reviewer Rating			Teacher Score
4. Did the author use the **correct term** to describe his or her investigation (e.g., experiment, observations, interpretation of a data set)?	☐ No	☐ Partially	☐ Yes	0 1 2
Reviewers: If your group made any "No" or "Partially" marks in this section, please **explain how the author could improve** this part of his or her report.	**Author:** What revisions did you make in your report? Is there anything you decided to keep the same even though the reviewers suggested otherwise? Be sure to explain why.			

Section 3: The Argument	Reviewer Rating			Teacher Score
1. Did the author provide a **claim** that answers the guiding question?	☐ No	☐ Partially	☐ Yes	0 1 2
2. Did the author include **high-quality evidence** in his/her argument?				
• Were the data collected in an appropriate manner?	☐ No	☐ Partially	☐ Yes	0 1 2
• Is the analysis of the data appropriate and free from errors?	☐ No	☐ Partially	☐ Yes	0 1 2
• Is the author's interpretation of the analysis (what it means) valid?	☐ No	☐ Partially	☐ Yes	0 1 2
3. Did the author **present the evidence** in an appropriate manner by				
• using a correctly formatted and labeled graph (or table);	☐ No	☐ Partially	☐ Yes	0 1 2
• including correct metric units (e.g., m/s, g, ml); and	☐ No	☐ Partially	☐ Yes	0 1 2
• referencing the graph or table in the body of the text?	☐ No	☐ Partially	☐ Yes	0 1 2
4. Is the claim **consistent with the evidence**?	☐ No	☐ Partially	☐ Yes	0 1 2
5. Did the author include a **justification of the evidence** that				
• explains why the evidence is important (why it matters) and	☐ No	☐ Partially	☐ Yes	0 1 2
• defends the inclusion of the evidence with a specific science concept or by discussing his/her underlying assumptions?	☐ No	☐ Partially	☐ Yes	0 1 2
6. Is the **justification of the evidence** acceptable?	☐ No	☐ Partially	☐ Yes	0 1 2
7. Did the author discuss **how well his/her claim agrees with the claims made by other groups** and explain any disagreements?	☐ No	☐ Partially	☐ Yes	0 1 2
8. Did the author **use scientific terms correctly** (e.g., *hypothesis* vs. *prediction, data* vs. *evidence*) and **reference the evidence in an appropriate manner** (e.g., *supports* or *suggests* vs. *proves*)?	☐ No	☐ Partially	☐ Yes	0 1 2

Section 3: The Argument *(continued)*	Reviewer Rating	Teacher Score
Reviewers: If your group made any "No" or "Partially" marks in this section, please *explain how the author could improve* this part of his or her report.	**Author:** What revisions did you make in your report? Is there anything you decided to keep the same even though the reviewers suggested otherwise? Be sure to explain why.	

Mechanics	Reviewer Rating			Teacher Score
1. *Organization:* Is each section easy to follow? Do paragraphs include multiple sentences? Do paragraphs begin with a topic sentence?	☐ No	☐ Partially	☐ Yes	0 1 2
2. *Grammar:* Are the sentences complete? Is there proper subject-verb agreement in each sentence? Are there no run-on sentences?	☐ No	☐ Partially	☐ Yes	0 1 2
3. *Conventions:* Did the author use appropriate spelling, punctuation, and capitalization?	☐ No	☐ Partially	☐ Yes	0 1 2
4. *Word Choice:* Did the author use the appropriate word (e.g., *there* vs. *their, to* vs. *too, than* vs. *then*)?	☐ No	☐ Partially	☐ Yes	0 1 2

Teacher Comments:

Total: _____ /50

Investigation Report Peer-Review Guide: Advanced Placement Version

Report By: _____   Author: Did the reviewers do a good job?   1   2   3   4   5

ID Number

Rate the overall quality of the peer review

Reviewed By: _____   _____   _____   _____

ID Number   ID Number   ID Number   ID Number

Section 1: Introduction and Guiding Question	Reviewer Rating			Teacher Score
1. Did the author provide enough **background information**?	☐ No	☐ Partially	☐ Yes	0 1 2
2. Is the background information **accurate**?	☐ No	☐ Partially	☐ Yes	0 1 2
3. Did the author **describe the goal** of the study?	☐ No	☐ Partially	☐ Yes	0 1 2
4. Did the author make the **guiding question** explicit and explain how the guiding question is related to the background information?	☐ No	☐ Partially	☐ Yes	0 1 2
Reviewers: If your group made any "No" or "Partially" marks in this section, please **explain how the author could improve** this part of his or her report.	**Author:** What revisions did you make in your report? Is there anything you decided to keep the same even though the reviewers suggested otherwise? Be sure to explain why.			

Section 2: Method	Reviewer Rating			Teacher Score
1. Did the author describe **the procedure** he or she used to gather data and then explain why he or she used this procedure?	☐ No	☐ Partially	☐ Yes	0 1 2
2. Did the author explain **what data** were collected (or used) during the investigation and why they were collected (or used)?	☐ No	☐ Partially	☐ Yes	0 1 2
3. Did the author describe **how he or she analyzed the data** and explain why the analysis helped him or her answer the guiding question?	☐ No	☐ Partially	☐ Yes	0 1 2

Section 2: Method *(continued)*	Reviewer Rating			Teacher Score
4. Did the author use the **correct term** to describe his or her investigation (e.g., experiment, observations, interpretation of a data set)?	☐ No	☐ Partially	☐ Yes	0 1 2

Reviewers: If your group made any "No" or "Partially" marks in this section, please **explain how the author could improve** this part of his or her report.	**Author:** What revisions did you make in your report? Is there anything you decided to keep the same even though the reviewers suggested otherwise? Be sure to explain why.

Section 3: The Argument	Reviewer Rating			Teacher Score
1. Did the author provide a **claim** that answers the guiding question?	☐ No	☐ Partially	☐ Yes	0 1 2
2. Did the author include **high-quality evidence** in his/her argument? • Were the data collected in an appropriate manner?	☐ No	☐ Partially	☐ Yes	0 1 2
• Is the analysis of the data appropriate and free from errors?	☐ No	☐ Partially	☐ Yes	0 1 2
• Is the author's interpretation of the analysis (what it means) valid?	☐ No	☐ Partially	☐ Yes	0 1 2
3. Did the author **present the evidence** in an appropriate manner by • using a correctly formatted and labeled graph (or table);	☐ No	☐ Partially	☐ Yes	0 1 2
• including correct metric units (e.g., m/s, g, ml); and	☐ No	☐ Partially	☐ Yes	0 1 2
• referencing the graph or table in the body of the text?	☐ No	☐ Partially	☐ Yes	0 1 2
4. Is the claim **consistent with the evidence**?	☐ No	☐ Partially	☐ Yes	0 1 2
5. Did the author include a **justification of the evidence** that • explains why the evidence is important (why it matters) and	☐ No	☐ Partially	☐ Yes	0 1 2
• defends the inclusion of the evidence with a specific science concept or by discussing his/her underlying assumptions?	☐ No	☐ Partially	☐ Yes	0 1 2
6. Is the **justification of the evidence** acceptable?	☐ No	☐ Partially	☐ Yes	0 1 2
7. Did the author discuss **how well his/her claim agrees with the claims made by other groups** and explain any disagreements?	☐ No	☐ Partially	☐ Yes	0 1 2
8. Did the author **use scientific terms correctly** (e.g., *hypothesis* vs. *prediction*, *data* vs. *evidence*) and **reference the evidence in an appropriate manner** (e.g., *supports* or *suggests* vs. *proves*)?	☐ No	☐ Partially	☐ Yes	0 1 2

Section 3: The Argument *(continued)*

Reviewers: If your group made any "No" or "Partially" marks in this section, please ***explain how the author could improve*** this part of his or her report.	**Author:** What revisions did you make in your report? Is there anything you decided to keep the same even though the reviewers suggested otherwise? Be sure to explain why.

Section 4: Limitations and Implications	Reviewer Rating			Teacher Score
1. Did the author discuss the ***limitations of the study*** and what he or she could have done in order to ***increase the rigor*** of the investigation?	☐ No	☐ Partially	☐ Yes	0 1 2
2. Did the author discuss ***sources of error*** that were unavoidable in the collection of the data?	☐ No	☐ Partially	☐ Yes	0 1 2
3. Did the author discuss ***new questions*** to explore?	☐ No	☐ Partially	☐ Yes	0 1 2

Reviewers: If your group made any "No" or "Partially" marks in this section, please ***explain how the author could improve*** this part of his or her report.	**Author:** What revisions did you make in your report? Is there anything you decided to keep the same even though the reviewers suggested otherwise? Be sure to explain why.

Mechanics	Reviewer Rating			Teacher Score
1. **Organization:** Is each section easy to follow? Do paragraphs include multiple sentences? Do paragraphs begin with a topic sentence?	☐ No	☐ Partially	☐ Yes	0 1 2
2. **Grammar:** Are the sentences complete? Is there proper subject-verb agreement in each sentence? Are there no run-on sentences?	☐ No	☐ Partially	☐ Yes	0 1 2
3. **Conventions:** Did the author use appropriate spelling, punctuation, and capitalization?	☐ No	☐ Partially	☐ Yes	0 1 2
4. **Word Choice:** Did the author use the appropriate word (e.g., *there* vs. *their, to* vs. *too, than* vs. *then*)?	☐ No	☐ Partially	☐ Yes	0 1 2
Teacher Comments:				

Total: _____/56

IMAGE CREDITS

All images in this book are stock photographs or courtesy of the authors unless otherwise noted below.

Lab 4

Figure L4.1: Martin23230, Wikimedia Commons, CC BY-SA 3.0. *https://commons.wikimedia.org/wiki/File:Americas_(orthographic_projection).svg*

Map of North America: Wikimedia Commons, GFDL 1.2. *https://commons.wikimedia.org/wiki/File:North_America_(orthographic_projection).svg*

Lab 6

Figure L6.1: U.S. Patent and Trademark Office, Public domain. *https://tinyurl.com/zvwyrcm*

Lab 8

Figure L8.1: Mj-bird, Wikimedia Commons, CC BY-SA 3.0. *https://commons.wikimedia.org/wiki/File:Internal_combustion_engine_pistons_of_partial_cross-sectional_view.jpg*

Lab 9

Figure L9.1: U.S. Air Force, Flickr, *https://c1.staticflickr.com/5/4068/4317655660_61a60f6576_b.jpg*

Lab 11

Figure L11.1: FaceMePLS, Wikimedia Commons, CC BY-SA 2.0. *https://commons.wikimedia.org/wiki/File:Kermis_Malieveld_Den_Haag.jpg*

Lab 13

Figures 13.1 and L13.2: Chetvorno, Wikimedia Commons, Public domain. *https://commons.wikimedia.org/wiki/Category:Diagrams_of_pendulums#/media/File:Simple_gravity_pendulum.svg*

Figure L13.1: Sjoerd22, Wikimedia Commons, Public domain. *https://commons.wikimedia.org/wiki/File:H6_clock.jpg*

Lab 14

Figure L14.1: Svjo, Wikimedia Commons, CC BY-SA 3.0. *https://commons.wikimedia.org/wiki/File:Mass-spring-system.png*

Lab 16

Figure L16.1: Max Andrews, Wikimedia Commons, CC BY-SA 3.0. *https://upload.wikimedia.org/wikipedia/commons/6/67/Concussion_Anatomy.png*

Figure L16.2: Damon J. Moritz, Public domain. *https://commons.wikimedia.org/wiki/File:US_Navy_031230-N-9693M-004_Navy_linebacker_Bobby_McClarin_sacks_Texas_Tech_quarterback_B.J._Symons_during_the_EV1.Net_Houston_Bowl_at_Reliant_Stadium_in_Houston,_Texas.jpg*

Lab 22

Figure L22.1: Warren K. Leffler, Wikimedia Commons, Public domain. *https://en.wikipedia.org/wiki/1979_energy_crisis#/media/File:Line_at_a_gas_station,_June_15,_1979.jpg*

Lab 23

Figure L23.1: Federal Highway Administration, U.S. Department of Transportation, Public domain. *www.fhwa.dot.gov/rakeman/1830.htm*

INDEX

Page numbers printed in **boldface** type refer to figures or tables.

Index

Index